MACROPHAGES IN THE HUMAN BODY

MACROPHAGES IN THE HUMAN BODY

A Tissue Level Approach

Edited by

NIELS OLSEN SARAIVA CAMARA

Professor, Department of Immunology, Institute of Biomedical Sciences, University of Sao Paulo, Brazil

TÁRCIO TEODORO BRAGA

Professor, Department of Pathology, Federal University of Parana, Parana, Brazil

ELSEVIER

ACADEMIC PRESS

An imprint of Elsevier

Academic Press is an imprint of Elsevier
125 London Wall, London EC2Y 5AS, United Kingdom
525 B Street, Suite 1650, San Diego, CA 92101, United States
50 Hampshire Street, 5th Floor, Cambridge, MA 02139, United States
The Boulevard, Langford Lane, Kidlington, Oxford OX5 1GB, United Kingdom

ISBN: 978-0-12-821385-8

For information on all Academic Press publications
visit our website at https://www.elsevier.com/books-and-journals

Publisher: Stacy Masucci
Acquisitions Editor: Linda Versteeg-Buschman
Editorial Project Manager: Susan E. Ikeda
Production Project Manager: Kiruthika Govindaraju
Cover Designer: Miles Hitchen

Typeset by STRAIVE, India

Contents

Contributors

Renan Willian Alves
Federal University of ABC, Center for Human and Natural Sciences, São Paulo, Brazil

Hans-Joachim Anders
LMU Hospital, Department of Medicine IV, Division of Nephrology, Munich, Germany

Luís Eduardo Coelho Andrade
Rheumatology Division, Medicine Department, Escola Paulista de Medicina, Universidade Federal de São Paulo, São Paulo, Brazil

Vinicius Andrade-Oliveira
Federal University of ABC, Center for Human and Natural Sciences, São Paulo, Brazil

Maria Christina W. Avellar
Department of Pharmacology, Federal University of São Paulo, Paulista School of Medicine, São Paulo, São Paulo, Brazil

Emiliano Barreto
Laboratory of Cellular Biology, Biological Sciences and Health Institute, Federal University of Alagoas, Maceio, Alagoas, Brazil

Tárcio Teodoro Braga
Department of Basic Pathology, Federal University of Parana, Curitiba; Postgraduate Program in Biosciences and Biotechnology, Instituto Carlos Chagas, Fiocruz-Parana, Brazil

Niels Olsen Saraiva Camara
Department of Immunology, Institute of Biomedical Sciences, University of São Paulo; Nephrology Division, Federal University of São Paulo, São Paulo, Brazil

Nicholas Collins
Metaorganism Immunity Section, Laboratory of Immune System Biology, National Institute of Allergy and Infectious Diseases, National Institutes of Health, Bethesda, MD, United States

Eloisa Martins da Silva
Paulista Medical School, Federal University of São Paulo, São Paulo, Brazil

Wilson de Melo Cruvinel
Pontifícia Universidade Católica de Goiás, Goiânia, Brazil

M.G. de Oliveira
Neuroimmune Interactions Laboratory, Department of Immunology, Institute of Biomedical Sciences, University of São Paulo, São Paulo, SP, Brazil

Alexandre Wagner Silva de Souza
Rheumatology Division, Medicine Department, Escola Paulista de Medicina, Universidade Federal de São Paulo, São Paulo, Brazil

Lorena Doretto-Silva
Federal University of ABC, Center for Human and Natural Sciences, São Paulo, Brazil

Rebeca Bosso dos Santos Luz
Department of Basic Pathology, Federal University of Parana, Curitiba, Brazil

L.C. Faria
Neuroimmune Interactions Laboratory, Department of Immunology, Institute of Biomedical Sciences, University of São Paulo, São Paulo, SP, Brazil

Lucas D. Faustino
Division of Rheumatology, Allergy and Immunology, Center for Immunology and Inflammatory Diseases, Massachusetts General Hospital, Harvard Medical School, Boston, MA, United States

Sandra Y. Fukada
School of Pharmaceutical Sciences of RibeirãoPreto, Department of BioMolecular Sciences, University of São Paulo, Ribeirão Preto, SP, Brazil

Siamon Gordon
Graduate Institute of Biomedical Sciences, College of Medicine, Chang Gung University, Taoyuan City, Taiwan; Sir William Dunn School of Pathology, University of Oxford, Oxford, United Kingdom

Clive M. Gray
Division of Immunology, Institute of Infectious Disease and Molecular Medicine and Department of Pathology, Faculty of Health Sciences, University of Cape Town; Division of Molecular Biology and Human Genetics, Department of Biomedical Sciences, Stellenbosch University, Cape Town, South Africa

Seong-Ji Han
Metaorganism Immunity Section, Laboratory of Immune System Biology, National Institute of Allergy and Infectious Diseases, National Institutes of Health, Bethesda, MD, United States

Araceli Aparecida Hastreiter
Department of Immunology, Institute of Biomedical Sciences, Universidade de São Paulo, São Paulo, Brazil

Nathan Klopfenstein
Department of Medicine, Division of Infectious Diseases; Department of Pathology, Microbiology, and Immunology, Vanderbilt University Medical Center, Nashville, TN, United States

Julia Lichtnekert
LMU Hospital, Department of Medicine IV, Division of Nephrology, Munich, Germany

Djalma S. Lima-Junior
Metaorganism Immunity Section, Laboratory of Immune System Biology, National Institute of Allergy and Infectious Diseases, National Institutes of Health, Bethesda, MD, United States

Fernando O. Martinez
Faculty of Health and Medical Sciences, University of Surrey, Guildford, United Kingdom

Laura Helena Cherem Netto Nicolazzi
Department of Basic Pathology, Federal University of Parana, Curitiba, Brazil

L.G. Oliveira
Laboratory of Neuroimmunology of Arboviruses, Scientific Platform Pasteur-USP (SPPUSP); Neuroimmune Interactions Laboratory, Department of Immunology, Institute of Biomedical Sciences, University of São Paulo, São Paulo, SP, Brazil

Lais Cavalieri Paredes
Department of Immunology, Institute of Biomedical Sciences, Universidade de São Paulo, São Paulo, Brazil

J.P.S. Peron
Laboratory of Neuroimmunology of Arboviruses, Scientific Platform Pasteur-USP (SPPUSP); Immunopathology and Allergy Post Graduate Program, School of Medicine; Neuroimmune Interactions Laboratory, Department of Immunology, Institute of Biomedical Sciences, University of São Paulo, São Paulo, SP, Brazil

Jacqueline Pinon
Department of Medicine, Division of Infectious Diseases, Vanderbilt University Medical Center, Nashville, TN, United States

C.M. Polonio
Laboratory of Neuroimmunology of Arboviruses, Scientific Platform Pasteur-USP (SPPUSP); Neuroimmune Interactions Laboratory, Department of Immunology, Institute of Biomedical Sciences, University of São Paulo, São Paulo, SP, Brazil

Rodrigo Nalio Ramos
INSERM Laboratory of Integrative Cancer Immunology, Sorbonne Université, Sorbonne Paris Cité, Université Paris Descartes, Université Paris Diderot, Centre de Recherche des Cordeliers, Paris, France

Erivan S. Ramos-Junior
Department of Oral Biology & Diagnostic Science, The Dental College of Georgia, Augusta University, Augusta, GA, United States

Ana Carolina Guerta Salina
Department of Cellular, Molecular Biology and Pathogenic Bioagents, School of Medicine of Ribeirão Preto, University of São Paulo, Ribeirão Preto, São Paulo, Brazil

C. Henrique Serezani
Department of Medicine, Division of Infectious Diseases; Department of Pathology, Microbiology, and Immunology; Vanderbilt Institute of Infection, Immunology, and Inflammation; Vanderbilt Center for Immunobiology, Vanderbilt University Medical Center, Nashville, TN, United States

Stefanie Steiger
LMU Hospital, Department of Medicine IV, Division of Nephrology, Munich, Germany

Thaise M. Taira
School of Pharmaceutical Sciences of RibeirãoPreto, Department of BioMolecular Sciences, University of São Paulo, Ribeirão Preto, SP, Brazil

Eleonora Timperi
INSERM U932, Institut Curie, PSL Research University, Paris, France

N.G. Zanluqui
Laboratory of Neuroimmunology of Arboviruses, Scientific Platform Pasteur-USP (SPPUSP); Immunopathology and Allergy Post Graduate Program, School of Medicine; Neuroimmune Interactions Laboratory, Department of Immunology, Institute of Biomedical Sciences, University of São Paulo, São Paulo, SP, Brazil

Michael Z. Zulu
Division of Microbiology and Immunology, Yerkes National Primate Research Center, Emory University, Atlanta, GA, United States

CHAPTER 1

Macrophages: From Metchnikoff to 2020 and ahead

Rebeca Bosso dos Santos Luz[a], Laura Helena Cherem Netto Nicolazzi[a], Niels Olsen Saraiva Camara[b,c], and Tárcio Teodoro Braga[a,d]
[a]Department of Basic Pathology, Federal University of Parana, Curitiba, Brazil
[b]Department of Immunology, Institute of Biomedical Sciences, University of São Paulo, São Paulo, Brazil
[c]Nephrology Division, Federal University of São Paulo, São Paulo, Brazil
[d]Postgraduate Program in Biosciences and Biotechnology, Instituto Carlos Chagas, Fiocruz-Parana, Brazil

Macrophages are cells of great interest for immunology and other biology-related fields. The study of these cells is old, and, initially, only their microbicide functions were elicited. More functions were discovered with time, like their role in maintaining the physiology of tissues and clearing cellular debris and apoptotic cells. Also, it was discovered that macrophages are tissue-resident cells, unlike it was thought in the beginning, when they were thought to be only migrating cells. They were first discovered due to their fundamental role in phagocytosis and clearance of microorganisms by the researcher Ilya Ilyich Metchnikoff, also known as Elie Metchnikoff, in 1882. His efforts were an important key for the initial understanding about the innate immune response.

Metchnikoff was born in 1845 in a Ukrainian middle-class family. He was a zoologist, comparative embryologist, biologist, and an evolution enthusiastic when he was 16 years old [1]. At that time, he started acquiring knowledge about the biology of living beings, having a great admiration for Virchow's cellular theory. Because of his incredible genius, he was able to anticipate 2 years of his graduation in natural sciences at Kharkiv University. Thanks to that advance, with only 22 years old, he managed to become a teacher at Odessa University, where he remained for 15 years [1].

For many years, as a teacher, his main interest was comparative and evolutionary embryology. Metchnikoff thought that, in order to understand the physiological process that regulates the metazoan cellular differentiation, it could be essential to understand the morphology behind embryology. Before Metchnikoff left Odessa, he started to study embryonic development in some invertebrates, and in this field, there was a certain competition

Macrophages in the Human Body
https://doi.org/10.1016/B978-0-12-821385-8.00004-4

between Metchnikoff and Ernst Haeckel, also an evolutionist. Both researchers wanted to explain the cellular complexity of metazoan. Haeckel postulated that the organism originated from an invagination of the outer layer of the gastrula. On the other hand, Metchnikoff observed another pattern; he saw undifferentiated cells filling the empty space of the gastrula in a less organized manner [2].

These different views on the gastrula's development yielded Metchnikoff more ideas to research. He wanted to understand how those filling cells were maintained together and were later differentiated. So, here lays the essential link between the evolutionary embryologist and the immunologist. Metchnikoff thought that, in complex organisms, the cells needed to find harmony inside the chaotic growing body, and he thought that phagocytes were the key [2].

Because of his beliefs, Metchnikoff continued his observations on the so-called wandering cells, the cells that fill the gastrula. They have eating as their primary function, in animals without guts. He observed that in more complex animals, the mesoderm, originated from the gastrula, maintained their cells in the same manner that occurred in lower animals. Metchnikoff saw an opportunity in the phagocytes to better understand the phylogeny.

It was during those years of Metchnikoff's life that he performed the famous "eureka" experiment, in 1882, at Messina. As described by him, 30 years after the event and later published by his wife: "while I had remained alone at my microscope and was following the life of motile cells in a transparent starfish larva, I was struck by a novel idea. I began to imagine that similar cells could serve the defense of an organism against dangerous intruders" [1]. He ran to his garden to pick up a rose thorn, then injected it into the starfish larva, and performed the famous experiment that gave the initial idea for what we have known as macrophages. Although there were records about phagocytosis before Metchnikoff, he was responsible for the posterior experiments and defense of the phagocytosis theory [3].

This experiment brought light to his understanding of metazoan cell differentiation. He hypothesized that, in complex animals, phagocytes were cells that did not belong necessarily to any specific tissue and were always free in the organism, searching for microorganisms and cellular debris [2]. Besides the notorious phagocyte discovery, that experiment also helped to explain cellular immunity and the after called innate immunity. For almost the rest of his life, Metchnikoff researched to prove this theory. In one of these experiments, while observing tadpole's metamorphosis, he noted that phagocytic cells were involved directly in the removal of the animal's tail and its repair. He then concluded that the phagocytic functions had been evolutionarily conserved [2].

Although his work established the basis for cellular immunology studies, he faced many obstacles in order to consolidate his theory in the scientific community. Because of the emergence of positivism and its application in science, Metchnikoff's idea was thought to be theological [2], due to his lack of molecular mechanisms and for basing his theory on a broad manner. Additionally, humoral immunologists also tried to refute Metchnikoff's theory. They claimed that cellular defense was either insufficient or only happened because of the existence of a humoral response [4]. Despite Metchnikoff's efforts, there were various experiments that tried to overthrow the phagocytosis theory. Furthermore, Behring's and Kitasato's discovery of antibodies against the tetanus and diphtheria exotoxins and Paul Ehrlich's work on "magic bullet" were significant to overcome the so-called cellularist defensors of phagocytes [4]. The phagocytic theory was even claimed to be wrong by Robert Koch at a congress in 1891. Paul Ehrlich and Elie Metchnikoff, in 1908, won the Nobel Prize in Physiology or Medicine, in an attempt to end their duel and give proper recognition to both theories [4].

Nevertheless, the "humoralists" prevailed over the "cellularists" for a long time in the field of science. They casted cellular immunology aside for a while, and although the focus was on humoral immunology, some immunologists debated about the macrophages' classification. But in the 1960s, an enormous number of studies about the cellular basis of immunology emerged [4]. Categorizing similar cells by morphology, function, and origin was a crucial step to facilitate its understanding and study. Hence, the first attempt to classify macrophages was made by Metchnikoff himself, naming the "macrophage system," composed by the macrophages and the microphages (today known as polymorphonuclear leukocyte) [5]. With more knowledge being acquired, Aschoff brought a new proposition in 1924. He showed the idea of a reticuloendothelial system, initially composed of endothelial cells, fibroblasts, reticular cells of spleen and lymph nodes, reticuloendothelial cells, histiocytes, splenocytes, and monocytes. This system was unsuitable because of the low phagocytic capacity of the endothelial cells and the fibroblasts. They also had significant differences of morphology and origin; therefore, these cells left the classification group [5].

Despite the large flaws in that classification, it prevailed for 20 years, until an old organization system, proposed by Voltera, was made to reborn by J.A. Thomas in 1949. This system only included a few more histiocyte-like cells; therefore, the lack of specification remained [5]. So, in 1969, at a conference at Leiden, the Netherlands, the mononuclear phagocyte system was proposed. This was only possible because of the new discoveries by the scientific community about the phagocytes; thus, this new organization system is composed of macrophages, monocytes, and pro-monocytes. All these cells share the same origin in the bone marrow [6], similar high phagocytic ability, and similar immune receptors, indicating that they play a role in host defense, perform pinocytosis, and have a remarkable high adherence on surfaces. It is worth noting that the term "mononuclear phagocyte" was already used by Metchnikoff [5].

Since the beginning of the studies about phagocytosis as a part of host defense against microorganisms, Metchnikoff and some contemporaries, a number of them his students, were engaged in unveiling what was behind this new immunology area. Metchnikoff had discovered that the microorganism elimination could occur inside phagocytic cell and that acid chemical processes occurred inside these cells (macrophages and microphages) [7]. He also proposed the existence of certain enzymes produced by these cells, the macrocytosis, produced by macrophages, and the microcytosis, produced by microcytes. These enzymes were grouped together as the cytosis, and they were responsible for bacterial digestion. Furthermore, Metchnikoff discovered that the macrophages' interaction with microorganisms was crucial for phagocytosis, as well as macrophage membrane's amoebic expansion [7].

Although his contributions were great, mainly in the intracellular digestion field, one of his hypotheses was posteriorly refuted. Metchnikoff faithfully believed that macrophages possessed a certain mobility independence and that these features allowed macrophages to reach microorganism and perform digestion without needing any assistance. But, in 1975, Griffin and colleagues discovered what they called the zipper mechanism, which went against their initial thought. They were looking to understand if only the antibody attachment to the Fc macrophage's receptor was essential for complete phagocytosis. They performed various experiments, which had the principle to block the interaction with macrophage's free receptors, using an anti-macrophage antibody [8]. The study showed that when these receptors were blocked, there was no phagocytosis. Due to this, Griffin and colleagues proposed that phagocytosis occurs through macrophage membrane's interaction with other molecules, like a zipper, concluding that it was not just one signal responsible for the event.

This study was only possible because, previously, the idea that antibodies are bound to macrophage was proposed and tested. In 1964, there were already records of antibodies recognizing specifically macrophages [9], and in 1965, Berken and Benacerraf already had performed studies trying to understand this interaction [10]. They characterized optimal temperature, pH, and ligation time and realized that the interaction was reversible and weak. Additionally, they perceived that it was not all kinds of gamma globulins that were capable to interact with macrophages; their affinity happened only with the γ2-globulin, and they also discovered that these interactions were crucial for opsonization. Furthermore, they also managed, through pepsin digestion, to digest only the antibody's Fc portion. This prevented opsonization; thus, they concluded that antibody's recognition occurred through Fc portion; that is, macrophages have an Fc receptor [10].

Besides the great importance of how macrophages can perform phagocytosis, the discovery of a specific macrophage's antibody helped to narrow their study. On that context, in 1979, Timothy Springer and colleagues managed to discover an antigen (the integrin Mac-1) [11] recognized by a monoclonal antibody produced by his group [12]. In the study, M1/70 antibody was utilized to search antigens recognized by it. They looked on lymphoid and blood cells, the P388D macrophage lineage, and other cells. Through antigen title, a great antigen presence in the P388D cells was observed. The same test was realized with bacteria-infected murine peritoneum cells, divided into adherent and non-adherent ones. They got a similar result: the antigens prevailed on the adherent cells, which is a strong macrophage characteristic. On the same line, in 1981, Jonathan Austyn and Siamon Gordon produced the F4/80 monoclonal antibody [13] that recognizes specifically macrophages from the peritoneum, bone marrow, and other macrophages and does not utilize macrophage's Fc receptor; instead, the antibody identifies integrins. Hence, it was an enormous mark in macrophage's history.

A great amount of knowledge about the function and signal's transduction of the Fc receptor was developed about 20 years ago. More sophisticated experiments confirmed what Berken and Benacerraf found: the macrophage Fc importance for the opsonization. Such data allowed researchers to go further. It was found that the lack of gamma subunits on these receptors blocked phagocytosis effectiveness, which happens due to the location of ITAM domains at these subunits; therefore, the absence of these domains impairs proper signaling [14]. Besides, it was also possible to classify them [14] thanks to the capacity to isolate those receptors, a technology that did not exist at Berken's time (Fig. 1).

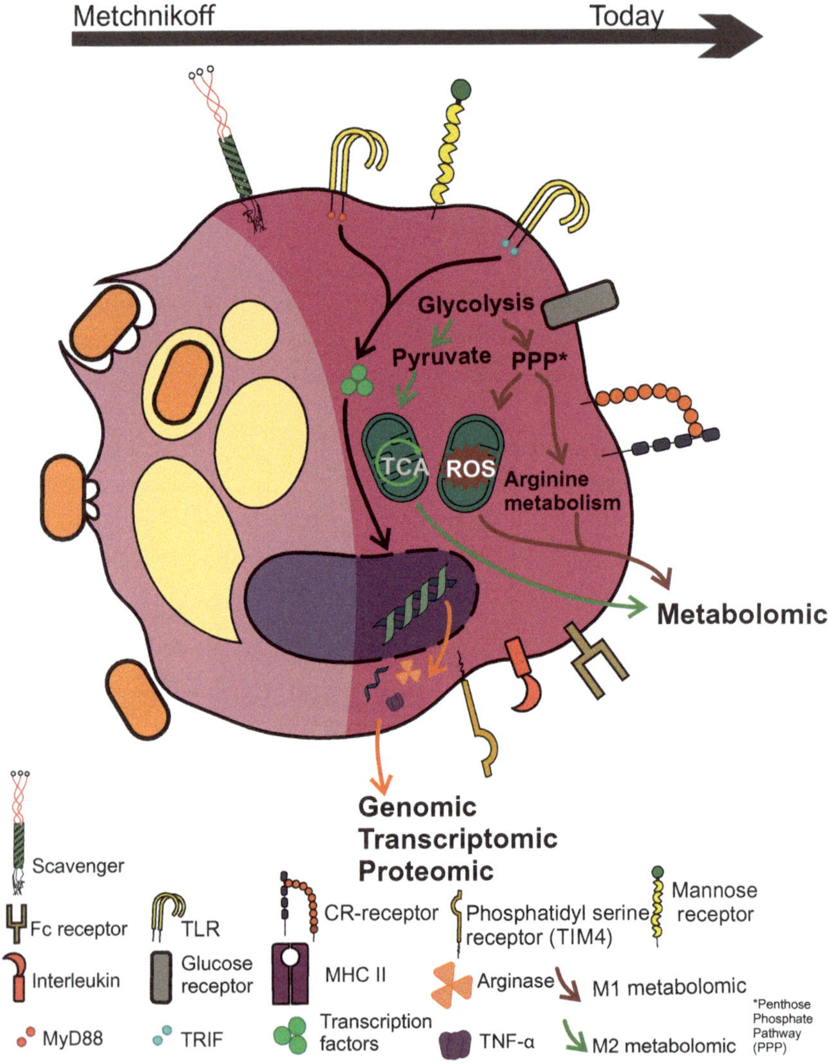

Fig. 1 From Metchnikoff to nowadays. The figure represents the progression of the macrophage's knowledge over time. At Metchnikoff's time, the phagocytic capacity of these cells was described. On the left part of the figure, the membrane projections and the internal digestion corpuscles demonstrate the characteristic phagocytosis knowledge acquired at that time. On the right part, the current information we have about the macrophages is shown. From the 1960s, the identification of phenotypic markers, demonstrated by the membrane receptors, revolutionized the macrophage's research. In the last decades, with the OMICs era (and demonstrated by genes, transcripts, proteins, and metabolic pathways), better categorization of macrophages was possible together with the knowledge of their phenotypic changes.

Just like macrophage's Fc receptors, the complement receptor was one of the first to be studied and verified to interact with those cells. In 1975, Griffin and colleagues effected one of the first characterizations of this interaction. Through the macrophage's incubation with C3b-opsonized erythrocytes, they noted a great interaction between these cells [15]. They also could perceive that the C3d form did not bind to the macrophages once the C3b cleavage inhibits the interaction. Besides, they observed that the engulfment only occurred when the cells were previously activated. Therefore, they could conclude that in phagocytosis promotion, the receptors are not independent and membrane contact with the particle in question is necessary, as proposed by Metchnikoff [7]. So, the phagocytosis activation by one particle does not cause the engulfment of neighbor cell. As studies advanced, it was possible to identify and classify the receptors for the complement system, on CR1, CR2, CR3, and CR4, as well as know that CR1 is the responsible receptor to interact with C3b [14].

The mannose receptor, together with the Fc receptor and complement system receptor, was one of the first to have their involvement in phagocytosis discovered in 1976 [16] and characterized in 1990 by Alan and colleagues [17]. These receptors were involved in the recognition of oligosaccharides with mannose, fucose, *N*-acetylglucosamine, glucose, and galactose by affinity order [18]. This capacity of recognition allows mannose receptor to also function as a scavenger receptor and to be part of host/pathogen distinction. They also play a role in endosomes, as it is believed that this receptor participates in endosome and lysosome fusion during the phagocytosis process. This is verified because of the loss of affinity in low pH [18]. Many other receptors were found to be part of phagocytosis, like the dectin-1, scavenger receptors, and MARCO [19], but the exact phagocytosis mechanism and posterior pathogen elimination were not fully clarified. Later, the molecular mechanism that causes changes in the cell's body, which disengage phagocytosis events, would be discovered. As mentioned above, the recognition of particles by membrane receptor is crucial. Subsequentially, it would be discovered that the F-actin polymerization and the Arp2/3 complex are essential for those internal changes [20].

Rosenthal and Shevach, in 1973, demonstrated one important function of macrophages in the immune response against pathogens [21]. At that time, the lymphocyte's response and their subsequential antibody production had been consistently well described. However, they were looking for a more profound explanation on how lymphocyte activation occurred. They saw that only antigen binding with the membrane immunoglobulin receptors did not explain the whole mechanism. Therefore, with advances in the field, macrophage's role was clear, but not their cellular mechanism. Rosenthal and Shevach were able to demonstrate that lymphocyte–macrophage interaction occurred through antigen presentation by macrophages via the major histocompatibility complex (MHC), causing lymphocyte proliferation [21]. In the same year, a major discovery was accomplished in immunology, the identification of a new cell type: the dendritic cells [22]. They were seen as being different from macrophages, first in morphology and later in other aspects, for instance, different lysosome types [23]. Later in his research, Steinman found that dendritic cells also possessed the MHC complex [24], and he and colleagues were able to prove that these cells also present antigens to the T lymphocyte [25]. Thus, in this time of the innate immunity history, we knew the mechanisms of antigen presenting and their role in T lymphocyte activation and clonal expansion. But an important link between macrophages from innate to acquired immunity was proposed by Janeway, in 1989, at the Cold Spring Harbor Symposium. He criticized the theory proposed by the studies of Landsteiner, which demonstrated that immune response was equally, by intensity and form, independent of the antigen stimuli and that it occurred through interaction with clonal receptors, generated by genetic rearrangement [26].

Back then, clonal selection theory and acquired immunity were the main subjects researchers were studying. Janeway then said that, because of that negligence, a view more like Metchnikoff's was left aside, just like the innate immunity [26]. So, explanations for second signals, generated by activated macrophages, for example, were thought to be for avoiding self-reactiveness by lymphocytes, and a lot of the lymphocyte's surface molecule functions were unknown. Janeway then opposed that explanation, proposing that these second signals were responsible for the effectiveness of pathogen elimination by phagocytic cells [26]. He also suggested that the recognition of microbes was not only because of the clonal receptors, but also because of receptors that recognized pathogen patterns, which he named "pattern recognition receptors (PRRs)" [26]. Thus, he proposed that this mechanism was the most conserved within evolution and that these receptors were essential to eliminate pathogens in invertebrates [26].

In fact, these receptors were found later in *Drosophila*, the Toll-like receptor (TLR). They were known to be part of the embryonic development of flies, but Rosetto and colleagues [27] showed that they may play a role in the pathogen's elimination in these insects. Bruno Lemaitre and Jules Hoffmann discovered the function of the fruit fly *Toll* gene in fungal infection [28]. Later, Medzhitov and colleagues [29] found, through in silico analysis, that a homologue of this receptor exists in humans and uses the same transcription factor—NF-κB—for its signaling. They could then be characterized as a membrane receptor with an extracellular leucin-rich domain and a cytosolic domain TIR/IL-1 that makes the signal transduction and has a broader recognition aspect, being able to recognize patterns of all microbe classes [30].

Other PRRs had their recognition domains and signaling properties discovered. One of their classes is important in macrophages: the NOD1 and NOD2 receptors, first discovered by Girardin and colleagues [31,32]. These receptors, in contrast with the TLR, are cytosolic and activate transcription factors in different manners, including via inflammasomes [33]. The inflammasome complex, in turn, was discovered by Tschopp and colleagues, in 2002, as a mechanism by which pro-inflammatory caspases activated IL-1β maturation [33]. Along with the discovery of the receptor's functions and their effects on the macrophage, questions about their regulation started to be surveyed. In 1993, Dalton and colleagues found the importance of the cytokine interferon-γ (IFN-γ) in the macrophage's immune response, once these cells have IFN-γ receptors [34]. It was known that IFN-γ had a role in the macrophage activation through macrophage's microbicide activity, data confirmed by experiments demonstrating the increased number of deaths in mice lacking IFN-γ genes after infection with *Mycobacterium bovis*.

IFN-γ is just one example among various cytokines that had their role discovered in the 1990s. In the beginning of 2000s, these roles were well established, such as the macrophage's activation by IL-12 and IL-18, in addition to IFN-γ. This allowed the proposition of the classical classification of macrophages: the M1 and M2. This division was made by Mills and colleagues [35] based on the activation by either Th1 or Th2 lymphocytes. They proposed a classification supported by experiments, which demonstrated that Th1 lymphocyte promoted a classically activated macrophage profile, with secreted nitric oxide and posterior elimination of pathogens through action of the lymphocyte-derived IFN-γ [35]. On the other hand, M2 macrophages, activated by the Th2 through IL-10, IL-4, and IL-5, had an anti-inflammatory profile, with increased arginase metabolism. The authors also suggested an inflammation control role of macrophages and concluded saying that these phenotypes could represent a third macrophage's phenotype.

On the other hand, Gordon and colleagues questioned some characteristics of the pre-existing classification of the macrophages [36]. They pointed out the issue involved in classifying IL-4 and IL-13 together with the IL-10 as promoting M2 macrophage. These three interleukins have several differences in macrophage's phenotype. For example, the inhibition of respiratory burst and TNF-α production is higher with the IL-10 stimuli, whereas the class II MHC induction and the mannose receptor-mediated endocytosis are higher with the IL-4 and IL-13 stimuli [36]. Gordon also said that the previous kind of classification hides other macrophage's roles and adds that it is hard to group macrophages in one or another group, because of their high heterogeneity of functions and their ability to adapt into various different tissues and microenvironments [36].

As more information was being acquired, some ideas about the macrophage's heterogeneity emerged, which has also affected the understanding of apoptosis. Gordon and Plüddemann highlighted the importance to study different mechanisms of this process in different tissues, even in different species, and comparative studies to better understand the incredibly diverse macrophage's phenotype [37]. Noticing that macrophages have so many different functions, phenotypes, products, regulations, and flaws, in case of deregulation, Mosser and Edwards were against the classification that prevailed during that time: the classically activated macrophages (M1) and the alternatively activated macrophages (M2) [38]. Together with the high heterogeneity of these cells, with their ability to switch phenotype depending on the environment, or even to have characteristics of two phenotypes at the same time, they proposed another way to visualize them [38]. They said that macrophages could exhibit a phenotype that varied among three main characteristics: host defense, wound healing, and maintaining homeostasis, with the latter being part of the original "alternatively activated macrophages." They did it because the non-host defense macrophages had so many different functions that they could not belong to the same classification [38]. Consequently, they aimed to characterize this new macrophage population. The wound-healing class is characterized by high arginase metabolism, which will further implicate in the production and deposition of extracellular matrix components. The crucial difference between this class and the regulatory macrophage is that the latter does not have the ability to contribute with extracellular matrix production. Thus, they can respond to stress through the release of glucocorticoids and promote an anti-inflammatory profile, reducing the immune response in later stages of inflammation, and they also have the ability to present antigens [38].

Additionally, Mosser and Edwards pointed some considerations about classically activated macrophages: IFN-γ and TNF-α were not necessary for activation, and the TLR can sometimes overcome the lack of one of them.

Finally, one important consideration made by them about these three types of macrophages is that their deregulation is the origin of some pathologies. For example, the immune response, when not well controlled by the regulatory macrophages, can be harmful, as well as uncontrolled wound healing, which can lead to fibrosis [38].

The debate on the best macrophage characterization led to other unveiled macrophage functions, besides their role in the immune response against pathogens. There were researchers looking for answers related to questions that involved macrophages in other contexts, like tissue repair. The role of this cell in this process is still a relevant subject in modern research. In 1975, Leibovich and Ross found that these phagocytes were essential for debris clearance and possibly for fibroblast proliferation [39]. They got their results by depleting macrophages with hydrocortisone acetate and antimacrophage serum. This ablation technique is still used, and now it is easier and more reliable to apply, thanks to the experiments of van Rooijen and van Nieuwmegen [40]. They are responsible for the discovery and standardizing of clodronate liposome-encapsulated technique for macrophage ablation, and it is the most utilized technique today for wound repair studies (Fig. 2).

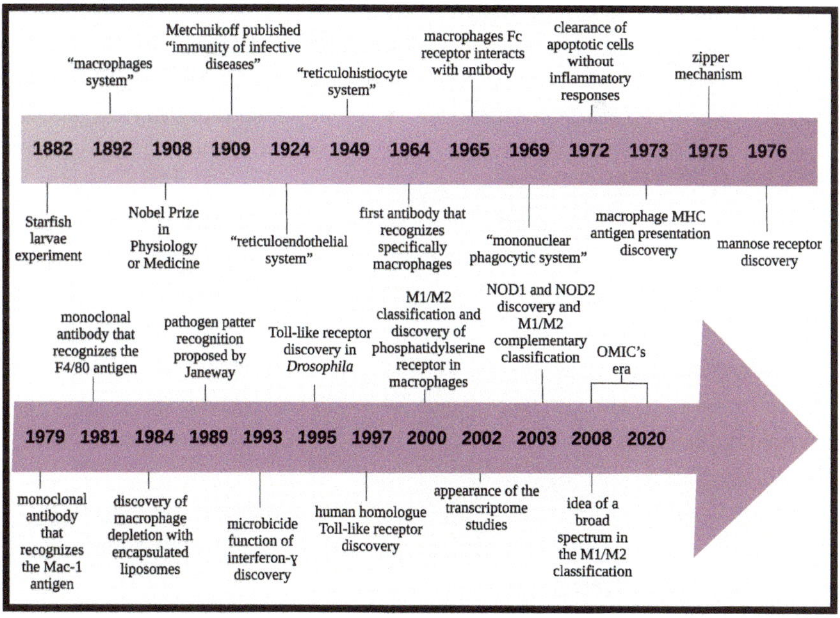

Fig. 2 Macrophage's timeline. The timeline shows the main discoveries about macrophage's function, identification, and phenotype, starting with the starfish larva experiment from Metchnikoff and ending with the demonstration of the OMICs nowadays.

Another macrophage function is their role in apoptosis. This event is involved in inflammation resolution, tissue remodeling, homeostasis, and even directly host defense. These studies started a long time ago, with Metchnikoff's observations about the importance of neutrophil clearance to inflammation's end [41], which is also important for macrophage's phenotype changes in tissue injuries. Although Metchnikoff's observations were beyond its time, it lacked molecular mechanisms. In 1972, studies about apoptosis were advancing, and there was a suggestion that changes in membrane configurations of apoptotic cells are responsible for triggering phagocytosis. It was also known that the apoptotic bodies were rapidly absorbed by phagocytic cells and that this process did not generate inflammatory responses [42], although the molecular mechanism was still not understood.

In the 1990s, the apoptosis' role in the inflammation was starting to be recognized, when apoptotic bodies' removal in that process was discovered [41]. Additionally, the non-inflammatory apoptosis' origin was confirmed by showing that macrophage and neutrophil's uptake did not cause the release of inflammatory cytokines. It was known at this time that a recognition mechanism was necessary for proper elimination of apoptotic cells by macrophages, although the exact receptor responsible for that was not known [41]. This lack of molecular information, however, did not last much. In the 2000s, it was discovered which molecules participated in the inhibition of inflammatory responses in macrophages, such as TGF-β, PGE_2, and PAF [43]. Another important discovery was the phosphatidylserine (PS) receptor in the macrophages, the first one to be found as specific for recognition of apoptotic cells, since they have the PS exposed on the external layer of their plasmatic membrane [43]. As all processes involving phagocytosis, the PS receptor is not the only one that can promote recognition of apoptotic bodies. The importance of the CD14 in this process was found out, as well as other receptors already known to act in phagocytosis, like scavenger receptors and lectin receptors [43]. On the other hand, macrophages can be induced to die for necroptosis or pyroptosis. These processes cause inflammation and occur after intense stress and can be induced by various danger-associated patterns (DAMPs) [44].

There is great interest nowadays in studying the ways that turn macrophages into the regulatory phenotype mainly because of their role in preventing autoimmune diseases. Helminths, for example, are demonstrated to stimulate this regulatory phenotype [45]. Another designation for these cells is "alternatively activated macrophages," mostly resident macrophages, because of their functions related with the regulatory counterpart, like the resolution of inflammation. This phenotype is very unique and also universal, and depending on their microenvironment, they have different origins and functions [46]. Resident macrophages can control the immune response through mechanisms of innate immune memory, also named "trained immunity" [47]. Such processes also occur with other innate immune cells, and this is a subject of actual interest [48]. Additionally, the endotoxin tolerance is a way of controlling the immune response, that is, a case when the macrophages do not respond to minimal concentrations of endotoxin. This behavior was observed long ago, in 1946, by Paul Beeson, and it is another current research interest because of its implications in some severe diseases related to bacterial infections [49].

Nowadays, several studies about macrophages have a great interest in understanding the mechanism underlying their rich heterogeneity. Gordon and Plüddemann say, "We need to explore modern technologies in our favor, using it to understand those mechanisms, and to search possible markers for each profile, which helps to find new therapies targets" [37]. In this line, there are transcriptome studies, which are mapping genetic changes in each macrophage's profile, as well as their chromatin changes, and one of the first studies about that subject is from Loke and colleagues [50]. Besides that, in 2010, the Human Immunology Project Consortium (HIPC) was established with the goal to provide a database about the immunological system. The aforementioned modern techniques are used, such as transcriptome, proteomic, multiplex, and other assays to evaluate immune response triggered by different stimuli, thus providing tools for immunologists around the world. Furthermore, this type of study is important to define more specific markers that may help to better characterize macrophages. Today, the surface markers that were largely used in the past to study these cells are known to be less reliable due to low specificity. The F4/80 and the CD68, for example, are now known to be present also in dendritic cells, which also express class II MHC. Because of their shared monocytic origin, this also happens with other surface markers [51]. Thus, genetic manipulations and the ongoing "OMICs" era may give a solution for these problems. Such source of techniques has the purpose to generate a profile for cells in different contexts.

In another study, Martinez and colleagues elucidated transcription profile under the influence of M-CSF, and in macrophage polarization in a more global manner [52]. They saw that M-CSF upregulated cell cycle genes, and during M1-to-M2 transition, there was differences in cellular metabolism, which reflected in those genes' expression [52]. The Immunological Genome

Project contributed to the macrophage's genetic profile. Considerable information about some genetic markers was demonstrated, like the lack of specificity of the traditional F4/80 macrophage marker, which was shown to be less expressed than it was known. Yet, transcription factors, surface markers, and other molecules were shown that have high macrophage specificity [53].

RNA sequencing (RNA-seq) and single-cell RNA sequencing (scRNA-seq) are recently developed techniques that show even more specificity, quality, and more information, when comparing with other genomics techniques [54]. They have the same purpose, although with different results, of offering an unbiased and a more complete data regarding the cell and even whole organ profile [55], like it is shown in MacParland's study, in which she applies, for the first time, the scRNA-seq technique in the whole human liver to search for distinct macrophage populations [56]. Thus, these methods brought a new way to study macrophages in their real environment, which gives us more trustworthy information.

With the intention to comprehend macrophage's transitions, metabolomic studies have also recently re-emerged. The metabolic profile of these cells changes according to the environment, stimuli, functions, and other causes. The well-established change in glycolysis metabolism, that happens in inflammatory macrophages, is triggered by IL-4 stimulation, which ultimately leads to an oxidative phosphorylation profile [57]. Metabolic pathways are incredibly dynamic and may change with some types of stimuli, ultimately altering macrophage's phenotype, although certain metabolic profiles are shared by different macrophage phenotypes, like that demonstrated in Suzuki's study, reporting that glycolysis also regulates macrophage differentiation into the regulatory phenotype [58]. Furthermore, Van den Bossche and colleagues suggest that studying these pathways, together with epigenetic analysis, is essential to a complete knowledge [57]. Thus, to give importance to these cells, their surprising functions, and their role in a variety of diseases, this book gives light to macrophage's characteristics within different tissues, contributing to today's researches to find a more complete understanding of these cells.

References

[1] Gordon S. Elie Metchnikoff, the man and the myth. J Innate Immun 2016;8:223–7. https://doi.org/10.1159/000443331.

[2] Tauber AI. Metchnikoff and the phagocytosis theory. Nat Rev Mol Cell Biol 2003;4:897–901. https://doi.org/10.1038/nrm1244.

[3] Gordon S. The macrophage: past, present and future. Eur J Immunol 2007;37:9–17. https://doi.org/10.1002/eji.200737638.
[4] Silverstein MA. A history of immunology. 2nd ed. Elsevier; 2009. p. 39–54.
[5] van Furth R, Cohn ZA, Hirsch JG, Humphrey JH, Spector WG, Langevoort HL. The mononuclear phagocyte system: a new classification of macrophages, monocytes, and their precursor cells. Bull World Health Organ 1972;46:845–52.
[6] Hume DA. The mononuclear phagocyte system. Curr Opin Immunol 2006;18:49–53. https://doi.org/10.1016/j.coi.2005.11.008.
[7] Metchnikoff E. Immunity in infective diseases. Cambridge University Press; 1905. p. 175–206.
[8] Griffin FM, Griffin JA, Leider JE, Silverstein SC. Studies on the mechanism of phagocytosis. I. Requirements for circumferential attachment of particle bound ligands to specific receptors on the macrophage plasma membrane. J Exp Med 1975;142:1263–82. https://doi.org/10.1084/jem.142.5.1263.
[9] Boyden SV. Cytophilic antibody in Guinea-pigs with delayed-type hypersensitivity. Immunology 1964;7:474–83.
[10] Berken A, Benacerraf B. Properties of antibodies cytophilic for macrophages. J Exp Med 1965;22.
[11] Springer T, Galfré G, Secher DS, Milstein C. Mac-1: a macrophage differentiation antigen identified by monoclonal antibody. Eur J Immunol 1979;9:301–6. https://doi.org/10.1002/eji.1830090410.
[12] Springer T, Galfrè G, Secher DS, Milstein C. Monoclonal xenogeneic antibodies to murine cell surface antigens: identification of novel leukocyte differentiation antigens. Eur J Immunol 1978;8:539–51. https://doi.org/10.1002/eji.1830080802.
[13] Austyn JM, Gordon S. F4/80, a monoclonal antibody directed specifically against the mouse macrophage. Eur J Immunol 1981;11:805–15. https://doi.org/10.1002/eji.1830111013.
[14] Aderem A, Underhill DM. Mechanisms of phagocytosis in macrophages. Annu Rev Immunol 1999;17:593–623. https://doi.org/10.1146/annurev.immunol.17.1.593.
[15] Griffin FM, Bianco C, Silverstein SC. Characterization of the macrophage receptor for complement and demonstration of its functional independence from the receptor for the Fc portion of immunoglobulin G. J Exp Med 1975;141(6):1269–77. https://doi.org/10.1084/jem.141.6.1269.
[16] Ezekowitz RA, Stahl PD. The structure and function of vertebrate mannose lectin-like proteins. J Cell Sci Suppl 1988;9:121–33. https://doi.org/10.1242/jcs.1988.supplement_9.6.
[17] Alan R, Ezekowitz B, Sastry K, Bailly P, Warner A. Molecular characterization of the human macrophage mannose receptor: Demonstration of multiple carbohydrate recognition-like domains and phagocytosis of yeasts in cos-1 cells. J Exp Med 1990;172:1785–94. https://doi.org/10.1084/jem.172.6.1785.
[18] Stahl PD. The macrophage mannose receptor: current status. Am J Respir Cell Mol Biol 1990;2:317–8. https://doi.org/10.1165/ajrcmb/2.4.317.
[19] Taylor PR, Martinez-Pomares L, Stacey M, Lin HH, Brown GD, Gordon S. Macrophage receptors and immune recognition. Annu Rev Immunol 2005;23:901–44. https://doi.org/10.1146/annurev.immunol.23.021704.115816.
[20] Rougerie P, Miskolci V, Cox D. Generation of membrane structures during phagocytosis and chemotaxis of macrophages: role and regulation of the actin cytoskeleton. Immunol Rev 2013;256:222–39. https://doi.org/10.1111/imr.12118.
[21] Rosenthal AS, Shevach EM. Function of macrophages in antigen recognition by guinea pig t lymphocytes: requirement for histocompatible macrophages and lymphocytes. J Exp Med 1973;138:1194–212. https://doi.org/10.1084/jem.138.5.1194.
[22] Steinman RM, Zanvil CA. Identification of a novel cell type in peripheral lymphoid organs of mice. J Exp Med 1973;137:1142–62.

[23] Steinman RM. Decisions about dendritic cells: past, present, and future. Annu Rev Immunol 2012;30:1–22. https://doi.org/10.1146/annurev-immunol-100311-102839.
[24] Steinman RM, Kaplan G, Witmer MD, Cohn ZA. Identification of a novel cell type in peripheral lymphoid organs of mice: V. Purification of spleen dendritic cells, new surface markers, and maintenance in vitro. J Exp Med 1979;149:1–16. https://doi.org/10.1084/jem.149.1.1.
[25] Nussenzweig MC, Steinman RM, Gutchinov B, Cohn ZA. Dendritic cells are accessory cells for the development of anti-trinitrophenyl cytotoxic T lymphocytes. J Exp Med 1980;152:1070–84. https://doi.org/10.1084/jem.152.4.1070.
[26] Janeway CA. Approaching the asymptote? Evolution and revolution in immunology. Cold Spring Harb Symp Quant Biol 1989;54:1–13. https://doi.org/10.1101/sqb.1989.054.01.003.
[27] Rosetto M, Engström Y, Baldari CT, Telford JL, Hultmark D. Signals from the IL-1 receptor homolog, toll, can activate an immune response in a drosophila hemocyte cell line. Biochem Biophys Res Commun 1995. https://doi.org/10.1006/bbrc.1995.1477.
[28] Lemaitre B, Nicolas E, Michaut L, Reichhart JM, Hoffmann JA. The dorsoventral regulatory gene cassette spätzle/toll/cactus controls the potent antifungal response in Drosophila adults. Cell 1996;86(6):973–83. https://doi.org/10.1016/s0092-8674(00)80172-5.
[29] Medzhitov R, Preston-Hurlburt P, Janeway CA. A human homologue of the Drosophila toll protein signals activation of adaptive immunity. Nature 1997;388:394–7. https://doi.org/10.1038/41131.
[30] Honda K, Taniguchi T. IRFs: master regulators of signalling by toll-like receptors and cytosolic pattern-recognition receptors. Nat Rev Immunol 2006;6:644–58. https://doi.org/10.1038/nri1900.
[31] Girardin SE, Boneca IG, Carneiro LAM, Antignac A, Jéhanno M, Viala J, Tedin K, Taha MK, Labigne A, Zähringer U, Coyle AJ, DiStefano PS, Bertin J, Sansonetti PJ, Philpott DJ. Nod1 detects a unique muropeptide from gram-negative bacterial peptidoglycan. Science 2003;300:1584–7. https://doi.org/10.1126/science.1084677.
[32] Girardin SE, Boneca IG, Viala J, Chamaillard M, Labigne A, Thomas G, Philpott DJ, Sansonetti PJ. Nod2 is a general sensor of peptidoglycan through muramyl dipeptide (MDP) detection. J Biol Chem 2003;278:8869–72. https://doi.org/10.1074/jbc.C200651200.
[33] Martinon F, Burns K, Tschopp J. The Inflammasome: a molecular platform triggering activation of inflammatory caspases and processing of proIL-β. Mol Cell 2002;10:417–26. https://doi.org/10.1016/S1097-2765(02)00599-3.
[34] Dalton DK, Pitts-Meek S, Keshav S, Figari IS, Bradley A, Stewart TA. Multiple defects of immune cell function in mice with disrupted interferon-γ genes. Science 1993;80-). 259:1739–42. https://doi.org/10.1126/science.8456300.
[35] Mills CD, Kincaid K, Alt JM, Heilman MJ, Hill AM. M-1/M-2 macrophages and the Th1/Th2 paradigm. J Immunol 2000;164:6166–73. https://doi.org/10.4049/jimmunol.164.12.6166.
[36] Gordon S. Alternative activation of macrophages. Nat Rev Immunol 2003;3:23–35. https://doi.org/10.1038/nri978.
[37] Gordon S, Plüddemann A. Macrophage clearance of apoptotic cells: a critical assessment. Front Immunol 2018;9:1–9. https://doi.org/10.3389/fimmu.2018.00127.
[38] Mosser DM, Edwards JP. Exploring the full spectrum of macrophage activation. Nat Rev Immunol 2008;8:958–69. https://doi.org/10.1038/nri2448.
[39] Leibovich SJ, Ross R. The role of the macrophage in wound repair. A study with hydrocortisone and antimacrophage serum. Am J Pathol 1975;78:71–99.
[40] van Rooijen N, van Nieuwmegen R. Elimination of phagocytic cells in the spleen after intravenous injection of liposome-encapsulated dichloromethylene diphosphonate—an enzyme-histochemical study. Cell Tissue Res 1984;238:355–8. https://doi.org/10.1007/BF00217308.

[41] Savill J. Apoptosis: a mechanism for regulation of the cell complement of inflamed glomeruli. Kidney Int 1992;41:607–12. https://doi.org/10.1038/ki.1992.92.
[42] Kerr JFR, Wyllie AH, Currie AR. Apoptosis: a basic biological phenomenon with wide-ranging implications in tissue kinetics. Br J Cancer 1972;26:239–57. https://doi.org/10.1038/bjc.1972.33.
[43] Messmer UK, Pfeilschifter J. New insights into the mechanism for clearance of apoptotic cells. BioEssays 2000;22:878–81. https://doi.org/10.1002/1521-1878(200010)22:10<878::AID-BIES2>3.0.CO;2-J.
[44] Chen Y, Hua Y, Li X, Arslan IM, Zhang W, Meng G. Distinct types of cell death and the implication in diabetic cardiomyopathy. Front Pharmacol 2020;11:42. https://doi.org/10.3389/fphar.2020.00042. eCollection 2020.
[45] Ziegler T, Rausch S, Steinfelder S, Klotz C, Hepworth MR, Kühl AA, et al. A novel regulatory macrophage induced by a helminth molecule instructs IL-10 in CD4+ T cells and protects against mucosal inflammation. J Immunol 2015;194(4):1555–64. https://doi.org/10.4049/jimmunol.1401217.
[46] Okabe Y. Molecular control of the identity of tissue-resident macrophages. Int Immunol 2018;30:485–91. https://doi.org/10.1093/intimm/dxy019.
[47] Netea MG, Domínguez-Andrés J, Barreiro LB, Chavakis T, Divangahi M, Fuchs E, Joosten LAB, van der Meer JWM, Mhlanga MM, Mulder WJM, Riksen NP, Schlitzer A, Schultze JL, Stabell Benn C, Sun JC, Xavier RJ, Latz E. Defining trained immunity and its role in health and disease. Nat Rev Immunol 2020;20(6):375–88. https://doi.org/10.1038/s41577-020-0285-6.
[48] Xing Z, Afkhami S, Bavananthasivam J, Fritz DK, D'Agostino MR, Vaseghi-Shanjani M, et al. Innate immune memory of tissue-resident macrophages and trained innate immunity: re-vamping vaccine concept and strategies. J Leukoc Biol 2020;108:825–34. https://doi.org/10.1002/JLB.4MR0220-446R.
[49] Biswas SK, Lopez-Collazo E. Endotoxin tolerance: new mechanisms, molecules and clinical significance. Trends Immunol 2009;30:475–87. https://doi.org/10.1016/j.it.2009.07.009.
[50] Loke P, Nair MG, Parkinson J, Guiliano D, Blaxter M, Allen JE. IL-4 dependent alternatively-activated macrophages have a distinctive in vivo gene expression phenotype. BMC Immunol 2002;3:1–11. https://doi.org/10.1186/1471-2172-3-7.
[51] Ferenbach D, Hughes J. Macrophages and dendritic cells: what is the difference? Kidney Int 2008;74:5–7. https://doi.org/10.1038/ki.2008.189.
[52] Martinez FO, Gordon S, Locati M, Mantovani A. Transcriptional profiling of the human monocyte-to-macrophage differentiation and polarization: new molecules and patterns of gene expression. J Immunol 2006;177:7303–11. https://doi.org/10.4049/jimmunol.177.10.7303.
[53] Gautier TS, Miller J, Melanie Greter C, Jakubzick SI, Helft J, Chow A, Elpek KG, Gordonov S, Mazloom AR, Ma'ayan A, Chua W-J, Hansen TH, Turley SJ, Merad M, Randolph GJ, The I.G., Consortium. Gene expression profiles and transcriptional regulatory pathways underlying mouse tissue macrophage identity and diversity. Nat Immunol 2013;13:1118–28. https://doi.org/10.1038/ni.2419.Gene.
[54] Beyer M, Mallmann MR, Xue J, Staratschek-Jox A, Vorholt D, Krebs W, et al. High-resolution transcriptome of human macrophages. PLoS One 2012;7. https://doi.org/10.1371/journal.pone.0045466.
[55] Cochain C, Vafadarnejad E, Arampatzi P, Pelisek J, Winkels H, Ley K, et al. Single-cell RNA-seq reveals the transcriptional landscape and heterogeneity of aortic macrophages in murine atherosclerosis. Circ Res 2018;122:1661–74. https://doi.org/10.1161/CIRCRESAHA.117.312509.
[56] MacParland SA, Liu JC, Ma XZ, Innes BT, Bartczak AM, Gage BK, et al. Single cell RNA sequencing of human liver reveals distinct intrahepatic macrophage populations. Nat Commun 2018;9:1–21. https://doi.org/10.1038/s41467-018-06318-7.

[57] Van den Bossche J, O'Neill LA, Menon D. Macrophage immunometabolism: where are we (going)? Trends Immunol 2017;38:395–406. https://doi.org/10.1016/j.it.2017.03.001.
[58] Suzuki H, Hisamatsu T, Chiba S, Mori K, Kitazume MT, Shimamura K, et al. Glycolytic pathway affects differentiation of human monocytes to regulatory macrophages. Immunol Lett 2016;176:18–27. https://doi.org/10.1016/j.imlet.2016.05.009.

CHAPTER 2

Macrophage differentiation

Eleonora Timperi[a] and Rodrigo Nalio Ramos[b]
[a]INSERM U932, Institut Curie, PSL Research University, Paris, France
[b]INSERM Laboratory of Integrative Cancer Immunology, Sorbonne Université, Sorbonne Paris Cité, Université Paris Descartes, Université Paris Diderot, Centre de Recherche des Cordeliers, Paris, France

Background

After macrophage (MΦ) discovery by Ilya Metchnikoff in the last century, many researchers have invested big efforts to better understand how human MΦ are dynamically differentiated in steady state and in a variety of diseases. Considering that MΦ have been evolutionarily conserved for millions of years [1,2], a better comprehension of their phenotype, plasticity, and functional features in human still represents an important topic for basic and applied sciences. Due to the fast-revolutionary technologies we are using nowadays, a series of important publications have dramatically re-oriented our view about MΦ origin and phenotype. One main change came from studies that have uncovered part of MΦ found in tissues established during embryogenesis and persisting during the life span with minor contribution of blood monocyte replacement [3–9]. In parallel, a series of studies have also reported the role of monocyte-derived MΦ in distinct inflammatory conditions, putting these cells as central for the modulation of immune properties and tissue homeostasis regulation, as reviewed in Ref. [10]. Studies reported in the last decades have deeply characterized human MΦ by a didactic oversimplification into two major entities: M1 versus M2 MΦ, mostly based on their profile associated with inflammation (pro-inflammatory M1 MΦ and alternative or anti-inflammatory M2 MΦ) [11–14]. Importantly, this bifunctional model of MΦ differentiation is hardly reproduced in ex vivo or in vivo analysis of human tissues. Also, tissue-resident macrophages (TRMs) or monocyte-derived MΦ, especially in chronic diseases, were hardly dissected before the advent of new technologies focused on the single-cell level definition.

Importantly, a series of reports have demonstrated the contribution of human circulating monocytes in MΦ differentiation and niche replacement in vivo in distinct organs, including skin, lung, heart, and intestine [15–18]. Bujko and collaborators showed that human MΦ subsets could

Macrophages in the Human Body
https://doi.org/10.1016/B978-0-12-821385-8.00005-6

be replaced within 3 weeks after duodenum transplantation by analyzing mismatch HLA class I chimerism [16]. The authors further confirmed that complete MΦ replacement in recipients may occur after 1 year of transplantation, where part of MΦ subsets were derived from either the own recipient circulating monocytes or transplanted tissues. Another example was reported by Bajpai and colleagues when analyzing endomyocardial specimens from men that received female heart transplantation (sex-mismatched heart transplant) [17]. Using in situ hybridization for Y chromosomes along with immunostaining on slides, the authors revealed that a great majority of $CD68^+CCR2^-$ TRM came from female donors, while about 30% of $CD68^+CCR2^+$ MΦ corresponded to recipient-derived MΦ (Y chromosome positive). Also, they established that $CCR2^-$ TRM subset was maintained through local proliferation, while $CCR2^+$ MΦ rely on both monocyte migration and local cell expansion.

Another newly established concept, the innate immunity training, has been currently explored in distinct context, and it is claimed that pathogen-derived factors, cytokines, and soluble elements, among others, could induce epigenetic and transcriptional modulation in innate cells [19]. It leads to a functional reprograming/imprinting in myeloid immune cells, probably of progenitors at the bone marrow as recently showed [20], prompting the innate immune system of the host to be resistant to infections.

Due to technical and ethical limitations, the study of human MΦ using lineage tracing or transgenic methods is not feasible. One alternative has been the use of bioinformatics and mathematics models to calculate interspecies conserved myeloid cell homologues between mice and humans. Interestingly, studies could efficiently find core genes that are evolutionarily conserved in distinct immune cells, allowing the establishment of translational studies [21,22]. Thus, the landscape of human MΦ spectrum has been drastically revised in the last years, mostly uncovered by new technological approaches.

Macrophage origin and plasticity

A series of studies have described two distinct ways of macrophage differentiation mainly based on their origin and progeny. It is well accepted nowadays that MΦ could be derived from both the embryonic phase—giving rise to a great part of TRM—and circulating blood monocytes—becoming monocyte-derived macrophages (Mo-Mac). Many studies concerning both TRM and Mo-Mac have been performed in murine models, due to the lack of key tools to their study in humans. Also, the lack of mutually exclusive markers able to functionally and phenotypically distinguish both subtypes in humans may represent an important limitation.

Mo-macrophage differentiation in vitro

Over the last years, the advent of new cutting-edge technologies has permitted to deeply explore the origin and diversity of human MΦ, changing the entire paradigm of bimodal M1 and M2 differentiation. Many observations indeed highlighted a profound heterogeneity of macrophages, mainly derived from healthy and pathological in vivo conditions [23–27], showing a larger transcriptional repertoire. In accordance, to understand the complexity of this scenario, some years ago, illustrious scientists have furnished a consensus nomenclature and guidelines to clarify the standard definitions about activation and differentiation of MΦ [28].

In vitro approaches of differentiation are still considered today a supportive method established according to the context of human MΦ findings (i.e., healthy, infections, cancers). The majority of human MΦ differentiation studies derive from in vitro assays, in which growth factors (i.e., macrophage-colony-stimulating factor [M-CSF], granulocyte-macrophage-colony-stimulating factor [GM-CSF]), cytokines (i.e., interferon gamma [IFN-γ], tumor necrosis factor-alfa [TNF-α], interleukin-4 [IL-4], IL-13), and microbial factors (i.e., Toll-like receptor [TLR], TLR agonist, lipopolysaccharide [LPS]) are used to generate Mo-Mac. Similarly, ex vivo conditioned media, obtained from several conditions (i.e., autoimmune disorders, solid tumors, tumor cell lines), have been largely used to recapitulate the complexity of ex vivo observations.

Although an oversimplification of human ex vivo interpretations, M1 versus M2 models have been widely explored in mice studies and similarly reproduced in humans [27]. The model is based on the similarities with T-helper 1 and T-helper 2 responses, associated with the differentiation of $CD4^+$ T cells in certain conditions [29]. M1-macrophages (M1-MΦ) obtained in vitro by culturing blood monocytes with IFN-γ (±TNF-α) were normally associated with interferon signaling and pro-inflammatory responses [30,31]. IFN-γ controls specific gene expression programs (interleukin-15 receptor alpha [IL15RA], interleukin 2 receptor alpha [IL2RA], interleukin 6 receptor [IL6R]), cell activation markers (cluster of differentiation 36 [CD36], CD38, CD97), and cell adhesion molecules (intercellular adhesion molecule 1 [ICAM1]) and acts as signal transducer and activator of transcription 1 (STAT1), IFN regulatory factor-1 (IRF-1) and suppressor of cytokine signaling-1 (SOCS1) in monocytes [32]. M1-MΦ may produce cytokines able to stimulate T-cell responses, resulting in an efficient phagocytosis of infected or tumor cells.

Due to their microbial killing activities, M1-MΦ show a nitric oxide (NO) metabolic phenotype [33]. M1-MΦ in vitro assays have been largely explored to study viral or bacterial infections [34,35]. Among different TLRs and TLR ligands (i.e., LPS, muramyl dipeptide, and lipoteichoic acid), LPS stimulates M1-like phenotype; even though LPS and IFN-γ show different gene expression profiles [27,36], the former leads to strong pro-inflammatory cytokine (interferon-β [IFN-β], interleukin-12 [IL-12], TNF-α, IL-6, and IL-1β) and chemokine (i.e., chemokine [C-C motif] ligand 2 [CCL2], chemokine [C-X-C motif] ligand 10 [CXCL10], and CXCL11) production and expression of antigen presentation molecules, such as major histocompatibility complex (MHC) members and co-stimulatory molecules [37].

In contrast, generation of M2-macrophages in vitro (M2-MΦ), furtherly subdivided into three distinct subsets (M2a-MΦ, M2b-MΦ, and M2c-MΦ) [38], is mainly induced by interleukin-4 (IL-4) (± interleukin-13 [IL-13]). Janus kinase inhibitor-1 (JAK1) and JAK3 activation, mediated by IL-4, induces STAT6 transcriptional program and SOCS3 module.

Arginase activity, mannose receptor (MRC1), prostaglandin-endoperoxide synthase PTGS1 (prostaglandin G/H synthase 1), and transcription factor IRF4 [14,39,40] are also included in the transcriptomic programs of IL-4 that leads to the decrease of phagocytosis and promotion of wound healing profiling, finally reducing the ability of antigen presentation for T cells. M2-MΦ are characterized by trophic polyamine/arginine metabolism [33], mainly observed in mouse studies [27], resulting as a consequence of an alternative phenotype. Although IL-13 and IL-4 effects on monocytes are not perfectly conserved [41]—in terms of transcriptional signatures—IL-13 is additionally used as a compound for M2-MΦ differentiation.

Among many molecules identified to modulate monocyte differentiation, certainly M-CSF and GM-CSF are the most studied. Both M-CSF and GM-CSF regulate survival, differentiation, and mobilization of myeloid cells. While the former is ubiquitously and constitutively produced at homeostatic conditions, the latter appears as part of host responses to injury or infections [42]. However, elevated levels of M-CSF characterize autoimmune disorders, bowel diseases, and cancer [43–46]. The opposing effects of M-CSF and GM-CSF on monocytes have been often associated with M2-like and M1-like phenotypes [11], respectively; however, this relationship should be applied with caution [28,47].

Human and mouse monocytes cultured with GM-CSF or M-CSF for 7 days show substantial differences in gene expression [11], also displaying distinguished metabolic profiles [48].

M-CSF signals include activation of ERK, phosphatidylinositol 3-kinase, and phospholipase C; induction of cell cycle genes (i.e., cyclins A2, B1, D1, and E1); and downregulation of human leukocyte antigen (HLA) members and complement C1QA/B/C subunits [27]. Additionally, it induces high levels of anti-inflammatory cytokines, such as IL-10, CCL2/MCP-1, and also activin A production [11], resulting in a macrophage polarization towards an immunosuppressive phenotype. IRF5 is also induced upon M-CSF monocyte treatment.

On the contrary, GM-CSF downstream signals involved JAK2/STAT5 and ERK [49,50], stimulating human monocyte survival in vivo [51]. Even though TNF-α, IL-8, IL-1β, and IL-6, at mRNA expression levels, were observed to be rapidly increased in GM-CSF-treated monocytes/macrophages in vivo, a significant production of these cytokines commonly required LPS stimulus [47,52]. Importantly, GM-CSF increases the levels of CCL17, the highly upregulated gene in GM-CSF-human monocyte culture [53]. GM-CSF transcriptome analysis additionally showed regulation of several surface molecules (i.e., CD14, FCγR1A, CD163) [54]. Summarizing, GM-CSF is able to induce microbicidal capacity, leukocyte chemotaxis, cell adhesion, complement- and antibody-mediated phagocytosis, and antigen presentation. Although GM-CSF, usually together with IL-4, is widely used in in vitro assays to generate mouse and human dendritic cell (DC) subpopulations, respectively, from bone marrow precursors and blood circulating monocytes [55], it has been demonstrated that the differentiation of monocyte-derived DCs from human circulating monocytes requires IRF4 and ARH activation [56]. In fact, according to many observations, the transcription factor IRF4 has been described as the key factor in regulating DC properties in GM-CSF-treated monocytes [57–59].

Although, as largely explained above, the bimodal MΦ differentiation models could simplify the human in vivo complexity, it has been observed that the addition of several compounds to the classical M1-M2 models revealed a wider spectrum of MΦ differentiation [22,28]; for example, when adding stimuli such as free fatty acid, high-density lipoprotein (HDL), oleic acid, or combinations of stimuli associated with chronic inflammation (i.e., PGE2, TNF-α, IFN-β, IFN-α), a varied spectrum of MΦ activation signatures compared to the initial bipolar axis has been detected [22].

Similar complex scenario is observed in MΦ obtained by culturing monocytes with ex vivo conditioned media, generated in peculiar pathological conditions [60–64]. For example, our recent work and other studies have developed the generation of monocyte-derived tumor-associated MΦ from tumor-conditioned media [60,64,65], recapitulating the human ex vivo observations.

Tissue imprinting and niche replacement

TRMs are present in all tissues of the body and play a key role in tissue defense and homeostasis [66,67]. These cells can derive from different progenitors: from yolk sac MΦ, fetal liver, and circulating monocytes [68,69]. Most of the findings established in the field come from mice studies, due to the possibility of conducting complex experimental models, not possible in human settings.

Despite many similarities, MΦ identity is unique for each tissue [70], resulting in specific MΦ programs, established during early development and largely adapted in a tissue-dependent manner [71]. Indeed, many signals, including cytokines, metabolites, and cell-cell contacts, drive the expression and activation of lineage-determining transcription factors (TFs) peculiar for each tissue. Both signal-dependent and lineage-determining TFs are critical to refine MΦ programs and imprint a transcriptional program encountering the tissue-specific necessities. Both ontogeny and microenvironmental signals can shape the cell identity of MΦ by epigenetic modifications or by regulating promoters and/or enhancers conferring cell-type specificity. In myeloid immune cells, PU.1 is a key transcriptional factor, remodeling the chromatin in a cell-specific manner, determining myeloid progenitor fate in a concentration-dependent manner [68]. One of the main targets of PU.1 is *Csfr1r*, which encodes for IL-34 receptor and M-CSF; PU.1, together with Csf1R, is essential for the formation of yolk sac MΦ. In addition, many other TFs play their functions interacting with PU.1, resulting in a regulation of TRM development [72]. After terminal differentiation in the tissue, *Mafb*, highly expressed in TRM, induces the block of cell cycle. Additionally, *Mafb*, together with c-*Maf*, desensitizes the MΦ from the proliferative effect of M-CSF by inhibiting genes such as *Klf4*, *Kfl2*, and *Myc* [73]. MAF and MAFB are indeed important for MΦ terminal differentiation. Therefore, PU.1, CEBP, MAFB, and MAF work together to shape common TRM function. Among many factors, M-CSF is certainly the key regulator of MΦ survival and proliferation in the tissues; other cytokines, like IL-34 and GM-CSF, are locally produced to maintain MΦ in specific tissues. For example, IL-34—the alternative ligand of CSFR1—is mainly expressed in the brain, produced by neurons in the cortex, hippocampus, and striatum; it binds with high-affinity CSFR1 and is essential for microglial cell maintenance. In the skin, the keratinocytes produce IL-34 and TGF-β. TGF-β acts by phosphorylating the transcription factor SMAD, downstream of TGFβR1, and by a SMAD-binding motif that is enriched in PU.1-bound enhancer sites in microglial cells [74]. The local production of TGF-β promotes retention of Langerhans cells in the tissue, by expression of the TF inhibitor of DNA-binding 2 (ID2) and runt-related TF 3 (RUNX3). Both human and mouse monocytes cultured in vitro with CSF1 (M-CSF) and TGF-β express microglia-specific genes [75], whereas they exhibit Langerhans cell-like characteristics when cultured with CSF2 (GM-CSF) and TGFβ [76].

Although TRMs are mainly generated in the early phases of embryogenesis, a good proportion of MΦ in the tissues can be derived from circulating monocytes. Classical monocytes are indeed equipped with special transcription programs, permitting them to migrate into the tissues under homeostatic conditions. Generally, classical monocytes are released by the bone marrow and remain in the circulation for approximately 1 day, where they can also traffic in the tissues to repopulate a small portion of TRM in distinct tissues, including dermis [77], heart [7,78], intestine [79–82], lung [6], pancreas [83], and testis [84]. Given the high self-renewal potential of the TRMs and the presence of blood-brain and epithelial barriers, epidermis [85,86], central nervous system [4,87,88], and alveolar space [3,6,8] are organs with exceptions for monocyte engraftment.

Mo-Mac acquire significant gene modifications compared to the circulating monocyte counterpart; they are able to adapt to the local tissue environment and acquire novel transcriptomic signatures showing similarity with TRMs of embryonic origin, even though epigenetic differences remain [71,89–91]. It has been demonstrated that Mo-Mac adopt their phenotype and functions of the TRMs they are replacing. The engraft at steady state and the acquisition of self-maintenance capacity of Mo-Mac in comparison with their embryonic counterparts are debated and are associated to the tissue. On the contrary, in inflammatory conditions, Mo-Mac show distinct gene signatures and often fail to self-maintain for prolonged periods. One hypothesis proposes that low-grade inflammation is required for the continuum monocyte recruitment. One example is the physical microtrauma induced by physiological muscle contraction in the heart, which promotes a special monocyte-derived niche in the organ [78].

Innate immune memory: A new concept of trained macrophage immunity

The concept of immunological memory entirely associated with adaptive immunity has been recently challenged [92]. The organisms lacking adaptive immunity (i.e., plants, invertebrates) may indeed mount immune responses generating resistance to second infections through metabolic and epigenetic rewiring of innate immune cells; this phenomenon is called trained immunity or innate immune memory. Trained immunity involves a set of innate immune cells—myeloid cells, natural killer (NK) cells, and innate lymphoid cells (ILC)—and relies on a trained functional state of these. Many observations suggest that trained immunity can persist for weeks to months [93,94], rather than years as in the adaptive immunity. In accordance, in vitro studies have demonstrated long-term memory effects on monocytes/macrophages [94,95]. Therefore, the changes observed in innate immune cells involve mainly TFs and epigenetic programs, not implicating however permanent genetic modifications (i.e., mutations, recombination) such as during rearrangement of B cells.

Importantly, the discrimination between trained immunity and immune cell activation and/or differentiation processes is subtle; for example, MΦ differentiation could be considered as a trained immunity event; however, MΦ differentiation occurs also during homeostatic conditions, while trained immunity is described as a reaction to a foreign insult. Hence, after MΦ differentiation processes, these cells can be "educated," in particular conditions (i.e., infection, vaccination) to be trained, acquiring capacities and functions differing from the physiological context.

In the past years, considerable studies have proved trained immunity in human and mice monocytes/macrophages. It has been demonstrated that monocytes/macrophages are able to build a long-lasting memory proinflammatory response upon a subsequent challenge, resulting in a protective effect against reinfection in an independent adaptive immune system manner.

Many discoveries in the field highlighted a robust participation of epigenetic variations in the establishment of trained immunity [93,94,96]. A partially disrupted chromatin structure and increased basal level expression of target genes, maintained for long periods, result as the key aspects in the protective functions acquired by innate cells, enabling to generate a sort of "memory" innate immunity.

Exploring wide commensal microorganism effects on functional reprogramming of monocytes, different pattern recognition receptors (PRRs) with various bacterial or viral ligands (i.e., β-glucan, glycosyl-phosphatidylinositols from *Toxoplasma gondii*, lipopolysaccharide, flagellin, poly (I:C), CpG DNA) have been analyzed in a recent study [96]. PRRs and bacterial/viral ligands have proficiently generated long-term effects on monocytes mainly mediated by epigenetic modifications, both involving histone methylation and acetylation.

It has also been demonstrated that the stress-response transcription factor ATF7 suppressed a group of genes, encoding factors involved in innate immunity in MΦ; the treatment with LPS, inducing phosphorylation of ATF7 via p38 kinase, led to the release of ATF7 from chromatin and a decrease in repressive histone H3K9me2, resulting in an "open" chromatin structure concurring to increased basal expression level of target genes [93].

Complementary discoveries have been described for β-glucan on human monocytes where it provokes epigenetic stable changes. Indeed, *Candida albicans* and β-glucan-induced functional reprogramming of monocytes leading to enhanced cytokine production in vivo and in vitro, resulting in a strong protection to secondary *Candida albicans* infection [94]. Similarly, β-glucan induced protective trained immunity in human monocytes infected with *Mycobacterium tuberculosis* in an IL-1-dependent manner [97]. As well, in several mice models, the key role of β-glucan in the secondary protection to *Staphylococcus aureus* infection has been demonstrated [98,99]. Additional examples of trained immunity in monocytes have been observed in *Saccharomyces cerevisiae* in vitro models. Enhancing TNF-α and IL-6 production, upon secondary stimulation with TLR ligands, *Saccharomyces cerevisiae* induced protective properties (i.e., phagosome acidification and internalization), lately facilitating microbial digestion [100].

Interestingly, trained immunity may result to be efficient in the vaccination processes. In agreement, several vaccines showed beneficial effects against second and diverse infections. Tuberculosis bacillus Calmette-Guérin (BCG) vaccination indeed provoked genome-wide epigenetic reprograming of monocytes and protected against experimental infection with an attenuated yellow fever virus vaccine strain [101]. Similarly, enhanced functions of circulating monocytes (persisting for at least 3months) after tuberculosis BCG vaccination, by the increased production of IFN-γ, TNF-α, and IL-1β in response to unrelated bacterial and fungal pathogens, have also been confirmed. These training effects were associated to NOD2 receptor and increased the methylation [102], suggesting that if on the one hand the vaccination process stimulates adaptive immunity, as well reported, on the other hand it may support trained immunity through a protection mediated by innate immune cells.

Together with epigenetic changes, cellular metabolic processes (i.e., glycolysis, cholesterol synthesis, fatty acid oxidation) have been largely reported during monocyte trained immunity. β-Glucan-trained monocytes displayed lactate production and glucose consumption reflecting a shift towards aerobic glycolytic metabolism dependent on the activation of mammalian target of rapamycin (mTOR) through the dectin-1/Akt/HIF1α pathway. The inhibition of Akt, mTOR, or HIF1α resulted in the blocking of trained immunity, since mice with myeloid cell-specific defect in HIF1α were unable to mount trained immunity against bacterial sepsis. This study suggested that Akt/mTOR/HIF1α-dependent induction of aerobic glycolysis represented the metabolic basis of trained immunity [103]. Similarly, it has been demonstrated that monocytes primed with β-glucan, BCG, and oxidized low-density lipoprotein (oxLDL) showed increased pro- and anti-inflammatory cytokine responses upon re-stimulation with non-related stimuli, showing a switch to glycolysis metabolism [104]. Importantly, ex vivo human observations highlighted that during the early stage of sepsis, ATP production is mainly provided by the glycolysis, while strong decrease of mitochondrial oxidative phosphorylation was observed [105]. As well, many in vitro studies have largely described M1-like MΦ as glycolytic cells with an important impairment of oxidative phosphorylation [106].

Collectively, these data suggest the beneficial effects of trained immunity as a host defense mechanism; however, a detrimental role of trained immunity has recently been proposed in the induction and/or maintenance of autoimmune and autoinflammatory diseases [107]. It has been reviewed that monocytes derived from patients with rheumatoid arthritis (RA), systemic lupus erythematosus (SLE), and chronic granulomatous disease (CGD) strongly shared metabolic, epigenetic, and inflammatory pathways of trained monocytes [107]. Indeed, circulating monocytes from RA and CGD patients have increased expression of pro-inflammatory cytokines, and ex vivo re-stimulation produced more IL-1β, IL-6, and TNF-α than in physiological conditions. Commonly, circulating monocytes derived from both RA and SLE showed activated mTOR pathway, and, interestingly, rapamycin inhibition improved SLE patient outcome, especially in CGD patients, and counterbalanced the pre-activation state [107].

In this context, trained immunity could operate as an essential trigger in the initiation of the diseases and in the maintenance or intensification of the symptoms; therefore, genetic or environmental factors could train monocytes/macrophages to be more predisposed to become activated, resulting in the amplification of symptoms. Although these interpretations underline the potential detrimental role for trained immunity, similarities between human ex vivo observations and the possibility to recapitulate by in vitro studies emphasize the importance of trained immunity as a novel tool for the study of different human diseases.

New macrophage spectral phenotype

The bifunctional model of human MΦ differentiation, M1 versus M2, has been drastically reoriented towards a large spectral phenotype according to tissue imprinting and microenvironmental elements. Many recent studies combining transcriptomic and proteomic analysis have detailed slight features of MΦ in distinct organs and tissues. Most of the knowledge emerged in the recent years has been due to new technological assessments using single-cell approaches in healthy and disease, such as multicolor flow cytometry, mass cytometry (CyTOF), single-cell RNA-sequencing (scRNA-seq), spatial single-cell RNA-sequencing, and multiplex immunofluorescence (Fig. 1).

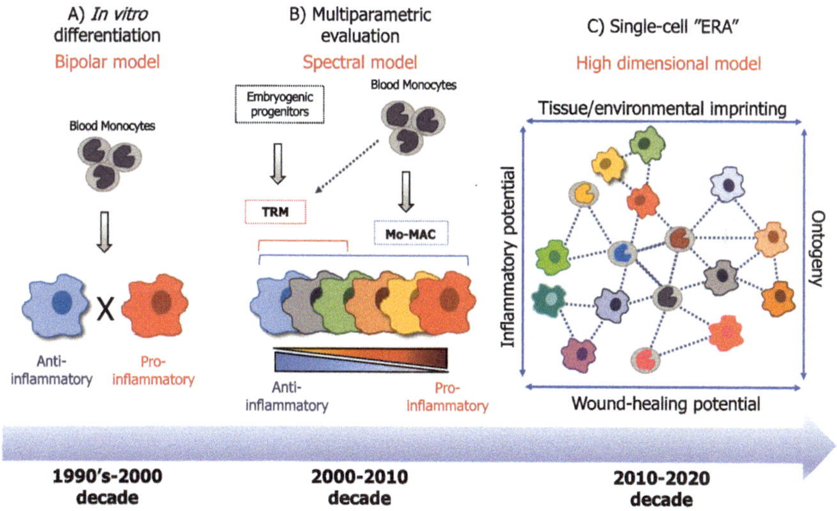

Fig. 1 Schematic overview of human macrophage differentiation during the last decades. (A) In vitro differentiation studies proposing a bipolar model of macrophage differentiation from circulating monocytes: M1 (pro-inflammatory) versus M2 (anti-inflammatory) cells. (B) With the advent of multiparametric technologies and elegant models, human macrophages were subsequently classified in a diverse spectrum of phenotype based on their ontogeny and phenotypical/functional characteristics considering anti-inflammatory to pro-inflammatory profiles. (C) Recent reports using single-cell and multidimensional features have opened an unlimited range of distinct functional, transcriptional, and epigenetic human macrophage profiles integrated to maps of trajectory. *Credit: figure provided by authors.*

Macrophage spectral phenotype in inflammation and diseases

Cancer

MΦ are one of the most abundant immune cells infiltrating human cancers [108]. These cells are normally called tumor-associated MΦ (TAMs) and have been characterized for years mostly based on the in vitro bimodal M1 vs. M2-like MΦ subsets. In fact, in vitro studies extensively showed that distinct stimuli could drive specific transcriptional and epigenetic programs in monocyte to MΦ differentiation, closely resembling to what is observed in ex vivo human tumors [22,109]. Importantly, in cancer patients, tumor systemic factors are also capable to modulate the transcriptional and functional profiles of monocytes at the bloodstream, pre-programing/sensitizing these cells to trigger distinct differentiation pathways, which impact MΦ or DC features in tissues [60,110–112].

Before the advent of the "single-cell era," markers such as CD163 and CD206 (*MRC1*), normally associated to anti-inflammatory M2 phenotype, were used in many retrospective studies for distinct tumor types to quantify TAMs and predict patients' outcome, as reviewed in Ref. [113]. In addition, few studies have uncovered specific markers associated to other TAMs' subtypes or pro-inflammatory M1-like MΦ; instead, phenotypes like $CD68^+CD86^+$ and $CD68^+HLADR^+$ have been used for this end [114,115]. Alternatively, expression of enzymes such as INOS and arginase-1 was also associated to M1- and M2-like MΦ phenotypes, respectively [116–118]. The development of unbiased multiparametric technologies in the recent years has modified the comprehension of TAMs from a bimodal profile towards a multifaced landscape in patients. Still, considering that TRMs are potentially present in all tissues of human body, single-cell technologies are critical tools for the proper discrimination of TAMs derived from monocytes or those originated from local tumor education of TRM.

In studies describing MΦ in brain tumors, distinct profiles of TAMs were revealed according to the tumor grade, mutational status, and tumor origin. Using combined scRNA-seq platforms in glioma cancer patients, Müller and collaborators [119] were pioneers in dissecting transcriptomic differences between blood-derived TAMs and resident microglia. The authors revealed a signature composed of 66 genes that discriminated brain MΦ clusters and described three putative subsets based on the protein expression: $CD11b^+P2RY12^-CX3CR1^-HLADR^+$ ($CX3CR1^-$ blood-derived TAM), $CD11b^+CX3CR1^+P2RY12^-HLADR^+$ ($CX3CR1^+$ blood-derived TAM), and $CD11b^+CX3CR1^+P2RY12^+HLADR^{low}$ (microglia).

Importantly, blood-derived TAM accumulation variated according to the glioma subtype and may co-express M1- and M2-like MΦ canonical markers, such as IL-10 and TNF-α, and CD206 and CD86, respectively. Using high-dimensional single-cell analysis via mass cytometry (CyTOF) combined with multicolor flow cytometry and immunofluorescence, another report by Friebel and collaborators [120] has uncovered MΦ heterogeneity in both primary tumor gliomas and brain-derived metastasis. Interestingly, the authors reported how monocyte-to-MΦ transition may happen across non-tumor brain samples. Tumor gliomas presenting *IDH1*-mutated ($IDH1^{mut}$) or wild-type ($IDH1^{wt}$) profiles, as well as melanoma and carcinoma metastasis derived from brain tumors, may display particular abundance/phenotype of TAMs. The authors demonstrated a continuous differentiation process from monocytes to TAM by the downregulation of CCR2 and CD33, along with CD163 and MERTK upregulation and emergence of three final branches of distinct monocyte-derived TAMs. One intermediate TAM phenotype expressing CD14, CD64, and HLADR seemed to give rise to the next three TAMs with mixed phenotype: two TAM subsets expressing CD163, CD206, and CD169 but characterized by $CD38^{+}CD209^{+}PD\text{-}L1^{high}$ and $CD38^{-}CD209^{-}PD\text{-}L1^{low}$ phenotypes, respectively, and a third TAM cluster characterized by higher CADM1 and CX3CR1 expression. Importantly, tumor type and tumor microenvironment (TME) site, rather than the tissue, may dictate the differentiation of distinct phagocytes in this context, since non-differentiated monocytes were accumulated in $IDH1^{mut}$ gliomas, $CADM1^{+}$ $CX3CR1^{+}$ TAMs were mostly found in $IDH1^{wt}$ gliomas, and TAMs expressing higher levels of CD38, PD-L1, and CD209 were equally infiltrated in melanoma and carcinoma brain metastasis. In another recent study, Leruste et al. [121] uncovered TAM clusters at a single-cell level in rhabdoid tumors, a pediatric cancer that presents one of the lowest mutational burdens among cancers—a biallelic inactivation of *SMARCB1*. In this report, authors found microglia cells and five transcriptionally distinct TAM subsets, including three clusters that express high levels of CD163, but also positive for matrix remodeling genes, IFN-type I responding genes, and genes linked to angiogenesis. Also, two other clusters presented a stressed/metabolic altered profile, by expressing various heat-shock proteins and ribosomal genes, respectively.

Using high-dimensional analysis by combining scRNA-seq with mass cytometry, Lavin and collaborators [122] analyzed the myeloid cell diversity in lung adenocarcinoma patients' cohort. The authors found four main MΦ subsets presenting distinct tissue localization and abundance between non-tumor and tumor lung lesions. One general MΦ cluster (expressing *HLA-DR*, *MAFB*, and *CSF1R*) was found equally distributed in tumor and non-tumor lung tissues. Besides, two lung-specific TRM subsets expressing high levels of CD86 and CD206 also depict similarities at transcriptomic and protein levels and were more abundant in non-tumor lung tissues. In contrast, one TAM subset was identified to be specifically accumulated in lung cancer lesions and was characterized by a higher expression of *APOE*, *TREM2*, *SPP1*, *MARCO*, and the transcriptional factor *PPAR-gamma*. Furthermore, in a more recent report evaluating non-small cell lung cancer (NSCLC) patients [123], the authors revealed nine transcriptionally distinct clusters of MΦ using scRNA-seq analysis, including subsets involved in immune cell recruitment defined as *CXCL5*$^+$ MΦ, *CXCL12*$^+$ MΦ, and CXCL9$^+$CXCL10$^+$ MΦ. Interestingly, by comparing these subsets to canonical M1- and M2-like states, none of the MΦ clusters exhibited exclusively a M1-like phenotype, while three subsets presented enrichment on M2-like gene signature. This profile is expected for TAMs, but it suggests that M2 phenotype is not a defined distinct state, rather a dynamic state of differentiation. In addition, by comparing MΦ from human and mice species, the authors described conserved transcriptomic signatures revealing enrichment of key genes, such as *APOE*, *C1QA*, *C1QB*, *C1QC*, *CTSB*, and *CTSD*. However, among the distinct clusters defined, MΦ showed both conserved and divergent subsets between mouse and human, representing much more a heterogeneous gradient of differentiation than a precisely defined state. For instance, alveolar MΦ closely resembled, while TAMs presented particular tissue and species signatures. In contrast, the three monocyte subsets from both human and mice depicted high similarities at a transcriptional level, reflecting much more the canonical distinctions between monocyte subsets than monocyte state of activation. These findings suggest that the differentiation signals that drive the transcriptional programing of monocyte towards TAM phenotype are well conserved between both human and mice species, but heterogeneity within TAM subsets is probably determined by species particularities.

Similar diversity of TAMs was also found in a cohort of clear-cell renal cell carcinoma evaluated by mass cytometry assay [124]. Using about third distinct markers, the authors described 17 subsets of TAMs in which most of the molecules depicted a dynamic continuum of expression, rather a well-defined profile. Some TAM subsets mutually exclusively expressed CD169, CD204, or CD206, while other subpopulations showed co-expression of these three markers concomitant with CD163. Subsequently, by applying diffusion mapping method, the authors were able to model the transition of differentiation from monocyte to MΦ across clusters. Data demonstrated two distinct differentiation paths for monocytes, respectively, to classical $CD14^+$ blood monocyte and patrolling inflammatory $CD16^+$ monocytes. Moreover, two final differentiated MΦ groups were found in one branch, corresponding to (I) TRM expressing $CD206^{high}$ (found in both non-tumor and tumor tissues) and (II) TAMs (present mostly in tumors) expressing high levels of HLA-DR, CD68, and CD64, but a diverse spectra of CD38, CD163, CD169, and CD204 molecules.

Another study performed in breast cancer patients explored the multifaceted spectrum of monocytes and MΦ using scRNA-seq in blood, lymph node, non-tumor breast tissue, and primary breast cancer [125]. By employing diffusion map analysis, the study revealed four major branches of monocyte/MΦ transcriptional states distributed in 12 monocyte clusters and three TAM subsets. At one side of the map, monocytes were actually stratified in blood monocytes (two clusters), monocyte precursor in tumor (four clusters), and activated monocytes in tumors (six clusters), by presenting increased expression of genes associated to activation, angiogenesis, chemokines, and IFN responses. At the other extremity of the branches, TAMs were characterized by a core gene signature including *CD68*, *APOE*, *TREM2*, and *CHIT1*, but gradual increased expression of genes associated to both M2-like MΦ (*MARCO*, *NRP2*, and *B7-H3*) and M1-like MΦ (*CCL3*) was observed in the three TAM subsets. These findings may reflect the differentiation and activation of either recruited monocyte-derived TAMs or TRMs.

In colorectal cancer (CRC) patients, the myeloid compartment seems to present a more marked dichotomy of final differentiated TAM subsets as observed for other cancer types. Evaluating blood, non-tumor colorectal tissue, and colorectal cancer samples using two distinct platforms of scRNA-seq, Zhang and colleagues [126] have described four transcriptionally distinct MΦ and two mature TAMs. Indeed, $NLRP3^+$ MΦ were found in both normal and tumor tissues, while MΦ subsets expressing *PLTP* and *IL1B* were much more enriched in normal tissues, indicating that these three subsets could belong to TRM subgroup. Another MΦ subset enriched at tumor sites presented a transitional monocyte-to-MΦ profiling, since expressed group of genes belong to both monocyte lineage (*CD52*, *S100A6*, and *FCN1*) and MΦ lineage (*JUN*, *CD68*, and *MAFB*). When exploiting inferred trajectory tool, authors revealed two well-defined TAM clusters that are enriched in CRC lesions and present distinct functions: (I) $C1QC^{high}$

TAMs, exhibiting upregulated gene pathways related to complement activation (*C1QB*), TCR signaling, and phagocytosis (*CD163*, *MRC1*), and (II) *SPP1*^{high} TAMs, depicting upregulated gene pathways associated to TGF-beta signaling and angiogenesis (*VEGFA*, *VCAN*, *CXCL8*). In addition, by comparing transcriptional profiles of myeloid subsets between two merged murine tumor models (Renca and MC38) and CRC patients, the authors found a great functional analogy between TAMs from both humans and mice.

Altogether, the majority of these studies do not correlate significant increasing of TAM subset abundance during cancer progression, suggesting that the events governing monocyte-to-MΦ education and differentiation could happen during the initial carcinogenic process probably both in mice and humans. One main hypothesis could be the effect that tumor-derived factors may promote systemically in MΦ progenitors both at the bone marrow and at circulation, by provoking transcriptional modifications in monocytes' programing. This also implies that strategies aiming to deplete/reprogram monocytes and MΦ should be implemented at very early phases of the cancer initiation.

Autoimmune diseases

Differently from cancer diseases, where M2-like MΦ were currently associated to an unfavorable prognosis, for autoimmune diseases, it is the increasing of the pro-inflammatory M1-like MΦ phenotype that has been associated to illness progression. For distinct types of autoimmune diseases, including SLE [127] and inflammatory bowel diseases (IBD) [128], the M1/M2-like MΦ imbalance has been associated to worse disease prognosis. Chronic inflammation, mostly driven by an accumulation of M1-like MΦ, high producers of reactive oxygen species, and pro-inflammatory cytokines, could also promote a consequent Th1 × Th2 imbalance [33]. In addition, the reduced presence of wound-healing/anti-inflammatory MΦ may contribute to a chronic inflammatory milieu, driving disease maintenance and a worse prognosis. Of importance, similarly to TAMs, the phenotypical and functional features of MΦ in autoimmunity are still ambiguous, considering that monocyte/MΦ may co-express many of M1- and M2-like MΦ markers concomitantly [129,130]. Therefore, local microenvironment could promote the dynamic epigenetic/transcriptomic reprograming of monocytes and MΦ towards pro-inflammatory or wound-healing profiles [109].

In a study of patients with lupus nephritis (LNE), Arazi and colleagues [131] have described the transition of monocyte to MΦ using scRNA-seq approaches. Results revealed five transcriptionally distinct clusters of myeloid cells in kidney, with one conventional DC and four monocyte/MΦ cells. By deeply characterizing the monocyte/MΦ clusters, authors found inflammatory CD16$^+$ MΦ (expressing *CX3CR1*, *TNF*, *S100A8*, *S100A9*, *NFKB1*, and *TCF7L2*) and two highly phagocytic MΦ called phagocytic CD16$^+$ (expressing *CD36*, *SCARB2*, *CD163*, *NR1H3*, and *GPNMB*) and M2-like MΦ (characterized by *VSIG4*, *MSR1*, *MERTK*, *STAB1*, *SLC40A1*, *FOLR2*, and *CD209*), which were both enriched in patients' kidney when compared to healthy donors. In contrast, a cluster of TRMs were found enriched in normal kidneys and were characterized by the expression of *MMP2*, *ADAMTS10*, *HTRA1*, and *BHLHE41*. Importantly, TRMs from LNE patients presented an altered transcriptional programing in comparison to those found in healthy donors by showing upregulation of interferon stimulated genes (ISGs), of anti-inflammatory genes (*GRN*, *TMSB4X*, *CREB5*), as well as of inhibitors of TLR signaling (*GIT2*, *TNFAIP8L2*). This suggests that lupus disease not only induces an accumulation of inflammatory monocytic cells, but may also drive a re-education of TRMs in situ. In addition, by applying diffusion map method, the authors found similarities between CD16$^+$ blood monocytes and cluster of inflammatory CD16$^+$ MΦ and proposed that the progression towards MΦ differentiation could happen from (I) inflammatory blood monocyte, (II) phagocytic MΦ, and then (III) an alternatively M2-like MΦ. Interestingly, this trajectory is mostly driven by gradual downregulation of *NFKB1* expression, a key regulator of myeloid cell differentiation/activation, and a continuous increasing of *CD36* and *MERTK* genes.

Another report conducted by Smillie and collaborators [132] has uncovered many modules of immune cell subgroups involved in the inflamed and non-inflamed profiles found in patients with ulcerative colitis (UC). The authors generated a human scRNA-seq atlas from patients' colon mucosa, revealing seven distinct myeloid cell clusters, including MΦ (*APOE*, *TREM2*, *C1QB*, *C1QC*, *FOLR2*), cycling monocytes (*AIF1*, *HLA-DRA*, *TYROBP*, *MS4A6A*), inflammatory monocytes (*FCN1*, *AIF1*, *LYZ*, *S100A8*, *HLA-DRB1*), and DC subsets. While abundance of MΦ subset was significantly increased in UC patients compared to healthy tissues, colon of patients presenting inflamed UC was strongly infiltrated by inflammatory monocytes. Using cell-cell interaction network analysis, the authors also revealed that inflammatory monocytes were the most central node, suggesting their role in the interaction with other subsets. Importantly, the modulation of the tissue microenvironment strongly contributes to changes in the dynamics of monocyte/MΦ differentiation and function. Furthermore, in a cohort of UC patients submitted to anti-TNF blockage, resistant gene signature to treatment was strongly enriched in inflammatory monocytes, a subset that significantly increased in non-responding patients. These cells also expressed high levels of *Oncostatin M* (OSM), a molecule that phenocopies TNF in vivo, indicating that inflammatory monocytes participate in anti-TNF resistance by a compensatory mechanism. Interestingly, the molecular profiling of monocytes was also dynamically altered under TNF blockage, since the upregulation of *PTGS2*, *GOS2*, *MNDA*, and *RILPL2* genes by inflammatory monocytes was strongly associated to anti-TNF resistance.

In another study focused on Crohn's disease (CD), a type of disease that is part of inflammatory bowel subtypes, Martin and colleagues [133] described an important immune cellular module in ileal CD lesions using single-cell approaches. High-dimensional analysis revealed MΦ cluster (defined by core genes *CD14*, *CD68*, *FCGR3A*, *MERTK*, *CTSC*, *CTSD*, and *GLUL*) as one of the most represented in CD lesions. In fact, by further exploring MΦ subsets, the authors stated two transcriptionally heterogenous clusters: gut-resident MΦ, displaying upregulation of *MRC1*, *MAFB*, *CSF1R*, *DNASE1L3*, and *C1Q* genes, and inflammatory MΦ, defined by increased levels of *NFKBIA* and various cytokines/chemokines such as *IL1B*, *TNF*, *CCL3*, *CCL4*, and *OSM*. In addition, gut-resident MΦ and inflammatory MΦ display regulatory and inflammatory programs, respectively, strongly suggesting that local microenvironmental signals are critical for the transcriptional and epigenetic modulation of both subsets. By combining mass cytometry and immunofluorescence assays, the authors revealed that inflammatory MΦ are much more abundant in inflamed CD lesions in comparison to uninflamed ones. Besides, by increasing the resolution of inflammatory monocyte cluster, deep mining analysis revealed two distinct transcriptional programing: one containing mostly MΦ genes *C1Qs*, *GLUL*, *APOE*, and *ACP5* and another including molecules associated to monocyte-like cells in IBD disease, such as *S100A8*, *S100A9*, *S100A12*, *VCAN*, *IL1B*, and *OSM*. Also, the study revealed that the presence of inflammatory MΦ in tissues was associated to reduced blood circulatory $CD14^{+}CD16^{neg}$ monocytes, suggesting a dynamic transition program of monocyte differentiation to inflammatory MΦ in patients. Importantly, by analyzing a cohort of CD patients under anti-TNF treatment, the authors found inflammatory monocytes as one of the major actors for the prediction of resistance in patients, as similarly found for UC patients. In this context, the authors proposed strategies to avoid the massive infiltration of inflammatory monocytes in patients, like anti-CCR2 blocking antibodies, preserving a more balanced ratio between inflammatory and regulatory gut-resident MΦ.

In multiple sclerosis (MS), a recent report also uncovered monocyte/MΦ landscape by analyzing cerebrospinal fluid and blood at single-cell transcriptomic resolution [134]. Among 17 identified clusters, two transcriptionally distinct subsets of monocytes were described: Mono1 *FCGR3A*$^{+}$(CD16) and Mono2 *CD14*$^{+}$ *FCGR3A*^{int}, with the latter almost exclusively found in the cerebrospinal fluid and rarely in blood. Although no differences in abundance of monocyte clusters were found in the cerebrospinal fluid between MS patients and healthy volunteers, Mono2 displays a unique transcriptomic profile, but indicating a much higher heterogeneity than Mono1. Mono2 from cerebrospinal fluid seems to be modulated by a series of distinct cues from the microenvironment, since these cells express a group of genes of great diversity: *CD9*, *CD163*, *EGR1*, and *BTG2*, from classical monocyte markers; *C1QA*, *C1QB*, *MAF*, and *CSF1R*, from typical MΦ markers; *LYVE1*, defined in perivascular MΦ; and *TREM2*, *TMEM119*, and *GPR34*, also expressed by microglia. In addition, both Mono1 and Mono2 clusters found in MS patients' cerebrospinal fluid showed differentially expressed genes when compared to healthy donors. Among upregulated genes, Mono1 showed *IL12RB1*, *CD74*, *ITGB1*, *STAT1*, *IRF3*, and *IRF8*, and Mono2 showed *HLAC*, *CD74*, *MUM1*, and *JUN*, while for downregulated genes Mono1 displayed *HLAC*, *IRF7*, and *STAT5A*, and Mono2 presented *IRF3* and *HIF1A*. This report highlighted how the development of MS created a unique microenvironment that shapes monocyte/MΦ' transcriptome, generating a greater heterogeneity than in other autoimmune diseases.

In RA patients, Zhang and collaborators [135] applied combined analyses of scRNA-seq, mass cytometry, flow cytometry, and bulk RNA-seq to dissect the immune cell compartment, revealing similarities in myeloid clusters. The monocyte/MΦ compartment was characterized by four transcriptionally distinct clusters: *IL1B*$^{+}$ cells (also expressing *ATF3*, *FOS*, *CD83*), *NUPR1*$^{+}$ cells (*MERTK*, *FOLR2*, *TGFBI*), *C1QA*$^{+}$ cells (*C1QB*, *RNASE1*, *MARCO*), and *SPP1*$^{+}$ cells (*CSTB*, *FN1*, *VCAN*). These clusters were further characterized according to their potential activation status. *IL1B*$^{+}$ cluster presents upregulated genes defining the LPS response in monocytes and MΦ, suggestive of pro-inflammatory monocytes, while *SPP1*$^{+}$ cells express ISGs, including IFITM3 and IFI6. *NUPR1*$^{+}$ and *C1QA*$^{+}$ subsets do not show activation states based on known public data, possibly indicating a homeostatic status or a TRM profile. To validate the molecular profiling of these clusters, mass cytometry was implied identifying five major clusters of $CD14^+$ cells. By integrating mass cytometry and bulk RNA-seq data, the authors found that clusters $IL1B^+$, $C1QA^+$, and $SPP1^+$ comprise $CD14^+CD11c^+CD38^+$ cells (most abundant in leukocyte-rich RA), while $NUPR1^+$ cluster corresponded to a $CD14^+CD11c^-CD38^{+/-}$ phenotype and were less represented in RA tissues. Even assuming that $NUPR1^+$ subset inversely correlated with tissue inflammation, these cells showed up-regulation of GPNMB (osteoactivin) and CTSK (cathepsin K), suggesting a potential role in bone remodeling. Together, data reported in this study indicated a crucial role of inflammatory cytokines and interferons in the modulation of MΦ diversity in RA. The authors support the use of TNF and JAK kinase inhibitors for disease treatment, since inflammatory monocyte/MΦ may represent one of the major contributors for RA progression.

In agreement, another elegant recent work from Alivernini and colleagues [136] has uncovered cellular diversity of MΦ subsets constituting an atlas in RA patients. By applying scRNA-seq, deep-phenotypic, spatial, and functional assays, the authors described two main populations of MΦ based on MERTK and CD206 expression, which were further distributed in a highly diverse spectrum of profiles. scRNA-seq analysis performed among $CD11b^+CD64^+$ cells revealed nine transcriptionally distinct clusters of cells classified into four main subgroups of tissue-synovial MΦ: $MERTK^+TREM2^+FOLR2^{neg}$, $MERTK^+TREM2^{neg}FOLR2^{high}$, $MERTK^{neg}HLA^{high}$, and $MERTK^{neg}CD48^+$ cells. The authors further quantified clusters among healthy donors, patients with active RA, and patients under remission of RA and reported that $MERTK^+$ clusters and $MERTK^{neg}$ clusters were much more increased in remission and active RA patients, respectively. Going deeply in their phenotype, MΦ cluster defined by $MERTK^{neg}CD48^+S100A12^+CD9^{neg}CLEC10A^{neg}$ profile was the most significantly increased in patients with active RA, in contrast to $MERTK^+TREM2^{neg}FOLR2^{high}LIVE1^+$ subset that was the most abundant in remission RA patients compared to all other groups of patients. Therefore, using functional assays, the study found that $MerTK^{neg}CD206^{neg}$ MΦ produce pro-inflammatory cytokines and may stimulate inflammatory responses in synovial fibroblasts. On the other hand, $MerTK^+CD206^+$ MΦ clusters secrete lipid mediators that are involved in inflammation modulation and tissue repair. Notably, among $MERTK^+$ subsets, $TREM2^+$ cluster was also described by displaying a broad regulatory transcriptomic profile, suggestive of their role in immune modulation. In addition, $MERTK^+$

clusters in remission RA patients are defined by an exclusive transcriptome programing, expressing *KLF2*, *KLF4*, *NR4A1*, *NR4A2*, and *ATF3* transcription factors *and DUSP1* (dual-specificity phosphatase 1), able to participate in myeloid cell re-education towards tissue homeostatic phenotype. Importantly, the balance of MΦ subsets in active versus remission RA patients strongly suggests that the dynamics of differentiation and specific clues from each microenvironment may dictate the final phenotype acquired, consequently impacting disease progression.

Concluding remarks

Although MΦ/TAMs showed high heterogeneity at both transcriptional and protein levels for distinct types of diseases, the traditional bipolar M1 × M2-MΦ paradigm was not precisely aligned to the description of human MΦ from tissues. Much effort was applied in the last decades for the human MΦ in vitro reproduction, where mechanisms of monocyte to MΦ or DCs were molecularly uncovered and targeted by distinct experimental approaches. The emerging recent data using combined high-dimensional and multi-parametric technologies have constituted a new level of complexity and diversity in the human macrophage field. Additionally, newly suggested scenarios, like innate immune training and trajectory maps, bring fresh inputs for the comprehensive understanding of human MΦ ontogeny, differentiation, and function both in steady state and during diseases. Most interestingly, many recent independent studies, irrespective of the context, have been unbiased in revealing a core gene signature for human MΦ, also found in TAMs, including *APOE, C1QA, C1QB, C1QC, CTSB, CTSC, CTSD, SPP1*, and *TREM2*. Some of those were also among the most upregulated genes during Alzheimer's and neurodegenerative disease progression [137,138].

Furthermore, two common profiles of MΦ frequently found in single-cell studies are clusters with upregulated ISGs and cell cycling genes. Human studies performed in infection, cancer, or autoimmune disease context have been unbiased in finding these common MΦ subsets; however, their contribution to disease control, tissue damage/reorganization, or pathogen clearance needs to be further addressed by functional validation. Tissue spatial localization (i.e., tumor margin or stroma), as well as their potential for self-renewing, plasticity, and survival life span, could bring new light for their better characterization.

The use of trajectories/diffusion and cellular maps performed at a single-cell level have brought crucial new insights concerning monocyte-to-macrophage differentiation. Nevertheless, it is still unclear how typically inflammatory monocytes, found across distinct reports, are educated/modulated by environmental cues contributing to macrophage differentiation and function. Interestingly, patients presenting autoimmune diseases submitted to anti-TNF blockage (i.e., CD and UC) treatment display a distinct modulatory outline of the monocyte-to-MΦ differentiation, indicating changes in transcriptomic subset profiling. Thus, the investigation of both local and systemic effects in patients under clinical therapies may reveal new pathways for MΦ activation and differentiation.

Altogether, these findings strongly suggest significant similarities in the transcriptional human MΦ programming across distinct tissues, individuals, and diseases. The events responsible for the triggering and modulation of human MΦ phenotype and function should be further evaluated in transversally comprehensive studies, aiming to dissect in detail how the signals derived from human body/tumors/pathogens may modulate macrophage differentiation and activation in vivo.

References

[1] Tauber AI. Metchnikoff and the phagocytosis theory. Nat Rev Mol Cell Biol 2003;4(11):897–901.

[2] Cooper MD, Alder MN. The evolution of adaptive immune systems. Cell 2006;124(4):815–22.

[3] Yona S, Kim K-W, Wolf Y, Mildner A, Varol D, Breker M, et al. Fate mapping reveals origins and dynamics of monocytes and tissue macrophages under homeostasis. Immunity 2013;38(1):79–91.

[4] Ginhoux F, Greter M, Leboeuf M, Nandi S, See P, Gokhan S, et al. Fate mapping analysis reveals that adult microglia derive from primitive macrophages. Science 2010;330(6005):841–5.

[5] Schulz C, Gomez Perdiguero E, Chorro L, Szabo-Rogers H, Cagnard N, Kierdorf K, et al. A lineage of myeloid cells independent of Myb and hematopoietic stem cells. Science 2012;336(6077):86–90.

[6] Hashimoto D, Chow A, Noizat C, Teo P, Beasley MB, Leboeuf M, et al. Tissue-resident macrophages self-maintain locally throughout adult life with minimal contribution from circulating monocytes. Immunity 2013;38(4):792–804.

[7] Epelman S, Lavine KJ, Beaudin AE, Sojka DK, Carrero JA, Calderon B, et al. Embryonic and adult-derived resident cardiac macrophages are maintained through distinct mechanisms at steady state and during inflammation. Immunity 2014;40(1):91–104.

[8] Guilliams M, De Kleer I, Henri S, Post S, Vanhoutte L, De Prijck S, et al. Alveolar macrophages develop from fetal monocytes that differentiate into long-lived cells in the first week of life via GM-CSF. J Exp Med 2013;210(10):1977–92.

[9] Jakubzick C, Gautier EL, Gibbings SL, Sojka DK, Schlitzer A, Johnson TE, et al. Minimal differentiation of classical monocytes as they survey steady-state tissues and transport antigen to lymph nodes. Immunity 2013;39(3):599–610.

[10] Coillard A, Segura E. In vivo differentiation of human monocytes. Front Immunol 2019;13(10):1907.

[11] Lacey DC, Achuthan A, Fleetwood AJ, Dinh H, Roiniotis J, Scholz GM, et al. Defining GM-CSF- and macrophage-CSF-dependent macrophage responses by in vitro models. J Immunol 2012;188(11):5752–65.
[12] Jaguin M, Houlbert N, Fardel O, Lecureur V. Polarization profiles of human M-CSF-generated macrophages and comparison of M1-markers in classically activated macrophages from GM-CSF and M-CSF origin. Cell Immunol 2013;281(1):51–61.
[13] Doyle AG, Herbein G, Montaner LJ, Minty AJ, Caput D, Ferrara P, et al. Interleukin-13 alters the activation state of murine macrophages in vitro: comparison with interleukin-4 and interferon-γ. Eur J Immunol 1994;24(6):1441–5.
[14] Stein M, Keshav S, Harris N, Gordon S. Interleukin 4 potently enhances murine macrophage mannose receptor activity: a marker of alternative immunologic macrophage activation. J Exp Med 1992;176(1):287–92.
[15] Eguíluz-Gracia I, Schultz HHL, Sikkeland LIB, Danilova E, Holm AM, Pronk CJH, et al. Long-term persistence of human donor alveolar macrophages in lung transplant recipients. Thorax 2016;71(11):1006–11.
[16] Bujko A, Atlasy N, Landsverk OJB, Richter L, Yaqub S, Horneland R, et al. Transcriptional and functional profiling defines human small intestinal macrophage subsets. J Exp Med 2018;215(2):441–58.
[17] Bajpai G, Schneider C, Wong N, Bredemeyer A, Hulsmans M, Nahrendorf M, et al. The human heart contains distinct macrophage subsets with divergent origins and functions. Nat Med 2018;24(8):1234–45.
[18] McGovern N, Schlitzer A, Gunawan M, Jardine L, Shin A, Poyner E, et al. Human dermal CD14 + cells are a transient population of monocyte-derived macrophages. Immunity 2014;41(3):465–77.
[19] Mantovani A, Netea MG. Trained innate immunity, epigenetics, and covid-19. Phimister EG, editor, N Engl J Med 2020;383(11):1078–80.
[20] Priem B, van Leent MMT, Teunissen AJP, Sofias AM, Mourits VP, Willemsen L, et al. Trained immunity-promoting nanobiologic therapy suppresses tumor growth and potentiates checkpoint inhibition. Cell 2020;183(3):786–801.e19.
[21] The Immunological Genome Consortium, Gautier EL, Shay T, Miller J, Greter M, Jakubzick C, et al. Gene-expression profiles and transcriptional regulatory pathways that underlie the identity and diversity of mouse tissue macrophages. Nat Immunol 2012;13(11):1118–28.
[22] Xue J, Schmidt SV, Sander J, Draffehn A, Krebs W, Quester I, et al. Transcriptome-based network analysis reveals a Spectrum model of human macrophage activation. Immunity 2014;40(2):274–88.
[23] Mosser DM, Edwards JP. Exploring the full spectrum of macrophage activation. Nat Rev Immunol 2008;8(12):958–69.
[24] Boorsma CE, Draijer C, Melgert BN. Macrophage heterogeneity in respiratory diseases. Mediat Inflamm 2013;2013:769214.
[25] Chinetti-Gbaguidi G, Staels B. Macrophage polarization in metabolic disorders: functions and regulation. Curr Opin Lipidol 2011;22(5):365–72.
[26] Lawrence T, Natoli G. Transcriptional regulation of macrophage polarization: enabling diversity with identity. Nat Rev Immunol 2011;11(11):750–61.
[27] Martinez FO, Gordon S, Locati M, Mantovani A. Transcriptional profiling of the human monocyte-to-macrophage differentiation and polarization: new molecules and patterns of gene expression. J Immunol 2006;177(10):7303–11.
[28] Murray PJ, Allen JE, Biswas SK, Fisher EA, Gilroy DW, Goerdt S, et al. Macrophage activation and polarization: nomenclature and experimental guidelines. Immunity 2014;41(1):14–20.
[29] Loke P, Gallagher I, Nair MG, Zang X, Brombacher F, Mohrs M, et al. Alternative activation is an innate response to injury that requires CD4+ T cells to be sustained during chronic infection. J Immunol 2007;179(6):3926–36.

[30] Alexander WS, Hilton DJ. The role of suppressors of cytokine signaling (SOCS) proteins in regulation of the immune response. Annu Rev Immunol 2004;22:503–29.
[31] O'Shea JJ, Murray PJ. Cytokine signaling modules in inflammatory responses. Immunity 2008;28(4):477–87.
[32] Waddell SJ, Popper SJ, Rubins KH, Griffiths MJ, Brown PO, Levin M, et al. Dissecting interferon-induced transcriptional programs in human peripheral blood cells. PLoS One 2010;5(3):e9753.
[33] Mills CD, Kincaid K, Alt JM, Heilman MJ, Hill AM. M-1/M-2 macrophages and the Th1/Th2 paradigm. J Immunol 2000;164(12):6166–73.
[34] Pai RK, Convery M, Hamilton TA, Boom WH, Harding CV. Inhibition of IFN-gamma-induced class II transactivator expression by a 19-kDa lipoprotein from *Mycobacterium tuberculosis*: a potential mechanism for immune evasion. J Immunol 2003;171(1):175–84.
[35] Nandan D, Reiner NE. Attenuation of gamma interferon-induced tyrosine phosphorylation in mononuclear phagocytes infected with Leishmania donovani: selective inhibition of signaling through Janus kinases and Stat1. Infect Immun 1995;63(11):4495–500.
[36] Nau GJ, Richmond JFL, Schlesinger A, Jennings EG, Lander ES, Young RA. Human macrophage activation programs induced by bacterial pathogens. Proc Natl Acad Sci USA 2002;99(3):1503–8.
[37] Martinez FO, Gordon S. The M1 and M2 paradigm of macrophage activation: time for reassessment. F1000Prime Rep 2014;6:13.
[38] Edwards JP, Zhang X, Frauwirth KA, Mosser DM. Biochemical and functional characterization of three activated macrophage populations. J Leukoc Biol 2006;80(6):1298–307.
[39] Kreider T, Anthony RM, Urban JF, Gause WC. Alternatively activated macrophages in helminth infections. Curr Opin Immunol 2007;19(4):448–53.
[40] Martinez FO, Helming L, Milde R, Varin A, Melgert BN, Draijer C, et al. Genetic programs expressed in resting and IL-4 alternatively activated mouse and human macrophages: similarities and differences. Blood 2013;121(9):e57–69.
[41] Scotton CJ, Martinez FO, Smelt MJ, Sironi M, Locati M, Mantovani A, et al. Transcriptional profiling reveals complex regulation of the monocyte IL-1 beta system by IL-13. J Immunol 2005;174(2):834–45.
[42] Hamilton JA. Colony-stimulating factors in inflammation and autoimmunity. Nat Rev Immunol 2008;8(7):533–44.
[43] Chitu V, Stanley ER. Colony-stimulating factor-1 in immunity and inflammation. Curr Opin Immunol 2006;18(1):39–48.
[44] Menke J, Iwata Y, Rabacal WA, Basu R, Stanley ER, Kelley VR. Distinct roles of CSF-1 isoforms in lupus nephritis. J Am Soc Nephrol 2011;22(10):1821–33.
[45] Toh M-L, Bonnefoy J-Y, Accart N, Cochin S, Pohle S, Haegel H, et al. Bone- and cartilage-protective effects of a monoclonal antibody against colony-stimulating factor 1 receptor in experimental arthritis. Arthritis Rheumatol 2014;66(11):2989–3000.
[46] Hume DA, MacDonald KPA. Therapeutic applications of macrophage colony-stimulating factor-1 (CSF-1) and antagonists of CSF-1 receptor (CSF-1R) signaling. Blood 2012;119(8):1810–20.
[47] Hamilton JA. GM-CSF-dependent inflammatory pathways. Front Immunol 2019;10:2055.
[48] Izquierdo E, Cuevas VD, Fernández-Arroyo S, Riera-Borrull M, Orta-Zavalza E, Joven J, et al. Reshaping of human macrophage polarization through modulation of glucose catabolic pathways. J Immunol 2015;195(5):2442–51.
[49] Lehtonen A, Matikainen S, Miettinen M, Julkunen I. Granulocyte-macrophage colony-stimulating factor (GM-CSF)-induced STAT5 activation and target-gene

expression during human monocyte/macrophage differentiation. J Leukoc Biol 2002;71(3):511–9.
[50] Hansen G, Hercus TR, McClure BJ, Stomski FC, Dottore M, Powell J, et al. The structure of the GM-CSF receptor complex reveals a distinct mode of cytokine receptor activation. Cell 2008;134(3):496–507.
[51] Achuthan A, Aslam ASM, Nguyen Q, Lam P-Y, Fleetwood AJ, Frye AT, et al. Glucocorticoids promote apoptosis of proinflammatory monocytes by inhibiting ERK activity. Cell Death Dis 2018;9(3):267.
[52] Wakefield PE, James WD, Samlaska CP, Meltzer MS. Colony-stimulating factors. J Am Acad Dermatol 1990;23(5 Pt 1):903–12.
[53] Achuthan A, Cook AD, Lee M-C, Saleh R, Khiew H-W, Chang MWN, et al. Granulocyte macrophage colony-stimulating factor induces CCL17 production via IRF4 to mediate inflammation. J Clin Invest 2016;126(9):3453–66.
[54] Lehtonen A, Ahlfors H, Veckman V, Miettinen M, Lahesmaa R, Julkunen I. Gene expression profiling during differentiation of human monocytes to macrophages or dendritic cells. J Leukoc Biol 2007;82(3):710–20.
[55] Chow KV, Delconte RB, Huntington ND, Tarlinton DM, Sutherland RM, Zhan Y, et al. Innate allorecognition results in rapid accumulation of monocyte-derived dendritic cells. J Immunol 2016;197(5):2000–8.
[56] Goudot C, Coillard A, Villani A-C, Gueguen P, Cros A, Sarkizova S, et al. Aryl hydrocarbon receptor controls monocyte differentiation into dendritic cells versus macrophages. Immunity 2017;47(3):582–596.e6.
[57] Gao Y, Nish SA, Jiang R, Hou L, Licona-Limón P, Weinstein JS, et al. Control of T helper 2 responses by transcription factor IRF4-dependent dendritic cells. Immunity 2013;39(4):722–32.
[58] Lehtonen A, Veckman V, Nikula T, Lahesmaa R, Kinnunen L, Matikainen S, et al. Differential expression of IFN regulatory factor 4 gene in human monocyte-derived dendritic cells and macrophages. J Immunol 2005;175(10):6570–9.
[59] Yashiro T, Yamaguchi M, Watanuki Y, Kasakura K, Nishiyama C. The transcription factors PU.1 and IRF4 determine dendritic cell-specific expression of RALDH2. J Immunol 2018;201(12):3677–82.
[60] Ramos RN, Rodriguez C, Hubert M, Ardin M, Treilleux I, Ries CH, et al. CD163+ tumor-associated macrophage accumulation in breast cancer patients reflects both local differentiation signals and systemic skewing of monocytes. Clin Transl Immunol 2020;9(2):e1108.
[61] Benner B, Scarberry L, Suarez-Kelly LP, Duggan MC, Campbell AR, Smith E, et al. Generation of monocyte-derived tumor-associated macrophages using tumor-conditioned media provides a novel method to study tumor-associated macrophages in vitro. J Immunother Cancer 2019;7(1):140.
[62] Saha B, Bruneau JC, Kodys K, Szabo G. Alcohol-induced miR-27a regulates differentiation and M2 macrophage polarization of normal human monocytes. J Immunol 2015;194(7):3079–87.
[63] Estko M, Baumgartner S, Urech K, Kunz M, Regueiro U, Heusser P, et al. Tumour cell derived effects on monocyte/macrophage polarization and function and modulatory potential of *Viscum album* lipophilic extract in vitro. BMC Complement Altern Med 2015;24(15):130.
[64] Sawa-Wejksza K, Dudek A, Lemieszek M, Kaławaj K, Kandefer-Szerszeń M. Colon cancer-derived conditioned medium induces differentiation of THP-1 monocytes into a mixed population of M1/M2 cells. Tumour Biol 2018;40(9). 1010428318797880.
[65] Kuang D-M, Wu Y, Chen N, Cheng J, Zhuang S-M, Zheng L. Tumor-derived hyaluronan induces formation of immunosuppressive macrophages through transient early activation of monocytes. Blood 2007;110(2):587–95.

[66] Gordon S, Taylor PR. Monocyte and macrophage heterogeneity. Nat Rev Immunol 2005;5(12):953–64.
[67] Wynn TA, Chawla A, Pollard JW. Macrophage biology in development, homeostasis and disease. Nature 2013;496(7446):445–55.
[68] T'Jonck W, Guilliams M, Bonnardel J. Niche signals and transcription factors involved in tissue-resident macrophage development. Cell Immunol 2018;330:43–53.
[69] Ginhoux F, Guilliams M. Tissue-resident macrophage ontogeny and homeostasis. Immunity 2016;44(3):439–49.
[70] Murray PJ, Wynn TA. Protective and pathogenic functions of macrophage subsets. Nat Rev Immunol 2011;11(11):723–37.
[71] Lavin Y, Mortha A, Rahman A, Merad M. Regulation of macrophage development and function in peripheral tissues. Nat Rev Immunol 2015;15(12):731–44.
[72] Behre G, Whitmarsh AJ, Coghlan MP, Hoang T, Carpenter CL, Zhang DE, et al. C-Jun is a JNK-independent coactivator of the PU.1 transcription factor. J Biol Chem 1999;274(8):4939–46.
[73] Soucie EL, Weng Z, Geirsdóttir L, Molawi K, Maurizio J, Fenouil R, et al. Lineage-specific enhancers activate self-renewal genes in macrophages and embryonic stem cells. Science 2016;351(6274):aad5510.
[74] Abutbul S, Shapiro J, Szaingurten-Solodkin I, Levy N, Carmy Y, Baron R, et al. TGF-β signaling through SMAD2/3 induces the quiescent microglial phenotype within the CNS environment. Glia 2012;60(7):1160–71.
[75] Butovsky O, Jedrychowski MP, Moore CS, Cialic R, Lanser AJ, Gabriely G, et al. Identification of a unique TGF-β-dependent molecular and functional signature in microglia. Nat Neurosci 2014;17(1):131–43.
[76] Chopin M, Seillet C, Chevrier S, Wu L, Wang H, Morse HC, et al. Langerhans cells are generated by two distinct PU.1-dependent transcriptional networks. J Exp Med 2013;210(13):2967–80.
[77] Tamoutounour S, Guilliams M, Montanana Sanchis F, Liu H, Terhorst D, Malosse C, et al. Origins and functional specialization of macrophages and of conventional and monocyte-derived dendritic cells in mouse skin. Immunity 2013;39(5):925–38.
[78] Molawi K, Wolf Y, Kandalla PK, Favret J, Hagemeyer N, Frenzel K, et al. Progressive replacement of embryo-derived cardiac macrophages with age. J Exp Med 2014;211(11):2151–8.
[79] Zigmond E, Varol C, Farache J, Elmaliah E, Satpathy AT, Friedlander G, et al. Ly6C hi monocytes in the inflamed colon give rise to proinflammatory effector cells and migratory antigen-presenting cells. Immunity 2012;37(6):1076–90.
[80] De Calisto J, Villablanca EJ, Mora JR. FcγRI (CD64): an identity card for intestinal macrophages. Eur J Immunol 2012;42(12):3136–40.
[81] Ridaura VK, Faith JJ, Rey FE, Cheng J, Duncan AE, Kau AL, et al. Gut microbiota from twins discordant for obesity modulate metabolism in mice. Science 2013;341(6150):1241214.
[82] Bain CC, Bravo-Blas A, Scott CL, Perdiguero EG, Geissmann F, Henri S, et al. Constant replenishment from circulating monocytes maintains the macrophage pool in the intestine of adult mice. Nat Immunol 2014;15(10):929–37.
[83] Calderon B, Carrero JA, Ferris ST, Sojka DK, Moore L, Epelman S, et al. The pancreas anatomy conditions the origin and properties of resident macrophages. J Exp Med 2015;212(10):1497–512.
[84] Mossadegh-Keller N, Gentek R, Gimenez G, Bigot S, Mailfert S, Sieweke MH. Developmental origin and maintenance of distinct testicular macrophage populations. J Exp Med 2017;214(10):2829–41.
[85] Chorro L, Sarde A, Li M, Woollard KJ, Chambon P, Malissen B, et al. Langerhans cell (LC) proliferation mediates neonatal development, homeostasis, and

inflammation-associated expansion of the epidermal LC network. J Exp Med 2009;206(13):3089–100.

[86] Merad M, Manz MG, Karsunky H, Wagers A, Peters W, Charo I, et al. Langerhans cells renew in the skin throughout life under steady-state conditions. Nat Immunol 2002;3(12):1135–41.

[87] Kassmann CM, Lappe-Siefke C, Baes M, Brügger B, Mildner A, Werner HB, et al. Axonal loss and neuroinflammation caused by peroxisome-deficient oligodendrocytes. Nat Genet 2007;39(8):969–76.

[88] Ajami B, Bennett JL, Krieger C, Tetzlaff W, Rossi FMV. Local self-renewal can sustain CNS microglia maintenance and function throughout adult life. Nat Neurosci 2007;10(12):1538–43.

[89] van de Laar L, Saelens W, De Prijck S, Martens L, Scott CL, Van Isterdael G, et al. Yolk sac macrophages, fetal liver, and adult monocytes can colonize an empty niche and develop into functional tissue-resident macrophages. Immunity 2016;44(4):755–68.

[90] Cronk JC, Filiano AJ, Louveau A, Marin I, Marsh R, Ji E, et al. Peripherally derived macrophages can engraft the brain independent of irradiation and maintain an identity distinct from microglia. J Exp Med 2018;215(6):1627–47.

[91] Scott CL, T'Jonck W, Martens L, Todorov H, Sichien D, Soen B, et al. The transcription factor ZEB2 is required to maintain the tissue-specific identities of macrophages. Immunity 2018;49(2):312–325.e5.

[92] Netea MG, Joosten LAB, Latz E, Mills KHG, Natoli G, Stunnenberg HG, et al. Trained immunity: a program of innate immune memory in health and disease. Science 2016;352(6284):aaf1098.

[93] Yoshida K, Ishii S. Innate immune memory via ATF7-dependent epigenetic changes. Cell Cycle 2016;15(1):3–4.

[94] Quintin J, Saeed S, Martens JHA, Giamarellos-Bourboulis EJ, Ifrim DC, Logie C, et al. *Candida albicans* infection affords protection against reinfection via functional reprogramming of monocytes. Cell Host Microbe 2012;12(2):223–32.

[95] Ostuni R, Piccolo V, Barozzi I, Polletti S, Termanini A, Bonifacio S, et al. Latent enhancers activated by stimulation in differentiated cells. Cell 2013;152(1–2):157–71.

[96] Ifrim DC, Quintin J, Joosten LAB, Jacobs C, Jansen T, Jacobs L, et al. Trained immunity or tolerance: opposing functional programs induced in human monocytes after engagement of various pattern recognition receptors. Clin Vaccine Immunol 2014;21(4):534–45.

[97] Moorlag SJCFM, Khan N, Novakovic B, Kaufmann E, Jansen T, van Crevel R, et al. β-Glucan induces protective trained immunity against *Mycobacterium tuberculosis* infection: a key role for IL-1. Cell Rep 2020;31(7):107634.

[98] Di Luzio NR, Williams DL. Protective effect of glucan against systemic *Staphylococcus aureus* septicemia in normal and leukemic mice. Infect Immun 1978;20(3):804–10.

[99] Marakalala MJ, Williams DL, Hoving JC, Engstad R, Netea MG, Brown GD. Dectin-1 plays a redundant role in the immunomodulatory activities of β-glucan-rich ligands in vivo. Microbes Infect 2013;15(6–7):511–5.

[100] Rizzetto L, Ifrim DC, Moretti S, Tocci N, Cheng S-C, Quintin J, et al. Fungal chitin induces trained immunity in human monocytes during cross-talk of the host with *Saccharomyces cerevisiae*. J Biol Chem 2016;291(15):7961–72.

[101] Arts RJW, Moorlag SJCFM, Novakovic B, Li Y, Wang S-Y, Oosting M, et al. BCG vaccination protects against experimental viral infection in humans through the induction of cytokines associated with trained immunity. Cell Host Microbe 2018;23(1):89–100.e5.

[102] Kleinnijenhuis J, Quintin J, Preijers F, Joosten LAB, Ifrim DC, Saeed S, et al. Bacille Calmette-Guerin induces NOD2-dependent nonspecific protection from reinfection via epigenetic reprogramming of monocytes. Proc Natl Acad Sci USA 2012;109(43):17537–42.

[103] Cheng S-C, Quintin J, Cramer RA, Shepardson KM, Saeed S, Kumar V, et al. mTOR- and HIF-1α-mediated aerobic glycolysis as metabolic basis for trained immunity. Science 2014;345(6204):1250684.
[104] Bekkering S, Blok BA, Joosten LAB, Riksen NP, van Crevel R, Netea MG. In vitro experimental model of trained innate immunity in human primary monocytes. Clin Vaccine Immunol 2016;23(12):926–33.
[105] Liu TF, Vachharajani VT, Yoza BK, McCall CE. NAD+-dependent sirtuin 1 and 6 proteins coordinate a switch from glucose to fatty acid oxidation during the acute inflammatory response. J Biol Chem 2012;287(31):25758–69.
[106] Jha AK, Huang SC-C, Sergushichev A, Lampropoulou V, Ivanova Y, Loginicheva E, et al. Network integration of parallel metabolic and transcriptional data reveals metabolic modules that regulate macrophage polarization. Immunity 2015;42(3):419–30.
[107] Arts RJW, Joosten LAB, Netea MG. The potential role of trained immunity in autoimmune and autoinflammatory disorders. Front Immunol 2018;9:298.
[108] Gentles AJ, Newman AM, Liu CL, Bratman SV, Feng W, Kim D, et al. The prognostic landscape of genes and infiltrating immune cells across human cancers. Nat Med 2015;21(8):938–45.
[109] Piccolo V, Curina A, Genua M, Ghisletti S, Simonatto M, Sabò A, et al. Opposing macrophage polarization programs show extensive epigenomic and transcriptional cross-talk. Nat Immunol 2017;18(5):530–40.
[110] Ramos RN, Chin LS, Dos Santos AP, Bergami-Santos PC, Laginha F, Barbuto JA. Monocyte-derived dendritic cells from breast cancer patients are biased to induce CD4+CD25+Foxp3+ regulatory T cells. J Leukoc Biol 2012;92(3):673–82.
[111] Cassetta L, Fragkogianni S, Sims AH, Swierczak A, Forrester LM, Zhang H, et al. Human tumor-associated macrophage and monocyte transcriptional landscapes reveal cancer-specific reprogramming, biomarkers, and therapeutic targets. Cancer Cell 2019;35(4):588–602.e10.
[112] Chittezhath M, Dhillon MK, Lim JY, Laoui D, Shalova IN, Teo YL, et al. Molecular profiling reveals a tumor-promoting phenotype of monocytes and macrophages in human cancer progression. Immunity 2014;41(5):815–29.
[113] Caux C, Ramos RN, Prendergast GC, Bendriss-Vermare N, Menetrier-Caux C. A milestone review on how macrophages affect tumor growth. Cancer Res 2016;76(22):6439–42.
[114] Rakaee M, Busund L-TR, Jamaly S, Paulsen E-E, Richardsen E, Andersen S, et al. Prognostic value of macrophage phenotypes in resectable non-small cell lung cancer assessed by multiplex immunohistochemistry. Neoplasia 2019;21(3):282–93.
[115] Dong P, Ma L, Liu L, Zhao G, Zhang S, Dong L, et al. CD86+/CD206+, diametrically polarized tumor-associated macrophages, predict hepatocellular carcinoma patient prognosis. Int J Mol Sci 2016;17(3):320.
[116] Almatroodi SA, McDonald CF, Darby IA, Pouniotis DS. Characterization of M1/M2 tumour-associated macrophages (TAMs) and Th1/Th2 cytokine profiles in patients with NSCLC. Cancer Microenviron 2016;9(1):1–11.
[117] Dumars C, Ngyuen J-M, Gaultier A, Lanel R, Corradini N, Gouin F, et al. Dysregulation of macrophage polarization is associated with the metastatic process in osteosarcoma. Oncotarget 2016;7(48):78343–54.
[118] Sica A, Schioppa T, Mantovani A, Allavena P. Tumour-associated macrophages are a distinct M2 polarised population promoting tumour progression: potential targets of anti-cancer therapy. Eur J Cancer 2006;42(6):717–27.
[119] Müller S, Kohanbash G, Liu SJ, Alvarado B, Carrera D, Bhaduri A, et al. Single-cell profiling of human gliomas reveals macrophage ontogeny as a basis for regional differences in macrophage activation in the tumor microenvironment. Genome Biol 2017;18(1):234.

[120] Friebel E, Kapolou K, Unger S, Núñez NG, Utz S, Rushing EJ, et al. Single-cell mapping of human brain cancer reveals tumor-specific instruction of tissue-invading leukocytes. Cell 2020;181(7):1626–1642.e20.
[121] Leruste A, Tosello J, Ramos RN, Tauziède-Espariat A, Brohard S, Han Z-Y, et al. Clonally expanded T cells reveal immunogenicity of rhabdoid tumors. Cancer Cell 2019;36(6):597–612.e8.
[122] Lavin Y, Kobayashi S, Leader A, Amir E-AD, Elefant N, Bigenwald C, et al. Innate immune landscape in early lung adenocarcinoma by paired single-cell analyses. Cell 2017;169(4):750–765.e17.
[123] Zilionis R, Engblom C, Pfirschke C, Savova V, Zemmour D, Saatcioglu HD, et al. Single-cell transcriptomics of human and mouse lung cancers reveals conserved myeloid populations across individuals and species. Immunity 2019;50(5):1317–1334.e10.
[124] Chevrier S, Levine JH, Zanotelli VRT, Silina K, Schulz D, Bacac M, et al. An immune atlas of clear cell renal cell carcinoma. Cell 2017;169(4):736–749.e18.
[125] Azizi E, Carr AJ, Plitas G, Cornish AE, Konopacki C, Prabhakaran S, et al. Single-cell map of diverse immune phenotypes in the breast tumor microenvironment. Cell 2018. Jun [Internet] [cited 2018 Aug 20]; Available from: https://linkinghub.elsevier.com/retrieve/pii/S0092867418307232.
[126] Zhang L, Li Z, Skrzypczynska KM, Fang Q, Zhang W, O'Brien SA, et al. Single-cell analyses inform mechanisms of myeloid-targeted therapies in colon cancer. Cell 2020;181(2):442–459.e29.
[127] Triantafyllopoulou A, Franzke C-W, Seshan SV, Perino G, Kalliolias GD, Ramanujam M, et al. Proliferative lesions and metalloproteinase activity in murine lupus nephritis mediated by type I interferons and macrophages. Proc Natl Acad Sci 2010;107(7):3012–7.
[128] Steinbach EC, Plevy SE. The role of macrophages and dendritic cells in the initiation of inflammation in IBD. Inflamm Bowel Dis 2014;20(1):166–75.
[129] Soldano S, Trombetta AC, Contini P, Tomatis V, Ruaro B, Brizzolara R, et al. Increase in circulating cells coexpressing M1 and M2 macrophage surface markers in patients with systemic sclerosis. Ann Rheum Dis 2018;77(12):1842–5.
[130] Vogel DY, Vereyken EJ, Glim JE, Heijnen PD, Moeton M, van der Valk P, et al. Macrophages in inflammatory multiple sclerosis lesions have an intermediate activation status. J Neuroinflammation 2013;10(1):809.
[131] The Accelerating Medicines Partnership in SLE Network, Arazi A, Rao DA, Berthier CC, Davidson A, Liu Y, et al. The immune cell landscape in kidneys of patients with lupus nephritis. Nat Immunol 2019;20(7):902–14.
[132] Smillie CS, Biton M, Ordovas-Montanes J, Sullivan KM, Burgin G, Graham DB, et al. Intra- and inter-cellular rewiring of the human colon during ulcerative colitis. Cell 2019;178(3):714–730.e22.
[133] Martin JC, Chang C, Boschetti G, Ungaro R, Giri M, Grout JA, et al. Single-cell analysis of Crohn's disease lesions identifies a pathogenic cellular module associated with resistance to anti-TNF therapy. Cell 2019;178(6):1493–1508.e20.
[134] Schafflick D, Xu CA, Hartlehnert M, Cole M, Schulte-Mecklenbeck A, Lautwein T, et al. Integrated single cell analysis of blood and cerebrospinal fluid leukocytes in multiple sclerosis. Nat Commun 2020;11(1):247.
[135] Accelerating Medicines Partnership Rheumatoid Arthritis and Systemic Lupus Erythematosus (AMP RA/SLE) Consortium, Zhang F, Wei K, Slowikowski K, Fonseka CY, Rao DA, et al. Defining inflammatory cell states in rheumatoid arthritis joint synovial tissues by integrating single-cell transcriptomics and mass cytometry. Nat Immunol 2019;20(7):928–42.

[136] Alivernini S, MacDonald L, Elmesmari A, Finlay S, Tolusso B, Gigante MR, et al. Distinct synovial tissue macrophage subsets regulate inflammation and remission in rheumatoid arthritis. Nat Med 2020;26(8):1295–306.
[137] Krasemann S, Madore C, Cialic R, Baufeld C, Calcagno N, El Fatimy R, et al. The TREM2-APOE pathway drives the transcriptional phenotype of dysfunctional microglia in neurodegenerative diseases. Immunity 2017;47(3):566–581.e9.
[138] Mathys H, Davila-Velderrain J, Peng Z, Gao F, Mohammadi S, Young JZ, et al. Single-cell transcriptomic analysis of Alzheimer's disease. Nature 2019;570(7761):332–7.

CHAPTER 3

Metabolic requirement for macrophages

Araceli Aparecida Hastreiter, Lais Cavalieri Paredes, and Niels Olsen Saraiva Camara
Department of Immunology, Institute of Biomedical Sciences, University of São Paulo, São Paulo, Brazil

Introduction

The immune response involves dynamic changes in the function of immune cells. The stimuli and responses of the innate and adaptive immune system are already well known, especially the role of cytokines that regulate and direct the distinct processes. Over the past years, the characterization of the metabolism in immune cells has highlighted that not only cytokines play an important role, but also nutrients and oxygen levels trigger the polarization of the immune response [1].

The regulation of immune cells is directly related to metabolic adaptations: during the immune response, for example, the activation of macrophages by toll-like receptors (TLRs), a family of pathogen recognition receptors (PRRs), leads to a metabolic reshuffle that favors changes in the fate, differentiation, and function of these cells. Activated cells have a large biosynthetic demand, and the metabolic shift aims to maintain the balance between supply and demand for energy [1,2].

Immune cells are plastic and can adapt their metabolism according to their energetic requirements [3]. Metabolism consists of a series of simultaneous intracellular biochemical reactions that occur in the cytoplasm and mitochondria and aim to provide the adenosine triphosphate (ATP) needed for the cells to maintain bioenergetic homeostasis and cell survival and perform their functions [2,4]. The metabolic process is organized into catabolic and anabolic metabolic pathways, in which there is, respectively, the degradation and generation of complex macromolecules such as carbohydrates, lipids, proteins, and nucleic acids [4].

Metabolic pathways and their specific enzymes and proteins have been known for a long time. Glycolysis, fatty acid oxidation, and oxidative deamination and transamination of amino acids comprise the processes of

Macrophages in the Human Body
https://doi.org/10.1016/B978-0-12-821385-8.00010-X

catabolism and are ubiquitous throughout eukaryotic cells. Several molecules derived from catabolic processes are substrates for the tricarboxylic acid (TCA) cycle, and thereafter to the electron transport chain (ETC) for oxidative phosphorylation (OXPHOS), as described below [5]. However, as the metabolism is an integrated network, the regulation of the cellular metabolism is complex and less understood, mainly due to the interactions between cytosolic and mitochondrial metabolism, as well as between the pathways themselves [4,6].

Glycolysis

Glycolysis, also called the Embden-Meyerhof pathway, is one of the main metabolic pathways in all cell types for the maintenance of energy supply and synthesis of essential macromolecules [7]. This process comprises the conversion of glucose into pyruvate, which can either enter the TCA cycle or be fermented to lactate, which can occur in both oxygen deprivation (anaerobic glycolysis) and normoxic situations (aerobic glycolysis) [8].

Glucose enters macrophages through the GLUT glucose membrane transporter family [9]. Once inside the cell, glucose is irreversibly phosphorylated by hexokinase to generate glucose-6-phosphate (G6P), which can continue the process of glycolysis, or enter the pentose phosphate pathway (PPP), as described below [4]. The glycolytic pathway continues with two more reactions catalyzed by the enzymes phosphoglucose isomerase and phosphofructokinase, which convert G6P into fructose 1,6-bisphosphate (FBP). This step is irreversible and, so far, promotes the spending of two molecules of ATP [10,11]. In the following step, FBP undergoes lysis and conversion into two molecules of glyceraldehyde 3-phosphate (G3P), which go through a series of enzymatic steps that produce phosphoenolpyruvate (PEP), which is finally converted into pyruvate by pyruvate kinase [12], as described in Fig. 1.

The final yield of the glycolytic pathway is 2 ATP, 2 NADH, and 2 pyruvate molecules per molecule of glucose. Nicotinamide adenine dinucleotide (NADH) must recycle continuously to NAD^+ by OXPHOS, and in the absence of NAD^+, glycolysis is interrupted [11]. Depending on the oxygen availability, pyruvate undergoes complete oxidation in the mitochondria by TCA cycle, or in the absence of oxygen, pyruvate is converted to lactate, which is secreted from cells [13]. Highly proliferative and cancer cells lean to convert most of the pyruvate into lactate regardless of the availability of oxygen and have low rates of OXPHOS, a process called *Warburg effect*. This metabolic alteration, despite providing a lower supply of ATP, is beneficial

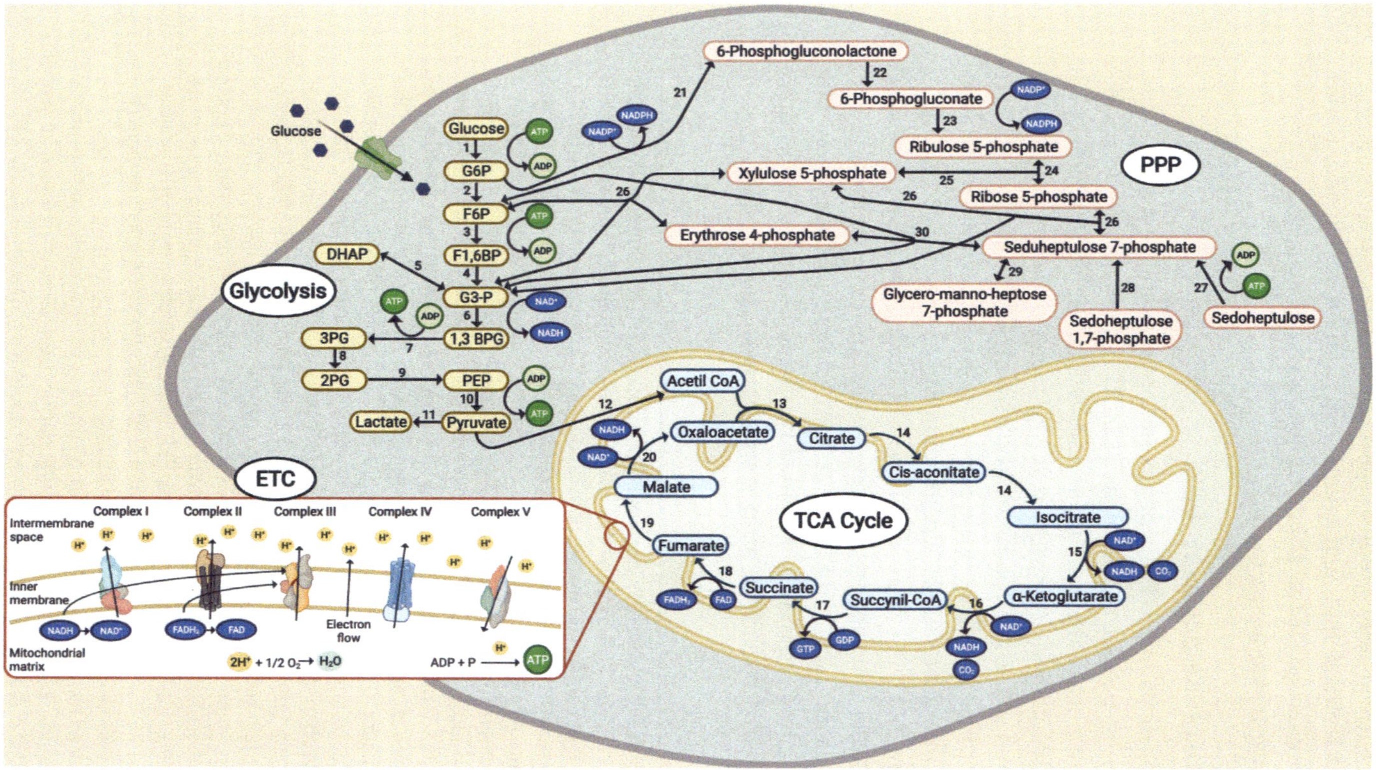

Fig. 1 Schematic figure of the main metabolic pathways of macrophages. *Glycolysis* is represented by a series of reactions catalyzed by the following enzymes, as numbered in the scheme: (1) hexokinase; (2) glucose phosphate isomerase; (3) phosphofructokinase; (4) fructose bisphosphate aldolase; (5) triosephosphate isomerase; (6) glyceraldehyde 3-phosphate dehydrogenase; (7) phosphoglycerate kinase; (8) phosphoglycerate mutase; (9) enolase; (10) pyruvate kinase; (11) lactate dehydrogenase; (12) pyruvate dehydrogenase complex. *Tricarboxylic acid (TCA) cycle* is represented by a series of reactions catalyzed by the following enzymes, as numbered in the scheme: (13) citrate synthase;

(Continued)

for these cells for two main reasons: it provides essential precursors for the synthesis of nucleotides, lipids, and amino acids required during mitosis, and intermediate molecules of the glycolytic pathway can be a carbon source for other cellular metabolic pathways [2].

Pentose phosphate pathway (PPP)

Pentose phosphate pathway (PPP), or hexose monophosphate shunt, occurs in the cytosol, in parallel to glycolysis [12]. PPP provides precursors for the synthesis of nucleotides and amino acids, which are important for cell proliferation. In addition, it provides reducing molecules, maintains carbon homeostasis, and acts in the control of oxidative stress [14].

PPP starts with an oxidative phase, in which G6P is converted to ribulose 5-phosphate, and nicotinamide adenine dinucleotide phosphate ($NADP^+$) is reduced into NADPH. This initial phase provides the maintenance of the redox balance, as it supplies reducing power for the generation of the antioxidant glutathione [12]. In addition, NADPH plays an important role in the inflammatory response, as it is used by several enzymes, for example NADPH oxidase, that generate reactive oxygen species (ROS) [15].

Fig. 1, Cont'd (14) aconitase; (15) isocitrate dehydrogenase; (16) alpha-ketoglutarate dehydrogenase complex; (17) succinyl coenzyme A (CoA) synthetase; (18) succinate dehydrogenase; (19) fumarase; (20) malate dehydrogenase. *Electron transport chain (ETC)* in mitochondria is represented by complexes I, II, III, IV, and V. The ETC utilizes NADH and $FADH_2$ to generate ATP. These reducing equivalents are produced during glycolysis and fatty acid breakdown and in TCA pathways. The transport starts from NADH binding to complex I or from succinate/FADH2 binding to complex II, initiating electron flow that ultimately results in ATP production in complex V. *Pentose phosphate pathway (PPP)* is represented by a series of reactions catalyzed by the following enzymes, as numbered in the scheme: (21) glucose 6-phosphate dehydrogenase; (22) 6-phosphogluconolactonase; (23) 6-phosphogluconate dehydrogenase; (24) ribose 5-phosphate isomerase; (25) ribulose 5-phosphate epimerase; (26) transketolase; (27) sedoheptulose conversion enzymes found in mammals; (28); sedoheptulose conversion enzymes found in fungi (*S. cerevisiae*) and plants; (29) sedoheptulose conversion enzymes found in bacteria; (30) transaldolase. Abbreviations: G6P, glucose 6-phosphate; F6P, fructose 6-phosphate; F1,6BP, fructose 1,6-bisphosphate; G3-P, glyceraldehyde 3-phosphate; DHAP, dihydroxyacetone phosphate; 1,3BPG, 1,3-bisphosphoglycerate; PEP, phosphoenolpyruvate; ATP, adenosine triphosphate; ADP, adenosine diphosphate; NAD^+, nicotinamide adenine dinucleotide; NADH, the reduced form of NAD^+; nicotinamide adenine dinucleotide phosphate; NADPH, the reduced form of $NADP^+$; CO_2, carbon dioxide; GDP, guanosine diphosphate; GTP, guanosine triphosphate; FAD, flavin adenine nucleotide; $FADH_2$, the reduced form of FAD. *(Figure created in BioRender.com. Credit: Araceli Aparecida Hastreiter and Lais Cavalieri Paredes.)*

Next, in the non-oxidative phase, ribulose 5-phosphate can be converted into ribose 5-phosphate, a precursor of nucleotides, or into erythrose 4-phosphate, a precursor for the biosynthesis of aromatic amino acids and histidine, interconnecting PPP to the Calvin cycle [12,16,17]. Depending on the biochemical demand, ribose 5-phosphate can be metabolized to sedoheptulose 7-phosphate, a connexon metabolite between PPP, glycolysis, amino acid biosynthesis, and metabolism of open-chain sugars [14].

The regulation of PPP is a key mechanism in the redox homeostasis, and it occurs through both the post-transcriptional and post-translational modulation of the enzymes and proteins involved in this pathway. Furthermore, the availability of intermediate substrates of connected pathways also modulates PPP. For example, the inhibition of glucose 6-phosphate dehydrogenase switches cellular metabolism from glycolysis to PPP and results in increased NADPH conversion [18].

Tricarboxylic acid cycle

The tricarboxylic acid (TCA) cycle, which is also known as the citric acid cycle or Krebs cycle, was first described in 1937 by Hans Krebs. This pathway is central to the oxidative metabolism and comprises eight reactions, which are initiated by the oxidation of acetyl-coenzyme A (CoA), a product obtained through decarboxylation of pyruvate, β-oxidation, and other processes [19,20]. This 8-step reaction results in the production of carbon dioxide, reducing equivalents in the form of nicotinamide adenine nucleotide (NADH), flavin adenine dinucleotide ($FADH_2$), and guanosine triphosphate (GTP). These reducing equivalents later on feed the electron transport chain, with the primary goal of production of adenosine triphosphate (ATP) in oxidative phosphorylation [21]. The TCA cycle also provides substrates for biosynthetic reactions in the cytosol via ancillary reactions [22]. The series of reactions mentioned are enzymatically catalyzed and are arranged in Fig. 1. It is important to highlight that the TCA cycle acts as a central hub in cellular metabolism, since numerous substrates can feed into it. The cycle starts with the reaction that combines acetyl-CoA, generated from amino acids, pyruvate oxidation, or fatty acids, with oxaloacetate to generate citrate, as showed in Fig. 1 [23].

As the cycle continues, the metabolites originated are transported to the cytosol to meet the demands of macromolecule synthesis. When these intermediates of the cycle are transported, the cycle needs replenishment to keep it functioning. This process is named anaplerosis, where various inputs can occur. One relevant mechanism is the conversion of glutamine to glutamate and subsequently to α-ketoglutarate, in a process called glutaminolysis. Another is the conversion of pyruvate to mitochondrial oxaloacetate by pyruvate decarboxylase. The function of TCA cycle is to oxidize acetyl-CoA into two molecules of CO_2, generating ATP and 3 NADH and 1 $FADH_2$ that later feed the ETC complexes I and II, respectively, as mentioned previously [23,24].

Oxidative phosphorylation and electron transport chain

Oxidative phosphorylation (OXPHOS) is the final step towards the generation of energy in the form of ATP in eukaryotic cells, and it occurs in the mitochondria. Each mitochondrion has a double-membrane structure, constituted of an inner and an outer membrane, with the first one being invaginated forming cristae. This heavily folded inner membrane, called cristae, contains many copies of the respiratory chain components (OXPHOS complexes) [25,26].

ETC is composed of two freely mobile electron transfer carrier's ubiquinone (CoQ) and cytochrome *c*. The ETC is also constituted of more than 80 polypeptides that are grouped together into four enzymatic complexes and ATP synthase (complex V). Usually, oxidative phosphorylation can be translated into the transfer of electrons from NADH and $FADH_2$ to complexes of ETC later captured by molecular oxygen, with this process associated to the pumping of protons across the inner membrane, from matrix into the mitochondria intermembrane space (IMS), creating an electrochemical gradient composed of pH gradient and membrane potential. This resulting proton motive force is then dissipated through ATP synthase, leading to phosphorylation of ADP [27].

Complex I (CI), also named as NADH-ubiquinone oxidoreductase, is one of the largest multisubunit complexes known, comprising 46 polypeptide chains. This complex transfers electrons from NADH to ubiquinone. Through conformational changes in the carrier ubiquinone, which is reduced to ubiquinol, a proton translocation channel is formed. There is the release of energy caused by the transfer of two electrons from NADH to CoQ in this complex, which further induces the pumping of four protons from the mitochondrial matrix into the IMS [28]. The complex II is independent of CI and is also named as succinate dehydrogenase. It is part of the Krebs cycle (TCA cycle) and of the ETC, which makes it a link between OXPHOS and metabolism [29]. This complex is an independent entry site for electrons in which they are donated from succinate to CoQ, by a similar process that happens in CI. In general terms, FAD is reduced to $FADH_2$

when it captures electrons from succinate and then transfers these electrons to FeS clusters. After that, CoQ is reduced to ubiquinol after acquiring the electrons from the FeS cluster. In this process, there is no coupling of electron transfer to translocation of protons [30,31].

The complex III (CIII), also named as cytochrome *bc*1 complex, participates by transferring electrons carried by ubiquinol to cytochrome *c*. The electron transfer process of CIII culminates in the displacement of four protons into the inner mitochondrial space [32]. Finally, cytochrome *c* oxidase, or complex IV (CIV), delivers electrons from cytochrome *c* to the terminal electron acceptor O_2 producing H_2O [30,33].

In conclusion, the ETC passes on two electrons at a time to mono-oxygen generating one water molecule, which results in the pumping of 4, 4, and 2 protons from the matrix to IMS through complexes I, III, and IV, respectively. Then, the protons pass to the matrix via F_0, one of the functional domains of complex V (CV), transferring the energy created by the proton electrochemical gradient to F_1, the other functional domain of CV, which culminates in a conformational change in the complex that results in the phosphorylation of ADP to form ATP [30,34].

Spectrum of macrophages and immunometabolism

Macrophages are present in virtually all tissues and are derived from the hematopoietic stem and myeloid progenitor cells in the bone marrow up to the monocyte stage [35,36]. Monocytes are released to the bloodstream and move to tissues, where they differentiate to macrophages. Tissue-specific macrophages establish themselves during embryogenesis and remain throughout life in their innate tissue, as a result of the self-renewal process [37,38]. However, circulating monocytes can be recruited to tissues in their steady state to repopulate long-lived specific tissue-resident macrophages, for example, peritoneal macrophages, osteoclasts, dust cells, microglia, histiocytes, Kupffer cells, and intraglomerular mesangial cells, among others. In addition, monocytes can migrate to the tissues in response to inflammation via gradients of inflammatory cytokines [35,37,38].

Initially only thought as immune effector cells, macrophages later gained attention due to their homeostatic roles. These cells are involved in the clearance of many erythrocytes each day. They also contribute to the removal of cellular debris generated during tissue remodeling and to the clearance of cells that have undergone apoptosis. The receptors that are responsible for these processes include scavenger, phosphatidyl serine, thrombospondin, and complement receptors and integrins [39]. Through their TLRs, they have become one of the primary sensors of danger in the organism [40]. The broad spectrum of macrophage functions reflects their heterogeneity and high plasticity in response to changes in environmental conditions and stimuli, such as those induced by inflammation and hormones. Thus, macrophages can reprogram their function and metabolism to polarize their phenotype, resulting in pro-inflammatory or pro-regenerative cells [41].

In order to mimic the T-cell literature, macrophages have also been classified according to a linear scale, with M1 and M2 as the terminals of two extremities. In this classification, M1 corresponds to classically activated macrophages and M2 to alternatively activated ones [42]. Later, M2 designation expanded to include all other types of macrophage [43]. However, it is well known that this M2 classification comprises cells with very distinct characteristics related to their physiology and biochemistry [44]. For the purpose of the explanations related to metabolism and treatment of diseases, these cells are referred to as pro-inflammatory (M1) and pro-regenerative (M2) in the following sections.

Pro-inflammatory macrophages

Pro-inflammatory macrophages, or M1 macrophages, are classically activated by interferon γ (IFN-γ), interleukin-1β (IL-1β), and/or pathogen-associated molecular patterns (PAMPs), such as lipopolysaccharide (LPS), and sustain inflammatory responses [45]. They compose the first line of defense against pathogens and induce the Th1 polarization of $CD4^+$ lymphocytes, through the production of pro-inflammatory mediators, such as nitric oxide synthase (iNOS), tumor necrosis factor α (TNF-α), IL-6, IL-12, and IL-23 [46–50].

The activation of pro-inflammatory macrophages requires a metabolic adaptation to meet the energy and precursor demands, which are required for phagocytosis and for the synthesis of inflammatory mediators. In the 1980s, the laboratory of Eric Newsholme developed fundamental studies on macrophage metabolism, by combining the cellular consumption and oxidation of glucose, glutamine, and fatty acids, with the activity of enzymes of the main metabolic pathways, and they demonstrated that glutamine and glucose are substantial fuels for murine macrophages; however, these molecules are poorly oxidized [51].

Pro-inflammatory macrophages increase their glycolytic and PPP activity and exhibit increased glucose consumption and lactate release whereas limit OXPHOS and fatty acid oxidation, especially by activating the transcription factor HIF-1α [12,41,52,53]. The signaling pathways TLR/NF-kB and AKT/mTOR are highlighted in the regulation of HIF-1α expression. The activation of NF-kB, the main regulator of macrophage functions, is due to several inflammatory signals, such as PRRs or pro-inflammatory cytokines, regulating HIF-1α gene expression, among others [53,54]. On the other hand, growth factors, such as GM-CSF, and pathogen detection receptors, as TLR4, activate the AKT/mTOR pathway [55]. HIF-1α is a regulator of macrophage metabolism by controlling the gene expression of GLUT1, glycolytic enzymes [53,54], and pyruvate conversion-related enzymes, which delimit pyruvate entry into the TCA cycle and promote the maintenance of NAD^+ levels [56,57].

Pro-inflammatory macrophages exhibit increased activity of the oxidative phase, but not of the non-oxidative phase of PPP, which occurs by the downregulation of sedoheptulose kinase and, consequently, decreases sedoheptulose 7-phosphate synthesis. This results in increased reduction of $NADP^+$ into NADPH. The NADPH oxidase complex produces ROS, including the superoxide that is converted into hydrogen peroxide by superoxide dismutase. This molecule promotes target cell death and inflammation, as well as affects several other macrophage functions [58,59]. Furthermore, the oxidative phase of PPP is crucial for the balance of redox mechanisms in activated macrophages, for the intense production of cytokines and prostaglandins, and also for the high turnover maintenance of plasma membrane and phagocytic activity, through fatty acid synthesis and incorporation into phospholipids and triacylglycerol [60].

The metabolic changes observed in pro-inflammatory macrophages also include a breakdown in the TCA cycle, which leads to the accumulation of regulatory intermediate metabolites, such as citrate, succinate, and itaconate [12]. The citrate accumulation is due to the downregulation of isocitrate dehydrogenase and upregulation of the mitochondrial citrate carrier that occurs in response to cellular activation by LPS, TNF-α, or IFN-γ; inhibits two important enzymes from the glycolytic pathway (phosphofructokinase and pyruvate kinase); and activates acetyl-CoA carboxylase [12,61]. On the other hand, itaconate enhances succinate accumulation, by inhibiting succinate dehydrogenase (SDH) in LPS-activated macrophages, and also helps to stabilize the levels of the anti-inflammatory transcription factor NRF2 [62]. Finally, succinate indirectly blocks HIF-1α degradation, through prolyl hydroxylase (PHD) inhibition, which culminates into the ROS production, pro-inflammatory mediators, and inflammasome activation [63,64].

Pro-regenerative macrophages

Pro-regenerative macrophages comprise a heterogeneous cell subtype linked to the inflammation resolution [65]. High phagocytic capacity, synthesis and secretion of pro-angiogenic factors, extracellular matrix components, chemotactic factors, and a large number of anti-inflammatory mediators, such as TGF-β, IL-10, and YM1, are some of the characteristics of pro-regenerative macrophages that reveal their key role in repair and tissue remodeling [66–68].

Pro-regenerative macrophage differentiation is induced by IL-4/IL-13 stimulus, and they maintain their polarization state by affecting the STAT6 and PGC-1β signaling pathways [69]. They also primarily adopt oxidative glucose metabolic pathways, having a complete TCA cycle, fatty acid oxidation, and mitochondrial energy metabolism [70,71]. It is noteworthy to mention that IL-4 induces OXPHOS and ATP citrate lyase activation and is responsible for the expansion of acetyl-CoA production that acts in histone acetylation process, an important epigenetic regulator of the expression of some M2 marker genes [72,73].

One of the intermediates of TCA cycle, citrate, can travel outside of mitochondria and be converted back to acetyl-CoA and oxaloacetate in the nucleus and in the cytosol. The activity and protein levels of ATP citrate lyase (ACLY), the enzyme responsible for this process, have been reported to be regulated by the Akt-mTORC1 signaling axis in macrophages activated by IL-4, one of the polarizing anti-inflammatory signals. A higher activity of ACLY is related to a subset of genes that are associated with production of chemokines and with cellular proliferation [73]. This indicates that a context-dependent metabolic control of histone acetylation is decisive to drive specific macrophage states. Studies in the field have demonstrated that acetyl-CoA is not just a passive acetyl donor group, but also acts as a key signaling molecule involved in the regulation of the activity of some specific transcription factors and histone marks that may dictate cellular functions, such as differentiation and proliferation, by regulating gene expression [74,75].

Also related to TCA cycle, α-ketoglutarate (α-KG) was demonstrated to have an essential role in favoring a pro-regenerative profile in macrophages while repressing pro-inflammatory responses. In IL-4-activated macrophages, glutaminolysis is elevated, resulting in higher production of α-KG, which contributes to the attainment of a pro-regenerative phenotype through the alteration of the activity of histone demethylase Jmjd3, which is another epigenetic process related to the metabolic state of these cells [76].

The metabolism of glucose, fatty acids, and also glutamine is deeply influenced by the macrophage activation state [77]. In relation to fatty acid (FA) utilization by these cells, it is known that FA could derive from the uptake and lipolysis of lipoprotein particles, for example the LDL and the VLDL through receptors such as CD36, a receptor that is induced in activated M2 macrophages, or by other lipoprotein receptors such as the scavenger receptor B-I [78,79]. Additionally, lipoproteins can modulate macrophage lipid metabolism by the presence of active mechanisms to efflux phospholipids and cholesterol that involve active ATP-dependent pumps, such as ABCA1 and ABCG1, which perform by interacting with extracellular lipid acceptors. The mobilization of cholesterol and lipids from these cells is crucial to modulate their function [80–82].

Other components that can affect macrophage metabolism are insulin and adrenaline. In macrophages, insulin was shown to affect the rate of phagocytosis and the levels of glucose utilization and to increase the maximum activities in vitro of citrate synthase, hexokinase, and glucose-6-phosphate dehydrogenase. Adrenaline, in physiological concentrations, had increased H_2O_2 synthesis in activated cells. In BCG-activated macrophages, glucose usage was high and was further increased by the addition of 5 nM adrenaline [77].

Mitochondria have been demonstrated to have a central role in macrophage polarization. These organelles have fundamental roles in the immune system through the regulation of macrophage responses to tissue damage, inflammation, and infection [83]. A study has shown that the interferon-stimulated gene 15 (IS15) regulates some mitochondrial processes related to immune and metabolic processes via mitophagy-dependent pathway. This gene promotes the synthesis of nitric oxide (NO) and arginase-1 activation, leading macrophages to polarize to a pro-regenerative phenotype [84].

Perspectives for therapies

Pro-inflammatory and pro-regenerative (M1/M2) macrophages are considered the opposite extremities of the full spectrum of macrophage activation. The transition between different macrophage phenotypes is responsible for the initiation, maintenance, and resolution of inflammatory diseases [85]. An example of the contribution of these cells to a disease can be described during an infection [86,87]. When the tissues are challenged by pathogenic bacteria, monocytes present in circulation are recruited and differentiated into macrophages and in general are polarized towards M1 phenotype at early stages of infection. Then, TLRs recognize the PAMPs present in bacteria, resulting in the activation of macrophages and production of pro-inflammatory mediators that are capable of killing the pathogens and activating the adaptive immune system [88]. If the macrophage-mediated inflammation cannot be controlled, a cytokine storm is formed, culminating in severe sepsis [89]. In order to counteract this inflammation, M1 macrophages may undergo apoptosis or be polarized towards an M2 phenotype to facilitate wound healing and protect the host. Other studies have demonstrated the importance of macrophage polarization for regulation of the immune system and protection against diseases, in the context of SARS-CoV and respiratory syncytial virus infections, atherosclerosis, cardiovascular diseases, etc. [90–94]. Furthermore, in patients with genetic syndromes or systemic disorders, such as obesity, autoimmune reactions, and tumors, these immune cells are responsible for the exacerbation of fibrosis, atherosclerosis, tumor growth, etc. It is noteworthy that the outcome of the inflammatory reaction towards regeneration or fibrosis is also correlated with the timely switching of M1 to M2 phenotype's polarization [95].

Taking this into consideration, there are a few strategies available for M1 cell inhibition. One of them is by injecting M2 cells, which are capable of inhibiting and eliminating M1-related inflammatory damage. Another strategy is exploring the use of chemokines and cytokines that helps in the shifting of M1 macrophages towards the M2 phenotype. Finally, another strategy that could be used is by blocking the signaling pathways associated with the inflammatory response promoted by M1 cells [96].

A transition of M1 towards M2 could be beneficial for the prevention and treatment of Alzheimer's disease. One potential regulator of mitochondrial metabolism is the curcumin, a natural polyphenol that is widely used to treat some conditions, such as atherosclerosis and acute kidney injury [97,98]. Curcumin has been demonstrated to interfere in mitochondrial function, increase the percentage of M2, and decrease the percentage of M1 macrophages, as well as improve the concentration of IL-10 and reduce pro-inflammatory cytokines [99]. In RAW264.7 cells after stimulation with LPS, Cur was capable of reducing the expression and secretion of a few important pro-inflammatory cytokines [100]. These and other studies highlight that this substance can regulate mitochondrial polarization and may have a protective role in inflammatory diseases.

The interventions may involve in situ or ex vivo reprogramming, with autologous transplantation of the ex vivo modified cells. In the case of proliferative diseases, as anticancer therapies, the opposite could be explored, that is, evoking M1 cells into the tissue, promoting the cell death and inflammation, while in inflammatory diseases, the increase in M2 population could stimulate the processes of repair or regeneration, depending on the extent of the lesion, extracellular matrix remodeling, and angiogenesis [101,102]. The relevance of this cell therapy is evident. However, the use of macrophages as a therapeutic tool is restricted due to lack of safety and efficiency in the approaches for their reprogramming. To overcome these problems, studies regarding the role of specific genes in macrophage functions are still needed, in a way that genetic modification of these cells results in the suppression or in the boost of the expression of genes in order to obtain stable M1 or M2 phenotypes. With these stable phenotypes, gene interaction and signaling pathway studies could culminate in reliable targets for drug development [103]. Finally, both in vivo and in vitro manageable activation of macrophages can provide powerful therapies for inflammatory and proliferative diseases.

References

[1] Huang N, Perl A. Metabolism as a target for modulation in autoimmune diseases. Trends Immunol 2018;39(7):562–76.

[2] Donnelly RP, Finlay DK. Glucose, glycolysis and lymphocyte responses. Mol Immunol 2015;68(2 Pt C):513–9.

[3] Menk AV, Scharping NE, Moreci RS, Zeng X, Guy C, Salvatore S, et al. Early TCR signaling induces rapid aerobic glycolysis enabling distinct acute T cell effector functions. Cell Rep 2018;22(6):1509–21.

[4] Judge A, Dodd MS. Metabolism. Essays Biochem 2020;64(4):607–47.

[5] Nolfi-Donegan D, Braganza A, Shiva S. Mitochondrial electron transport chain: oxidative phosphorylation, oxidant production, and methods of measurement. Redox Biol 2020;37, 101674.

[6] Fernie AR, Carrari F, Sweetlove LJ. Respiratory metabolism: glycolysis, the TCA cycle and mitochondrial electron transport. Curr Opin Plant Biol 2004;7(3):254–61.

[7] Akram M. Mini-review on glycolysis and cancer. J Cancer Educ 2013;28(3):454–7.

[8] Riksen NP, Netea MG. Immunometabolic control of trained immunity. Mol Asp Med 2021;77, 100897.

[9] Deng D, Yan N. GLUT, SGLT, and SWEET: structural and mechanistic investigations of the glucose transporters. Protein Sci 2016;25(3):546–58.

[10] Lenzen S. A fresh view of glycolysis and glucokinase regulation: history and current status. J Biol Chem 2014;289(18):12189–94.

[11] Xiong Y, Lei QY, Zhao S, Guan KL. Regulation of glycolysis and gluconeogenesis by acetylation of PKM and PEPCK. Cold Spring Harb Symp Quant Biol 2011;76:285–9.

[12] Viola A, Munari F, Sanchez-Rodriguez R, Scolaro T, Castegna A. The metabolic signature of macrophage responses. Front Immunol 2019;10:1462.

[13] Vander Heiden MG, Cantley LC, Thompson CB. Understanding the Warburg effect: the metabolic requirements of cell proliferation. Science 2009;324(5930):1029–33.

[14] Stincone A, Prigione A, Cramer T, Wamelink MM, Campbell K, Cheung E, et al. The return of metabolism: biochemistry and physiology of the pentose phosphate pathway. Biol Rev Camb Philos Soc 2015;90(3):927–63.

[15] Winkler BS, DeSantis N, Solomon F. Multiple NADPH-producing pathways control glutathione (GSH) content in retina. Exp Eye Res 1986;43(5):829–47.
[16] Zhao G, Pease AJ, Bharani N, Winkler ME. Biochemical characterization of gapB-encoded erythrose 4-phosphate dehydrogenase of *Escherichia coli* K-12 and its possible role in pyridoxal 5′-phosphate biosynthesis. J Bacteriol 1995;177(10):2804–12.
[17] Clasquin MF, Melamud E, Singer A, Gooding JR, Xu X, Dong A, et al. Riboneogenesis in yeast. Cell 2011;145(6):969–80.
[18] Fico A, Paglialunga F, Cigliano L, Abrescia P, Verde P, Martini G, et al. Glucose-6-phosphate dehydrogenase plays a crucial role in protection from redox-stress-induced apoptosis. Cell Death Differ 2004;11(8):823–31.
[19] Mailloux RJ, Bériault R, Lemire J, Singh R, Chénier DR, Hamel RD, et al. The tricarboxylic acid cycle, an ancient metabolic network with a novel twist. PLoS One 2007;2(8), e690.
[20] Shi L, Tu BP. Acetyl-CoA and the regulation of metabolism: mechanisms and consequences. Curr Opin Cell Biol 2015;33:125–31.
[21] Bowtell JL, Marwood S, Bruce M, Constantin-Teodosiu D, Greenhaff PL. Tricarboxylic acid cycle intermediate pool size: functional importance for oxidative metabolism in exercising human skeletal muscle. Sports Med 2007;37(12):1071–88.
[22] Williamson JR, Cooper RH. Regulation of the citric acid cycle in mammalian systems. FEBS Lett 1980;117(Suppl):K73–85.
[23] Martínez-Reyes I, Chandel NS. Mitochondrial TCA cycle metabolites control physiology and disease. Nat Commun 2020;11(1):102.
[24] DeBerardinis RJ, Chandel NS. Fundamentals of cancer metabolism. Sci Adv 2016;2(5), e1600200.
[25] Chaban Y, Boekema EJ, Dudkina NV. Structures of mitochondrial oxidative phosphorylation supercomplexes and mechanisms for their stabilisation. Biochim Biophys Acta 2014;1837(4):418–26.
[26] Kadenbach B, Ramzan R, Wen L, Vogt S. New extension of the Mitchell theory for oxidative phosphorylation in mitochondria of living organisms. Biochim Biophys Acta 2010;1800(3):205–12.
[27] Heales SJ, Gegg ME, Clark JB. Oxidative phosphorylation: structure, function, and intermediary metabolism. Int Rev Neurobiol 2002;53:25–56.
[28] Tan P, Feng Z, Zhang L, Hou T, Li Y. The mechanism of proton translocation in respiratory complex I from molecular dynamics. J Recept Signal Transduct Res 2015;35(2):170–9.
[29] Cecchini G. Function and structure of complex II of the respiratory chain. Annu Rev Biochem 2003;72:77–109.
[30] Zhao RZ, Jiang S, Zhang L, Yu ZB. Mitochondrial electron transport chain, ROS generation and uncoupling (review). Int J Mol Med 2019;44(1):3–15.
[31] Iverson TM. Catalytic mechanisms of complex II enzymes: a structural perspective. Biochim Biophys Acta 2013;1827(5):648–57.
[32] Trumpower BL. A concerted, alternating sites mechanism of ubiquinol oxidation by the dimeric cytochrome bc(1) complex. Biochim Biophys Acta 2002;1555(1–3):166–73.
[33] Wikstrom MK. Proton pump coupled to cytochrome c oxidase in mitochondria. Nature 1977;266(5599):271–3.
[34] Jonckheere AI, Smeitink JA, Rodenburg RJ. Mitochondrial ATP synthase: architecture, function and pathology. J Inherit Metab Dis 2012;35(2):211–25.
[35] Gordon S, Taylor PR. Monocyte and macrophage heterogeneity. Nat Rev Immunol 2005;5(12):953–64.
[36] Hastreiter AA, Galvao Dos Santos G, Cavalcante Santos EW, Makiyama EN, Borelli P, Fock RA. Protein malnutrition impairs bone marrow endothelial cells affecting hematopoiesis. Clin Nutr 2020;39(5):1551–9.

[37] Ginhoux F, Guilliams M. Tissue-resident macrophage ontogeny and homeostasis. Immunity 2016;44(3):439–49.
[38] Hachim D, LoPresti ST, Rege RD, Umeda Y, Iftikhar A, Nolfi AL, et al. Distinct macrophage populations and phenotypes associated with IL-4 mediated immunomodulation at the host implant interface. Biomater Sci 2020;8(20):5751–62.
[39] Erwig LP, Henson PM. Immunological consequences of apoptotic cell phagocytosis. Am J Pathol 2007;171(1):2–8.
[40] Kono H, Rock KL. How dying cells alert the immune system to danger. Nat Rev Immunol 2008;8(4):279–89.
[41] Curi R, de Siqueira MR, de Campos Crispin LA, Norata GD, Sampaio SC, Newsholme P. A past and present overview of macrophage metabolism and functional outcomes. Clin Sci (Lond) 2017;131(12):1329–42.
[42] Gordon S. Alternative activation of macrophages. Nat Rev Immunol 2003;3(1):23–35.
[43] Martinez FO, Sica A, Mantovani A, Locati M. Macrophage activation and polarization. Front Biosci 2008;13:453–61.
[44] Edwards JP, Zhang X, Frauwirth KA, Mosser DM. Biochemical and functional characterization of three activated macrophage populations. J Leukoc Biol 2006;80(6):1298–307.
[45] Castoldi A, Naffah de Souza C, Camara NO, Moraes-Vieira PM. The macrophage switch in obesity development. Front Immunol 2015;6:637.
[46] Goerdt S, Politz O, Schledzewski K, Birk R, Gratchev A, Guillot P, et al. Alternative versus classical activation of macrophages. Pathobiology 1999;67(5–6):222–6.
[47] Hao NB, Lu MH, Fan YH, Cao YL, Zhang ZR, Yang SM. Macrophages in tumor microenvironments and the progression of tumors. Clin Dev Immunol 2012;2012, 948098.
[48] Noel W, Raes G, Hassanzadeh Ghassabeh G, De Baetselier P, Beschin A. Alternatively activated macrophages during parasite infections. Trends Parasitol 2004;20(3):126–33.
[49] Moraes-Vieira PM, Yore MM, Dwyer PM, Syed I, Aryal P, Kahn BB. RBP4 activates antigen-presenting cells, leading to adipose tissue inflammation and systemic insulin resistance. Cell Metab 2014;19(3):512–26.
[50] Shapiro H, Lutaty A, Ariel A. Macrophages, meta-inflammation, and immuno-metabolism. ScientificWorldJournal 2011;11:2509–29.
[51] Newsholme P, Gordon S, Newsholme EA. Rates of utilization and fates of glucose, glutamine, pyruvate, fatty acids and ketone bodies by mouse macrophages. Biochem J 1987;242(3):631–6.
[52] Rodriguez-Prados JC, Traves PG, Cuenca J, Rico D, Aragones J, Martin-Sanz P, et al. Substrate fate in activated macrophages: a comparison between innate, classic, and alternative activation. J Immunol 2010;185(1):605–14.
[53] Wang T, Liu H, Lian G, Zhang SY, Wang X, Jiang C. HIF1α-induced glycolysis metabolism is essential to the activation of inflammatory macrophages. Mediat Inflamm 2017;2017:9029327.
[54] van Uden P, Kenneth NS, Rocha S. Regulation of hypoxia-inducible factor-1alpha by NF-kappaB. Biochem J 2008;412(3):477–84.
[55] Kelley TW, Graham MM, Doseff AI, Pomerantz RW, Lau SM, Ostrowski MC, et al. Macrophage colony-stimulating factor promotes cell survival through Akt/protein kinase B. J Biol Chem 1999;274(37):26393–8.
[56] Semenza GL, Jiang BH, Leung SW, Passantino R, Concordet JP, Maire P, et al. Hypoxia response elements in the aldolase A, enolase 1, and lactate dehydrogenase A gene promoters contain essential binding sites for hypoxia-inducible factor 1. J Biol Chem 1996;271(51):32529–37.
[57] Kim JW, Tchernyshyov I, Semenza GL, Dang CV. HIF-1-mediated expression of pyruvate dehydrogenase kinase: a metabolic switch required for cellular adaptation to hypoxia. Cell Metab 2006;3(3):177–85.

[58] Newsholme EA, Newsholme P, Curi R. The role of the citric acid cycle in cells of the immune system and its importance in sepsis, trauma and burns. Biochem Soc Symp 1987;54:145–62.
[59] Newsholme P, Curi R, Gordon S, Newsholme EA. Metabolism of glucose, glutamine, long-chain fatty acids and ketone bodies by murine macrophages. Biochem J 1986;239(1):121–5.
[60] Haschemi A, Kosma P, Gille L, Evans CR, Burant CF, Starkl P, et al. The sedoheptulose kinase CARKL directs macrophage polarization through control of glucose metabolism. Cell Metab 2012;15(6):813–26.
[61] Yalcin A, Telang S, Clem B, Chesney J. Regulation of glucose metabolism by 6-phosphofructo-2-kinase/fructose-2,6-bisphosphatases in cancer. Exp Mol Pathol 2009;86(3):174–9.
[62] Mills EL, Ryan DG, Prag HA, Dikovskaya D, Menon D, Zaslona Z, et al. Itaconate is an anti-inflammatory metabolite that activates Nrf2 via alkylation of KEAP1. Nature 2018;556(7699):113–7.
[63] Lampropoulou V, Sergushichev A, Bambouskova M, Nair S, Vincent EE, Loginicheva E, et al. Itaconate links inhibition of succinate dehydrogenase with macrophage metabolic remodeling and regulation of inflammation. Cell Metab 2016;24(1):158–66.
[64] Tannahill GM, Curtis AM, Adamik J, Palsson-McDermott EM, McGettrick AF, Goel G, et al. Succinate is an inflammatory signal that induces IL-1β through HIF-1α. Nature 2013;496(7444):238–42.
[65] Mantovani A, Biswas SK, Galdiero MR, Sica A, Locati M. Macrophage plasticity and polarization in tissue repair and remodelling. J Pathol 2013;229(2):176–85.
[66] Shin TH, Kim HS, Kang TW, Lee BC, Lee HY, Kim YJ, et al. Human umbilical cord blood-stem cells direct macrophage polarization and block inflammasome activation to alleviate rheumatoid arthritis. Cell Death Dis 2016;7(12), e2524.
[67] Fuentes L, Roszer T, Ricote M. Inflammatory mediators and insulin resistance in obesity: role of nuclear receptor signaling in macrophages. Mediat Inflamm 2010;2010, 219583.
[68] Bohlson SS, O'Conner SD, Hulsebus HJ, Ho MM, Fraser DA. Complement, c1q, and c1q-related molecules regulate macrophage polarization. Front Immunol 2014;5:402.
[69] Huang SC, Everts B, Ivanova Y, O'Sullivan D, Nascimento M, Smith AM, et al. Cell-intrinsic lysosomal lipolysis is essential for alternative activation of macrophages. Nat Immunol 2014;15(9):846–55.
[70] Wang LX, Zhang SX, Wu HJ, Rong XL, Guo J. M2b macrophage polarization and its roles in diseases. J Leukoc Biol 2019;106(2):345–58.
[71] Liu PS, Ho PC. Mitochondria: a master regulator in macrophage and T cell immunity. Mitochondrion 2018;41:45–50.
[72] Dias C, Barbosa RM, Laranjinha J, Ledo A. Evaluation of mitochondrial function in the CNS of rodent models of Alzheimer's disease—high resolution respirometry applied to acute hippocampal slices. Free Radic Biol Med 2014;75(Suppl 1):S37.
[73] Covarrubias AJ, Aksoylar HI, Yu J, Snyder NW, Worth AJ, Iyer SS, et al. Akt-mTORC1 signaling regulates Acly to integrate metabolic input to control of macrophage activation. elife 2016;5.
[74] Ansó E, Weinberg SE, Diebold LP, Thompson BJ, Malinge S, Schumacker PT, et al. The mitochondrial respiratory chain is essential for haematopoietic stem cell function. Nat Cell Biol 2017;19(6):614–25.
[75] Moussaieff A, Rouleau M, Kitsberg D, Cohen M, Levy G, Barasch D, et al. Glycolysis-mediated changes in acetyl-CoA and histone acetylation control the early differentiation of embryonic stem cells. Cell Metab 2015;21(3):392–402.
[76] Liu PS, Wang H, Li X, Chao T, Teav T, Christen S, et al. α-Ketoglutarate orchestrates macrophage activation through metabolic and epigenetic reprogramming. Nat Immunol 2017;18(9):985–94.

[77] Newsholme P, Costa Rosa LF, Newsholme EA, Curi R. The importance of fuel metabolism to macrophage function. Cell Biochem Funct 1996;14(1):1–10.
[78] Su X, Abumrad NA. Cellular fatty acid uptake: a pathway under construction. Trends Endocrinol Metab 2009;20(2):72–7.
[79] Feng J, Han J, Pearce SF, Silverstein RL, Gotto AM, Hajjar DP, et al. Induction of CD36 expression by oxidized LDL and IL-4 by a common signaling pathway dependent on protein kinase C and PPAR-gamma. J Lipid Res 2000;41(5):688–96.
[80] Norata GD, Pirillo A, Ammirati E, Catapano AL. Emerging role of high density lipoproteins as a player in the immune system. Atherosclerosis 2012;220(1):11–21.
[81] Catapano AL, Pirillo A, Bonacina F, Norata GD. HDL in innate and adaptive immunity. Cardiovasc Res 2014;103(3):372–83.
[82] Larbi A, Fortin C, Dupuis G, Berrougui H, Khalil A, Fulop T. Immunomodulatory role of high-density lipoproteins: impact on immunosenescence. Age (Dordr) 2014;36(5):9712.
[83] Miller VJ, Villamena FA, Volek JS. Nutritional ketosis and mitohormesis: potential implications for mitochondrial function and human health. J Nutr Metab 2018;2018:5157645.
[84] Baldanta S, Fernández-Escobar M, Acín-Perez R, Albert M, Camafeita E, Jorge I, et al. ISG15 governs mitochondrial function in macrophages following vaccinia virus infection. PLoS Pathog 2017;13(10), e1006651.
[85] Liu YC, Zou XB, Chai YF, Yao YM. Macrophage polarization in inflammatory diseases. Int J Biol Sci 2014;10(5):520–9.
[86] Shaughnessy LM, Swanson JA. The role of the activated macrophage in clearing *Listeria monocytogenes* infection. Front Biosci 2007;12:2683–92.
[87] Sica A, Mantovani A. Macrophage plasticity and polarization: in vivo veritas. J Clin Invest 2012;122(3):787–95.
[88] Wynn TA, Chawla A, Pollard JW. Macrophage biology in development, homeostasis and disease. Nature 2013;496(7446):445–55.
[89] Stearns-Kurosawa DJ, Osuchowski MF, Valentine C, Kurosawa S, Remick DG. The pathogenesis of sepsis. Annu Rev Pathol 2011;6:19–48.
[90] Page C, Goicochea L, Matthews K, Zhang Y, Klover P, Holtzman MJ, et al. Induction of alternatively activated macrophages enhances pathogenesis during severe acute respiratory syndrome coronavirus infection. J Virol 2012;86(24):13334–49.
[91] Shirey KA, Pletneva LM, Puche AC, Keegan AD, Prince GA, Blanco JC, et al. Control of RSV-induced lung injury by alternatively activated macrophages is IL-4R alpha-, TLR4-, and IFN-beta-dependent. Mucosal Immunol 2010;3(3):291–300.
[92] Swirski FK, Nahrendorf M. Leukocyte behavior in atherosclerosis, myocardial infarction, and heart failure. Science 2013;339(6116):161–6.
[93] Khallou-Laschet J, Varthaman A, Fornasa G, Compain C, Gaston AT, Clement M, et al. Macrophage plasticity in experimental atherosclerosis. PLoS One 2010;5(1), e8852.
[94] Kirbis S, Breskvar UD, Sabovic M, Zupan I, Sinkovic A. Inflammation markers in patients with coronary artery disease—comparison of intracoronary and systemic levels. Wien Klin Wochenschr 2010;122(Suppl. 2):31–4.
[95] Chung L, Maestas DR, Housseau F, Elisseeff JH. Key players in the immune response to biomaterial scaffolds for regenerative medicine. Adv Drug Deliv Rev 2017;114:184–92.
[96] Qing J, Zhang Z, Novák P, Zhao G, Yin K. Mitochondrial metabolism in regulating macrophage polarization: an emerging regulator of metabolic inflammatory diseases. Acta Biochim Biophys Sin Shanghai 2020;52(9):917–26.
[97] Karuppagounder V, Arumugam S, Thandavarayan RA, Sreedhar R, Giridharan VV, Afrin R, et al. Curcumin alleviates renal dysfunction and suppresses inflammation by shifting from M1 to M2 macrophage polarization in daunorubicin induced nephrotoxicity in rats. Cytokine 2016;84:1–9.

[98] Chen F, Guo N, Cao G, Zhou J, Yuan Z. Molecular analysis of curcumin-induced polarization of murine RAW264.7 macrophages. J Cardiovasc Pharmacol 2014;63(6):544–52.
[99] Li B, Hu Y, Zhao Y, Cheng M, Qin H, Cheng T, et al. Curcumin attenuates titanium particle-induced inflammation by regulating macrophage polarization. Front Immunol 2017;8:55.
[100] Tan RZ, Liu J, Zhang YY, Wang HL, Li JC, Liu YH, et al. Curcumin relieved cisplatin-induced kidney inflammation through inhibiting mincle-maintained M1 macrophage phenotype. Phytomedicine 2019;52:284–94.
[101] Pan XQ. The mechanism of the anticancer function of M1 macrophages and their use in the clinic. Chin J Cancer 2012;31(12):557–63.
[102] Bi Y, Chen J, Hu F, Liu J, Li M, Zhao L. M2 macrophages as a potential target for antiatherosclerosis treatment. Neural Plast 2019;2019:6724903.
[103] Poltavets AS, Vishnyakova PA, Elchaninov AV, Sukhikh GT, Fatkhudinov TK. Macrophage modification strategies for efficient cell therapy. Cell 2020;9(6).

CHAPTER 4

Spleen, lymph nodes, and lymphoid tissues—Inflammatory system

Djalma S. Lima-Junior, Nicholas Collins, and Seong-Ji Han
Metaorganism Immunity Section, Laboratory of Immune System Biology, National Institute of Allergy and Infectious Diseases, National Institutes of Health, Bethesda, MD, United States

Introduction

During the course of evolution, higher vertebrates have developed sophisticated mechanisms that support the induction of adaptive immune responses, which are critical for host protection against infection and cancer, as well as the establishment of immunological memory. The complex process of initiating an immune response requires coordinated action by numerous cells of the innate and adaptive arms of the immune system. The innate immune system represents the first line of defense used by the host within hours of encountering an antigen. Innate immunity is mainly mediated by phagocytes such as dendritic cells and macrophages. These cells can respond to a broad range of antigens and are directly involved in regulating the initiation, development, and resolution of many inflammatory processes. Unlike innate mechanisms of host defense, the adaptive immune system is antigen-dependent and antigen-specific, in which responses are based primarily on the antigen-specific receptors expressed on the surfaces of T- and B lymphocytes. Cells from the innate and adaptive immune system are in constant interaction and the cross talk between these cells leads to highly efficient recognition and clearance of pathogens. However, defective interactions between these systems can result in harmful immunologic responses that promote the disruption of homeostatic immunity and the development of autoimmune diseases, as well as allograft rejection.

Several innate cells, especially macrophages, have been shown to be crucial for homeostatic and defense functions after local and systemic perturbation within tissues. Macrophages are professional phagocytes that internalize and degrade extracellular particles, cellular debris, and pathogens. These cells can also produce a wide array of mediators that alert and

Macrophages in the Human Body
https://doi.org/10.1016/B978-0-12-821385-8.00001-9

trigger the adaptive immune system to invaders. As such, they play critical roles in regulating host immunity and participate in many other biological processes such as tissue repair, clearance of apoptotic cells, and regulation of metabolism [1–3]. Notably, based on their anatomical location, functional specialization, and expression of surface markers, macrophages present a high degree of heterogeneity even within the same tissue. They can either derive from circulating monocytes or the embryonic yolk sac and then self-renewing during adult life independently of monocytes [4,5]. Most are dependent on colony-stimulating factor-1 (CSF1) for their development [6,7]. All tissues in the body contain macrophages and it is likely that local signals derived from the environment in which they are residing play key roles in promoting specialized properties that allow them to support unique functions required by the strict relationship between macrophages in their tissue of residence [5]. For instance, many different subsets of macrophages populate secondary lymphoid organs such as lymph nodes and spleen, under both steady-state and inflammatory conditions.

Secondary lymphoid organs (SLOs) are key sites of the immune system dedicated to the concentration of antigens as well as Ag-specific lymphocytes fundamental to initiate an efficient adaptive immune response. These organs possess mechanisms for the recruitment of lymphocytes and for the acquisition and retention of innate cells that present antigens to lymphocytes, thereby promoting the differentiation of naïve T cells into effector and memory T cells that eliminate the inciting antigen in peripheral tissues, leading to protective immunity [8–10]. The spleen and lymph nodes are primordial SLOs that provide the structural framework for cellular interaction between antigen-presenting cells (such as dendritic cells and macrophages), B and T cells, which is a critical process that orchestrates humoral and cellular immune responses. Strategically, spleen and lymph nodes are located to efficiently acquire foreign antigens entering via bloodstream and peripheral tissues, respectively. SLOs play an essential role in the immune system since a large number of antigen-presenting cells are present at these sites, which are capable of initiating a rapid immune response to foreign antigens. The microanatomy of these tissues, as well as the presence of specific chemokines (chemotactic factors that direct the migration of cell of the immune system), promote the positioning of lymphoid and non-lymphoid cells (e.g., stromal cells) to create a microenvironment that supports the development and activation of innate and adaptive immune responses. However, despite common features between the spleen and lymph node, differences in routes of antigen transport, lymphocyte trafficking, and the presence of unique cell populations determine the role of a specific secondary lymphoid tissue in promoting immune responses.

Spleen

The spleen is the largest secondary lymphoid organ in our body and plays an important role in filtering blood. The spleen also plays a crucial role in promoting immune responses against infectious agents and foreign material, as well as in removing defective or senescent erythrocytes from the circulation. Furthermore, the spleen is a site in which immune responses that are deleterious to the host are regulated. The physical organization of the spleen allows for the effective screening of pathogens and antigens in the blood, enabling immediate innate or long-lasting adaptive immune responses. Unlike lymph nodes, the spleen lacks afferent lymphatic vessels and HEVs, and therefore all cells and antigens enter the spleen via the blood through the splenic artery. The spleen contains a unique microvasculature that allows for the trafficking of antigens and cells. The blood leaves open capillaries, filters through the reticular connective tissue, and drains into the venous sinus without passing through the endothelial barrier [11]. The spleen is functionally and structurally divided into two distinct compartments: the blood containing red pulp, which is responsible for the filtering function of the spleen; and white pulp, which is dense with lymphoid cells that are dedicated to adaptive immunity. Between these two regions is the marginal zone (MZ) in rodents and the perifollicular zone in humans (Fig. 1) [12,13]. The MZ or perifollicular zone that surrounds the white pulp represents an important transition between the innate and acquired immune systems; it is the first region after the red pulp encountered by blood-borne antigens and is richly supplied with specialized phagocytic cells, especially distinct subsets of macrophages.

The splenic red pulp region

The red pulp makes up the majority of the stromal tissue of the spleen and mediates an immune function distinct from that of the white pulp. The red pulp region of the spleen functions as a blood filter that removes toxins, foreign material, and aged or dead cells from the circulation, while simultaneously surveying for pathogens and tissue damage, as well as recycling iron from aging red blood cells. Arterial blood is delivered to the red pulp by splenic cords, which consist of fibroblasts and reticular fibers, and release

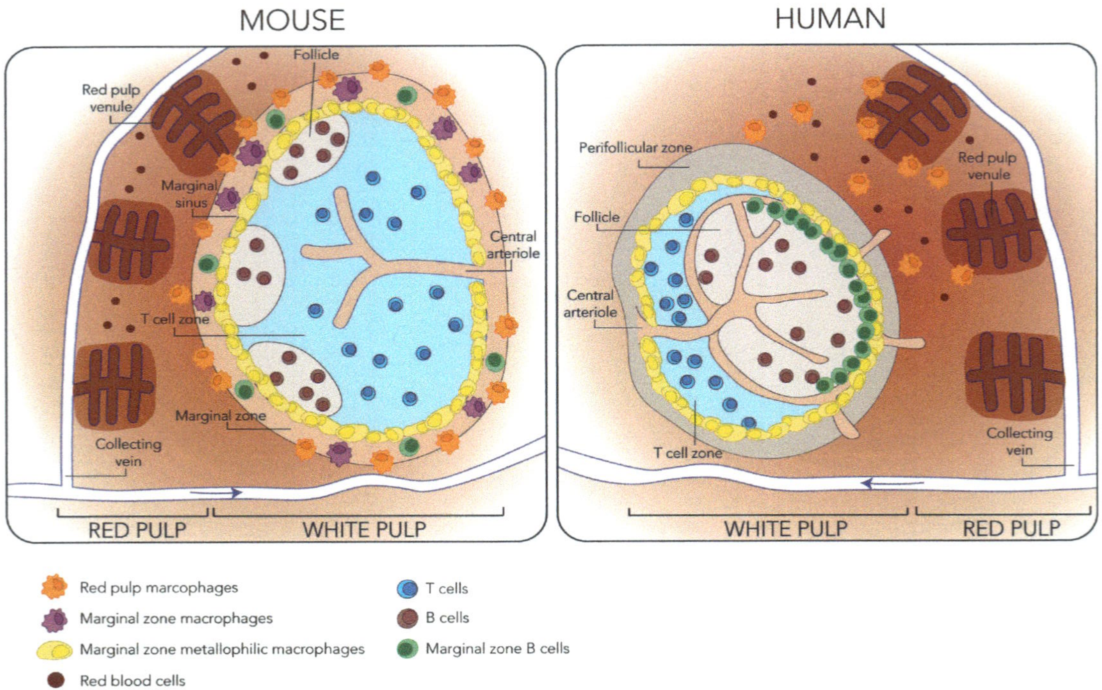

Lima-Junior, Collins & Han. Figure 1

Fig. 1 Mouse and human splenic architecture and cellular composition. The spleen can be divided into three main compartments: the red pulp (RP), the white pulp (WP), and the marginal zone (MZ) in rodents and perifollicular zone (PFZ) in humans. The MZ and RP are separated by the marginal sinus in rodent's spleen. The blood arrives at the splenic red pulp after entering the central arteriole, reaching the marginal sinuses and flowing passively through the MZ. The red pulp macrophages (RPMs) are typically found within the splenic cords, in close association with red blood cells and other blood cells/particles passing through red pulp venous sinuses. In mouse, but not in human, marginal zone metallophilic macrophages (MMMs) are located under the marginal sinuses, forming a concentric ring around the WP with MZMs and MZB cells. Marginal zone macrophages (MZMs) are located in the marginal zone above the marginal sinuses and are found in close proximity with marginal zone B cells (MZB). Murine MZB cells are found predominantly in the marginal zone but are constantly migrating between MZ and follicular compartments. In humans, MZB cells surround activated B cells in the follicle.

their contents into an open blood system without traditional endothelial linings [14]. In these cords, many macrophages are found. The blood passes into venous sinuses of the red pulp, which are located directly adjacent to the marginal zone. These sinuses are lined by endothelium that sits on a basement membrane comprised of endothelial cells and reticular fibers of the red pulp [15]. Blood flowing through the sinusoids is exposed to red pulp macrophages located within the splenic cords that are in close association with red blood cells and plasma cells (Fig. 1). These macrophages are responsible for removing aged, dead, or opsonized cells from the circulation, as well as for initiating immune responses. Other leukocytes with innate functions such as monocytes, dendritic cells, neutrophils, and gamma delta (gd) T cells also reside in the red pulp region [16]. Inflammatory processes can alter both the location and numbers of these cell populations to elicit an efficient and rapid adaptive immune response to foreign antigens, shaping the immune function and dynamic interaction of these cells. Plasmablasts and plasma cells, after antigen-specific differentiation in the follicles of the white pulp, migrate into red pulp toward a gradient of CXCL12 (which is higher in the RP) to produce antibodies that are carried into the circulatory system [16]. Effector $CD8^+$ T cells leave the white pulp to emigrate to the red pulp to clear the infection in the periphery of the spleen [17].

Red pulp macrophages

Red pulp macrophages (RPMs) form a vast network inside the red pulp and are characterized by high expression of F4/80, CD68, VCAM-1, CD31, MR, CD163, and Treml4 and by low expression of CD11b [18,19]. There is no clear consensus about the origin of RPMs, however, a recent study indicates that these macrophages are maintained by local proliferation at physiological conditions [20]. Conversely, after macrophage depletion or in some pathological conditions, bone marrow monocytes are able to differentiate into RPMs [21,22]. Their generation is dependent on the transcription factor Spi-C, which directly regulates VCAM-1 expression [19], while other studies have shown the importance of colony-stimulating factor (M-CSF/CSF-1) for this process. Interestingly, in the absence of M-CFS, there is a 50% reduction in red pulp macrophages indicating that the RPM population is composed of M-CSF-dependent and independent cells [23–25]. Consistent with this notion, reports have shown that the M-CSF-dependent RPM population has the potential to regulate immune responses by presenting a greater phagocytic capacity, higher response to prostaglandin

E2 (PGE2), and higher production of proinflammatory cytokines such as TNF-α and type I IFNs compared to the M-CSF-independent RPMs, which were less efficient at phagocytosis and produced higher amounts of PGE2 [26]. Also, it was shown in a model of graft-versus-host disease that mice lacking M-CSF-dependent macrophages had a decreased number of Foxp3 + regulatory T cells (Tregs), which resulted in accelerated pathology and exaggerated T-cell activation [27]. Similarly, another study showed that M-CSF-dependent RPMs suppressed $CD4^+$ T-cell responses via production of the anti-inflammatory cytokine TGF-b, and that they induced the generation of regulatory $CD4^+$ T cells that influenced the outcome of autoimmunity [28].

Red pulp macrophages are constantly monitoring the blood for bacteria and other foreign materials and play an important function in the phagocytosis of old and damaged erythrocytes and platelets. The phagocytosis of old and damaged red blood cells is mainly regulated by the ligation of CD47 to Signal-Regulatory Protein alpha (SIRPα) [29]. CD47 is a self-molecule ubiquitously expressed on many cell types, including red blood cells, which is important for avoiding clearance by phagocytes. The presence of normal CD47 expression on RBCs is an inhibitory signal for phagocytosis [30], nevertheless, aging RBCs express a modified isoform of this molecule that instead promotes phagocytosis by macrophages through mechanisms dependent on SIRPα binding [29]. Additionally, red pulp macrophages express the receptor Trem-like 4 that binds to apoptotic cells, improving its removal capacity of damaged cells [31].

Once an erythrocyte is phagocytosed by RPMs, the RBC-containing phagosome fuses with lysosomal vesicles, forming the erythrophagolysosome, which is accompanied by erythrocyte degradation by proteases and enzymatic catabolism of RBC components, resulting in the breakdown of hemoglobin and the release of heme [32]. The enzyme heme-oxygenase 1 (HO-1) catabolizes free heme, which is transported into the cytoplasm by the heme transporter HRG1 [32–34]. The catabolism of heme also releases ferrous iron that can be stored or reused. Red pulp macrophages are able to store larger amounts of iron in the form of ferritin aggregate haemosiderin, which is an insoluble complex of partially degraded ferritin (a cytosolic protein) [35]. These cells express several proteins involved in iron metabolism, such as Nramp1, ferroportin, heme oxygenase-1, and the scavenger receptor CD163 [36–39]. The iron transporter Nramp1, located on the phagolysosomal membrane, plays an important role in the iron transport into the cytosol [32,33]. The iron liberated from heme that is not stored as ferritin can be transported back to the circulation by the iron exporter ferroportin [33].

Red pulp macrophages have also been shown to perform immunological functions such as clearing blood-borne pathogens. The most well-characterized mechanism involved in this is the phagocytosis and elimination of pathogens from circulation. Also, iron uptake by red pulp macrophages has been shown to be important in limiting the growth of intraphagosomal pathogens by limiting their supply of iron [33]. Iron is one of the most important nutrients for several pathogens, but due to its scarcity under physiological conditions, various microorganisms are constantly competing for iron with the host. Some microorganisms, especially bacteria, have evolved a wide range of strategies to overcome iron shortage and to ensure sufficient uptake. One of these strategies relies on the synthesis and excretion of siderophores, which are molecules with a high affinity for iron that either bind free iron or sequester it from iron-binding proteins that are transported back into bacteria by specific receptors [33]. After pathogens encounter and stimulate Toll-like receptors, macrophages can secrete molecules such as lipocalin-2, which binds to siderophores, limiting pathogen iron uptake and consequently, controlling its growth.

Another way RPMs control iron availability and pathogen replication involves the intraphagosomal iron transporter Nramp1. Notably, Nramp1 is associated with protection against intracellular pathogens, such as *Mycobacterium bovis* BCG, *Leishmania donovani*, and *Salmonella typhimurium*, due to its modification of the intraphagosomal environment, which directly affects the replication of intracellular parasites [40,41]. Nramp1 synthesis can be also upregulated by stimulation with IFN-γ and LPS [42], a condition likely to occur during acute blood-borne infections. In addition, after IFN-γ or LPS stimulation, Nramp1-expressing cells exhibited greater enzymatic activity of inducible nitrous oxide synthase (iNOS), and increased amounts of NO, which is directly involved in protection against infections with intracellular pathogens [42]. Red pulp macrophages also express the mannose receptor (MR), a type I membrane molecule with two lectin activities, which recognizes mannose and sulfated ligands through its C-type lectin-like carbohydrate and cysteine-rich domains, respectively. MR is an endocytic pattern recognition receptor (PRR) and its expression in red pulp macrophages is likely to be involved in the uptake and clearance of endogenous and microbe-associated mannosylated ligands. Of note, MR expression in RPMs contributes to antiviral immunity by generating a rapid $CD8^+$ T-cell response [43]. Such activation of $CD8^+$ T cells contributes to immunity early after viral infection, which can reduce the spread of the virus. In addition, the mannose receptor has been shown to participate in the uptake of pathogens, such as *Leishmania* and *Candida albicans*, which can act in synergism with other PRRs to enhance the ability of macrophages to sense *Leishmania* or *Candida* infections [44,45]. Thus, in addition to the important functions in regulating erythrocyte homeostasis and iron metabolism, red pulp macrophages can also mediate immunity in the spleen by regulating uptake and elimination of pathogens through several mechanisms, such as expression of receptors involved in microbial binding (TLRs, MR) and iron hijacking.

The splenic white pulp

Blood is delivered to the spleen by the afferent splenic artery through several branches of arterioles into the marginal sinus [14]. These branches further divide in the spleen, which eventually ends in the venous sinuses of the red pulp (Fig. 1). The venous sinuses then leave the trabeculae to become central arteries that lack accompanying veins and are surrounded by a layer of lymphoid tissue as a continuous sheath, known as the white pulp. The white pulp, distinct from red pulp, is excluded from the bloodstream, and specific signals are required for the entry of blood.

The white pulp is organized as lymphoid sheaths around the arterial branches of the spleen and is subdivided into two distinct regions: an area of B-cell follicles and another area containing T cells (Fig. 1). It has a very similar organization to what is found in the lymph nodes, except for the apparent absence of high endothelial venules. The T-cell zone in the splenic white pulp is located in a central arteriole that is surrounded by a periarteriolar lymphoid sheath (PALS), whereas B cells are arranged in one or more follicles, all of which are bound by the marginal zone (MZ), which separates each individual WP from the RP (Fig. 1). Comparable to the situation of lymph nodes, the organization and maintenance of these compartments are controlled by specific chemokines that are able to attract B cells and T cells to their respective zones. For instance, chemokines such as CXCL13 is required to attract B cells expressing the chemokine receptor CXCR5 to follicular B-cell zones [46], while CCL19 and CCL21 are crucial to attract and retain $CCR7^+$ T cells and antigen-presenting DCs into T-cell zones of the white pulp [47]. CXCL13 is produced by follicular dendritic cells (FDCs) and stromal cells present in the follicles adjacent to the FDCs, while CCL19 and CCL21 are predominantly produced by stromal cells, in particular fibroblastic reticular cells (FRC). Furthermore, FRC produce the T-cell survival factor cytokine IL-7 [48]. CCL19 is also produced, albeit to a lesser extent, by $CD8^+$ DC in the T-cell zone [49]. CXCR5 signaling induces the expression of lymphotoxin (LT)$\alpha 1\beta 2$ on B cells, which in turn induces and maintains the maturation and differentiation of FDCs and their expression of CXCL13, thus mediating a positive feedback loop which is necessary for the establishment and maintenance of lymphoid follicle integrity [46,50]. Also, plasmacytoid DCs, which are important for innate and adaptive immune responses to viruses, constitutively migrate from the blood into the splenic white pulp in a CXCR4- and CCR7-dependent manner, toward the chemokines CXCL12 and CCL21 on fibroblastic reticular cells in the T-cell zone and in the marginal zone bridging channels [51].

White pulp macrophages

Both B-cell follicles and the T-cell zone of the white pulp contain discrete macrophage populations. Not much is known about the origin and life span of these populations. White pulp macrophages are characterized by the expression of the pan-macrophage marker CD68 and are found closely associated with follicular dendritic cells [52]. White pulp macrophages are specialized phagocytes that can regulate immune responses against pathogens. These cells express little or no F4/80 antigen. After antigen uptake and stimulation, CD68 is strikingly upregulated and these antigen-processing and presenting cells migrate into splenic white pulp, in particular to the germinal center region [53]. Germinal centers (GCs) are transient antigen-driven structures that form within the center of the follicle. GCs are anatomically divided into two distinct zones, a light zone that faces proximally to the site of antigen entry and a dark zone toward the T-cell area [54]. The light zone is clearly distinguishable by the presence of B cells interspersed with specialized stromal cells known as follicular dendritic cells (FDCs), which give this zone its "lighter" appearance [54]. In addition to antigen-specific GC B cells and FDCs, LZ also contains populations of naive IgD^+ B cells, T_{FH} cells, T cells, and macrophages. The dark zone is identified by the relative absence of FDCs and by the higher density of proliferating GC B cells. FDCs in the light zone express high amounts of Fc and complement receptors that sequester antigen-containing immune complexes, which lead to a selection process for high-affinity B cells followed by further editing of their B-cell receptors through hypermutation [54]. B cells expressing receptors with a high affinity for antigens will survive, and those with low-affinity receptors will undergo apoptosis and phagocytosis by a white pulp macrophage group named tingible body (TB) macrophages. These macrophages are a specialized group of phagocytes that engulf and eliminate apoptotic B cells that are frequent products of GC selection. This constant uptake of apoptotic cell debris and associated chromatin results in their tingible (stainable) staining pattern displayed by dark blue staining when using hematoxylin and eosin [55].

Recognition of apoptotic cells by these phagocytes involves the expression of several ligands and receptors, in particular, the expression of phosphatidylserine (PS) in the plasma membrane outer leaflet of apoptotic cells [56]. Tingible body macrophages have been shown to express various soluble molecules required for recognition of phosphatidylserine in apoptotic cells, such as the milk fat globule epidermal growth factor 8 (MFG-E8), growth arrest specific-6 protein (Gas6), T-cell immunoglobulin and mucin-domain-containing molecule (Tim4), and the tyrosine kinase Mer [57–59]. MFG-E8-deficient mice failed to remove apoptotic cells in the germinal center, which led to the production of autoantibodies and the development of autoimmune disease [60]. Interestingly, an elegant study from Kranich and colleagues showed that follicular dendritic cells deliver MFG-E8 to the TB macrophages, which mediates uptake of apoptotic cells [61]. Additionally, another report showed that Mer-deficient mice not only exhibited long-term accumulation of apoptotic cells in the germinal center but also showed a dysregulation of GC B-cell and $CD4^+$ helper T-cell responses and IFN-γ production, which promoted alteration of B-cell tolerance and the induction of autoimmunity [58]. These studies indicate that white pulp macrophages, in particular TB macrophages, primarily recognize and remove apoptotic cells from the germinal center reaction. As such, these cells play a fundamental role in the development and control of immune responses against antigens in the spleen.

The splenic marginal zone

Most of the arterial blood that enters the spleen ends in the venous sinusoidal system surrounding the white pulp, known as the marginal zone (MZ) [62,63]. Blood is delivered to the MZ by terminal arterioles, which release their contents into an open blood system without endothelial linings. The marginal zone is situated at the transition site between the scavenging red pulp and the lymphoid white pulp and it is distinguished from the RP by the absence of sinuses. Its specialized organization facilitates filtering of blood-borne pathogens, and initiation of innate and adaptive immune responses. In rodents, the MZ is well developed and separated from the WP by the marginal sinus, but in humans there is no clear MZ observed at the junction of white and red pulps [64]. Instead of an MZ, a specialized structure called the perifollicular zone has been identified in humans. This zone is observed as an extension of the sinuses of the red zone, as it is directly connected with this structure [65]. As most of the blood entering the spleen passes the marginal zone, it is together with the red pulp involved in the filtration of the blood, thereby constituting the primary site for the detection of blood-borne antigens and pathogens [33]. Because of their anatomical location at the border of the red pulp, they have a rich supply of open capillaries. Also, the marginal zone of the spleen is an important area for cells that are leaving the bloodstream and entering the white pulp. Its specialized organization facilitates filtering of blood-borne pathogens and the initiation of innate and adaptive immune responses.

The MZ contains a variety of different cell types including a distinct subset of B cells (marginal zone B (MZB) cells); a large number of innate-like lymphocytes; stromal fibroblasts; a special stromal endothelial cell called the marginal sinus lining cell that expresses mucosal vascular addressin cell-adhesion molecule 1 (MAdCAM1); and phagocytic populations such as DCs and neutrophils that are specialized to monitor the blood and help in the MZB cells activation, respectively [33,66–70]. Closely associated to MAdCAM1^{+} endothelial cells reside two specific populations of macrophages: the marginal zone metallophilic macrophages (MMMs); and the marginal zone macrophages (MZMs) which are in constant contact with MZB cells and dendritic cells, contributing to the uptake and clearance of blood-borne pathogens [66,69,71]. Remarkably, the marginal zone is the only microanatomical environment in which two phenotypically and functionally different macrophage populations coexist. These cells are firmly attached to a framework of reticular fibroblasts and sinus-lining endothelial cells in sinusoid spaces, which allows for continuous scanning of the blood for signs of infection, and serve a scavenging function to remove particulate material from the circulation [72]. Dendritic cells are thought to reside temporarily in the marginal zone and to migrate into the white pulp after stimulation and antigen uptake. Also, lymphocytes may reside in the marginal zone for longer periods during the process of transmigration into the white pulp. Whereas the white pulp is restricted to being involved in adaptive immune responses, including antibody responses to antigens in a T-cell-dependent manner, the MZ is involved in both innate and adaptive immunity to blood-borne pathogens (bacteria and viruses) and T-cell-independent antibody responses, through its specific macrophage populations and marginal zone B cells. [73]. This functional distinction between T-cell-dependent and T-cell-independent antibody responses correlates with the presence of two distinct B-cell populations: follicular and MZ B cells, respectively. To be fully efficient at mounting an antibody response, MZ B cells must establish, in a well-controlled spatiotemporal manner, interactions with specialized subsets of macrophages, DCs, and fibroblast-like cells and respond to environmental cytokines.

Reports have indicated that B cells play an important role in organizing and supporting the integrity of the marginal zone. Marginal zone B cells represent 5% of total splenic B cells and reside between the MZ and red pulp. Follicular and marginal zone B cells likely arise from a common progenitor but differentiate into cells with distinct roles within the spleen [33]. Marginal zone B cells are particularly well-equipped to capture antigens carried in the blood via complement receptors, may function as APCs, and play an important role in both T-cell-independent and dependent immune responses [12]. These cells are in a pre-activated state, which allows them to rapidly produce IgM after exposure to blood-borne antigens. The impact of marginal zone B cells on T-cell responses is not entirely clear but they may be prime naïve $CD4^{+}$ T cells and drive their differentiation to helper T-cell subsets (e.g., Th1, Th2), interact with NKT cells leading to antigen-specific regulatory T-cell generation [74], and can also serve as targets for viral infection which can result in targeted killing of marginal zone B cells by CD8+T cells [74].

Studies using various mouse models have shown that B cells are required for normal MZ development and maintenance, as well as for the presence and function of marginal zone macrophage subsets [69,75]. In lymph-node development and organization, LTβR signaling triggered by lymphotoxin LTα1β2 from lymphoid tissue inducer (LTi) cells leads to upregulation of expression of chemokines (CCL19, CCL21, CXCL12, and CXCL13) and adhesion molecules (ICAM1, VCAM1, and MAdCAM1) by stromal cells [76,77], and a similar role can be attributed for B cells in the organization of the spleen marginal zone. Splenic B cells express LTα1β2 that binds the LTβR receptor expressed by marginal zone endothelial and stromal cells, which promotes the growth and maturation of these cells and also the secretion of a range of chemokines that in turn influence the lodging and retention of several cellular subsets in this region. It is in agreement with the findings that mice lacking lymphoid CCL21 and CCL19 (two chemokines whose expression is dependent on LTα1β2), or in animals deficient to CCR7 (the common CCL21 and CCL19 receptor), there is a severe paucity of MZM next to the endothelial lining of the marginal sinus [78,79]. Another study showed that LTβR signaling on MZMs by LTα1β2-expressing MZ B cells was required for macrophage establishment in the marginal zone [80]. Remarkably, the authors found that loss of LTβR signaling in macrophages, but not endothelial cells or follicular dendritic cells, caused a reduction in the percentage of MZ with age, which was not observed until 1 month after birth. It was also observed that conditional depletion of LTβR signaling in MZMs provoked spontaneous germinal center formation, possibly due to a failure in their ability to perform phagocytosis and clear apoptotic cell-derived antigens in the marginal zone as a consequence of MZM dissipation. Altogether, these data suggest that a failure of LTβR signaling in MZMs can provoke a mislocalization of MZ B cells into the follicle, which disrupts cross talk between MZMs and B cells that subsequently interferes in the establishment and function of marginal zone macrophages in the spleen.

Macrophages in the marginal zone

The splenic marginal zone has a unique microanatomical environment containing specialized cells involved in evaluating the nature of arriving material and determining its fate. Within this region reside two phenotypically and functionally distinct macrophage populations: (1) the marginal zone macrophages (MZMs) that are located outside the follicles in the MZ compartment and form the outer ring closer to the red pulp and (2) the marginal metallophilic macrophages (MMMs) which line the B-cell follicles directly underneath the cells lining the sinus. Similar to tissue macrophages, they display receptors involved in recognition and phagocytosis of opsonized particles, such as Fc and complement receptors, but their outstanding capacity for binding and uptake of non-opsonized particles is based on the expression of particular sets of PRRs and scavenger receptors such as C-type lectin domain containing receptor SIGN-R1 (a mouse homolog of DC-SIGN) [81–83], the scavenger receptor MARCO [84,85], and the sialic-acid binding receptor CD169 (also termed Siglec-1 or sialoadhesin) [33,66,86]. MZMs are characterized by the expression of SIGN-R1 and MARCO, which recognize non-opsonized molecules, mainly blood-borne antigens [81]. Unlike MZMs, MMMs are defined, among other molecules, by high surface expression CD169, which recognizes sialic acids [33,66,86]. This combination of pathogen receptors, as well as their location relative to the blood flow in the spleen, makes these cells ideally suited to take up particulate antigens (Ag) present in the blood.

It is well known that the development of macrophages in the marginal zone occurs in the first weeks of life. However, little is known about the precise origin and specific factors that drive the differentiation of macrophages in the splenic marginal zone. It has been reported that the development of MZMs is dependent on the expression of the nuclear liver X receptor alpha (LXRα) [87]. Importantly, LXRα-deficient mice present with normal monocyte numbers and LXRα deficiency did not affect the development of macrophage subsets located in the red pulp or in other lymphoid tissues including liver, Peyer's patches, lymph nodes or intestine. Interestingly, adoptive transfer of wild-type monocytes into LXR-deficient mice leads to functional reestablishment of the splenic marginal zone macrophage population, indicating that circulating hematopoietic derived myeloid cells can function as precursors to MZMs within the spleen under conditions of LXRα deficiency, and that LXRα signaling in myeloid cells is required for the generation and maintenance of MZMs both during development and in adult life.

Despite the advances in studies involving MZ macrophages in recent years, there are limited tools for the study of these cells. In addition to the clodronate liposome treatment (used in vivo to remove macrophages from lymphoid tissues and other sites), several established genetic mouse models are commonly used to deplete MZ macrophages, including conditional depletion of MMMs or MZMs using injection of diphtheria toxin (DT) in transgenic mouse strains that express the DT receptor (DTR) under the control of macrophage promoters restricted to $CD11c^+$ or $CD169^+$ markers [88]. However, in both cases, administration of diphtheria toxin leads to depletion of MMMs as well as MZMs, which does not allow for the determination of which population of macrophage is responsible for specific functions. Another limitation in the study of these populations is the difficulty in their isolating due to their minor frequency in the spleen, their close contact with stromal and endothelial cells, and the particular properties of their surface markers that are quickly internalized by the macrophage during sample preparation.

Marginal zone macrophages and tissue immunity

Marginal zone macrophages, the second major macrophage population of the marginal zone, constitutively express a set of receptors with selective specificity for pathogens. This and their dispersed residence throughout the marginal zone make these cells important scavengers, acting as sentinels that screen for blood-borne antigens. Different from MMMs, MZMs are in close contact with MZ B cells and together these cell populations have developed efficient mechanisms for phagocytosis of antigens including immune complexes, systemic pathogens, and self-antigens, such as oxidized lipids and apoptotic cells [82,89,90]. The nature of the antigens phagocyted by MZMs as well as their close physical and mechanistic association with MZB cells determine the immunological role of these cells, which is involved in the development and differentiation of both cell populations [71,91]. It has been shown that MZMs quickly lost SIGN-R1 expression and phagocytic activity when MZ B cells are absent due to either genetic developmental defects or following transient migration of B cells out of the MZ by administration of LPS or FTY720 (a drug that blocks emigration from lymphoid organs into blood and lymph) [71,92]. Reciprocally, MZ B cells also were unable to capture antigens and produce IgM in response to polysaccharides when SIGN-R1-expressing MZMs were not present. Loss of MZMs (by treatment with clodronate-containing liposomes) also leads to MZ B-cell retention loss and impaired particulate antigen trapping [71]. In the absence of MZ B cells, $CD11c^+$ dendritic cells distribution is also altered, and these cells no longer concentrate in the MZ bridging channels and instead are redistributed circumferentially around the MZ, which can likely alter the adaptive immune response [92]. Thus, a constant dialog between MZ B cells and MZ macrophages is required for promoting MZ integrity and organization, as well as for efficient clearance of pathogens and immune responses, indicating that these cells can functionally and physiologically influence each other.

Marginal zone macrophages are particularly important in defense against pathogens, and delayed maturation of these cells in newborn mice and human infants, or splenectomy in adults, results in increased susceptibility to infection. Reports have shown that various pathogen receptors expressed by MZMs participate in the recognition and elimination of certain pathogens. SIGNR1 is an important C-type lectin present on marginal zone macrophages and responsible for recognition of polysaccharides antigens [81,93]. The C-type lectin domain of mouse SIGNR1 has high similarity to human DC-SIGN, and the amino acid residues important for both $Ca2^+$ and ligand binding are conserved in the murine homolog [81]. The ligand-binding specificities of SIGNR1 are quite similar to the ones described for DC-SIGN and have been shown to have binding specificity for mannose-, fucose-, and *N*-acetylglucosamine-terminating oligosaccharides [94]. SIGNR1 binds to yeasts and the yeast-derived particle zymosan, leading to efficient uptake of zymosan and production of tumor necrosis factor (TNF)-α [81,95]. SIGN-R1 also binds to bacteria *Streptococcus pneumoniae* [96], *Mycobacterium tuberculosis* [97], *Escherichia coli*, and *Salmonella typhimurium* [98], facilitating efficient phagocytosis and bacterial clearance, which lead to protection against septicemia. Furthermore, SIGNR1 has been shown to be involved in the uptake of viruses such as Marburg and Ebola virus and the S protein of severe acute respiratory syndrome coronavirus [99], as well as human immunodeficiency virus (HIV) [81]. This correlates with findings showing that marginal zone macrophages are essential for the clearance of viruses [100].

It has also been shown that SIGNR1 can associate with Toll-like receptors to capture antigens and increase activation of NF-κB [98,101,102]. Using lipopolysaccharide (LPS) from *E. coli* or *Salmonella minnesota*, it was observed that LPS binds to SIGN-R1 via oligosaccharides in the non-reductive end of the LPS core region. This binding entirely abrogated the interaction of SIGNR1 with FITC-dextran and *E. coli*. Also, LPS binding to SIGN-R1 enhanced oligomerization of TLR4 as well as degradation of IκB-α, and it was found that SIGN-R1 physically associates with the TLR4-MD-2 complex to promote cytokine production [98]. Additional studies showed that antigens from different pathogens such as *Mycobacterium leprae*, *Candida albicans*, *Mycobacterium tuberculosis*, measles virus, and human immunodeficiency virus-1 bind to DC-SIGN and modulate different TLR-mediated cytokine responses. The authors also found that the molecular mechanism(s) underlying the DC-SIGN-TLR responses depend on DC-SIGN-induced activation of the serine and threonine kinase Raf-1-acetylation, which modulates signaling of different TLRs. Taken together, these findings raise an important point, which is that TLRs-SIGNR1 interactions greatly enhance the efficiency by which the marginal zone macrophages can eliminate pathogens. In addition to interactions with TLRs, SIGNR1 was also demonstrated to interact with murine ICAM-2, which is widely expressed on lymphocytes [103] and could therefore function as an adhesion receptor mediating contact between SIGNR1^{+} cells and leukocytes in the marginal zone [81]. Strikingly, marginal zone macrophages and B cells are in close contact, indicating that marginal zone macrophages play a role in the migration and retention of B cells in the marginal zone through SIGNR1-ICAM-2 interaction.

The SIGNR1 molecule contains a triacid cluster in its cytoplasmic region that is thought to be responsible for both binding and internalization of particles, after which the SIGNR1-particle complex is targeted to lysosomes for antigen processing [66,86]. However, MZMs do not express MHC class II and it has been postulated that similar to classical complement pathway activation, but without the traditional requirement for antibody, degraded products bound to SIGNR1 are released from macrophages and directly bind the complement C1 subcomponent, C1q, and assembles the complex C4bC2a or C3 convertase that catalyzes C3b opsonin formation [104]. This phenomenon was shown to be important for resistance against blood-borne *Streptococcus pneumoniae* infection. Also, SIGNR1-bound particles released from the macrophages can be taken up by marginal zone B cells after opsonization by complement [105].

MARCO, another class of pathogen-biding receptor expressed in MZMs, belongs to class A scavenger receptors and its structure resembles the SR-A1 (Scavenger Receptor A1) molecule, consisting of a trimer of disulfide-bonded proteins with a collagenous structure. It is basically composed of short cytoplasmic and transmembrane domains, an extracellular spacer domain, a long extracellular collagenous domain, and a C-terminal fifth domain, forming a scavenger receptor cysteine-rich (SRCR) domain [84]. MARCO is constitutively expressed in several subsets of naive tissue-resident macrophages, especially in MZMs present in the spleen and lymph node [84]. Expression of MARCO can also be rapidly upregulated in other macrophage populations, such as alveolar and liver macrophages [84,85,106]. Consistent with that, it was shown that TLR engagement by LPS induced increased expression of MARCO, which stimulated MARCO-mediated phagocytic activity, contributing to efficient capturing and clearance of invading microbial pathogens [107,108]. Indeed, failure in this function caused increased antigen load in the MZ region of the spleen, contributing to aberrant activation of the MZ B cells [107].

MARCO binds to a broad range of self and pathogenic ligands, including acetylated and oxidized low-density lipoprotein and bacterial components [66,84,109,110]. It was found that gram-negative and gram-positive bacteria such as *E. coli* and *Staphylococcus aureus* are recognized by MARCO [66,109]. It has also been shown that MARCO is crucial for responses to viruses such as influenza, vaccinia, and herpes simplex virus [111–113].

Besides capture and phagocytosis function, MARCO activity is indispensable for the integrity and proper function of the MZ cells. Interestingly, it was demonstrated that MARCO expression is essential for retention and trafficking of the MZ B cells in the marginal zone since disruption of the cross talk between MZ B cells with MARCO results in migration of MZ B cells to the splenic follicles [91]. Additionally, ablation of MARCO leads to a defective microarchitecture of the splenic MZ and an impaired T-independent type 2 response when challenged with pneumococcal polysaccharide [114].

Marginal metallophilic macrophages and tissue immunity

Marginal metallophilic macrophages are essential for trapping microbial polysaccharides and blood-borne pathogens such as viruses and bacteria. In contrast to the position of the marginal zone macrophages that are located in an apparently random manner along the width of the marginal zone toward the red pulp, MMMs form a thin rim at the inner border of the marginal zone [115] and are more associated with the white pulp, found in a similar position surrounding the T-cell zone of lymph nodes directly underneath the subcapsular sinus of these organs [16]. This localization adjacent to T- and B-cell-rich splenic gives them an important role in transport of microbial antigens to B and T cells. Marginal metallophilic macrophages show a high affinity for silver (metallophilic) and are less phagocytic than MZMs [62,116]. The macrophage growth factor M-CSF (CSF-1) has been shown to be required for the development of MMMs, demonstrated by their absence in CSF-1-deficient op/op mutant mice [117].

These cells express the pattern recognition receptor CD169 that has been implicated in the binding and internalization of sialylated *Neisseria meningitidis* in an Sn- and sialic acid-dependent manner [118]. Further, CD169 has also been shown to function as an uptake receptor for *Campylobacter jejuni* [119]. Experimental models in which marginal zone macrophages and marginal metallophilic macrophages were selectively depleted by the uptake of clodronate liposomes show that trapping of *Listeria monocytogenes* in the spleen and early control of infection are lost in the absence of these cells [120]. Likewise, a study showed that although *Listeria monocytogenes* is captured by a wide range of phagocytes, macrophages in the marginal zone have a microbicidal activity that is required for initial control of bacteria replication [121]. Further, studies have shown that MMMs are critical to effective establishment of a type I interferon response and consequent antiviral protection in experimental models using infections with *Lymphocytic choriomeningitis* virus (LCMV) or herpes simplex virus (HSV) [100,122,123]. Interestingly, it was also observed in these studies that MMMs play a crucial role in capturing, degrading, and clearing *Listeria monocytogenes* or LCMV, although this was not coupled with direct Ag presentation [120,123]. Consistent with these observations, a study by Haan and colleagues showed that MMMs were essential for the development of T-cell-dependent immunity by mechanisms dependent on cross-presentation of blood-borne antigens by splenic DCs [124]. Using conjugated OVA antibodies to specifically target different macrophages in the spleen, the authors demonstrated that antigens targeted to MMMs as well as blood-borne adenoviruses are efficiently captured by MMMs, transferred to $CD8a^+$ dendritic cells, leading to efficient cross-presentation and activation of cytotoxic T lymphocytes. In addition to clearance of bacteria and viruses, MMMs are also involved in the uptake of parasites such as *Leishmania donovani* [125] and *Trypanosoma cruzi* [126].

Removal of apoptotic cells by macrophages in the marginal zone and immune tolerance

The efficient removal of apoptotic cells and immunologic tolerance toward apoptotic cell-associated antigens is fundamental for the maintenance of homeostasis and prevention of immunological disorders and development of autoimmune diseases. Several studies have demonstrated that alterations in the ability of phagocytes to capture or clear apoptotic cells result in autoimmunity [60,127–129]. Resident macrophages such as macrophages present in the marginal zone play an important role in this process by expressing a range of surface receptors that mediate the rapid recognition and uptake of apoptotic cells in various tissues.

Splenic marginal zone macrophages are located relative to the blood flow in the spleen, making these cells ideally suited to scavenge the blood for apoptotic material entering the spleen from circulation. This strategic location endows these cells with the capacity to rapidly capture and clear endogenous material. Both MZMs and MMMs express multiple receptors involved in the recognition and uptake of apoptotic cells. As mentioned earlier, MZMs express the scavenger receptor A (SR-A) and MARCO, and SIGNR-1 that are implicated in the uptake of apoptotic cells and in the prevention of autoimmune diseases [130–133]. MMMs highly express CD169, Tim-4, and Treml4 that are receptors involved in the capture and phagocytosis of early and late apoptotic cells [31,134,135].

In the spleen, deletion of both MZ macrophage subsets impairs systemic tolerance toward apoptotic cells and accelerates the development and severity of autoimmune diseases [88,136,137]. It showed that intravenous injection of apoptotic cells loaded with a fragment of the myelin oligodendrocyte glycoprotein peptide (MOG peptide) causes an accumulation of apoptotic cell contents in the marginal zone and suppression of immune response against this antigen, which was sufficient to prevent the development of experimental autoimmune encephalomyelitis [88]. This tolerance depended on selective engulfment of dying cells by $CD8a^+$ DCs regulated by macrophages in the marginal zone. Previous studies identified that this unique subset of $CD8\alpha^+$ DCs is able to capture dead cells in the splenic MZ, migrate to the T zone, and inhibit T-cell responses [138,139]. The elimination of both MZMs and MMMs resulted in increased phagocytosis of apoptotic cells by $CD8\alpha^-$ DCs instead of $CD8\alpha^+$ DCs, which enhanced priming of cytotoxic T cells and resulted in failure to suppress immune responses [88,136]. Another study showed that IDO expression is upregulated in splenic MZMs upon phagocytosis of apoptotic cells. Of note, genetic or pharmacologic inhibition of IDO enhanced innate and adaptive immunity to apoptotic cells, which promoted enhanced pathology associated with systemic autoimmune disease development [140]. Subsequent studies showed that MMMs exposed to apoptotic cells rapidly induced the expression of the chemokine CCL22, which lead to the migration and activation of regulatory T cells and dendritic cells within the white pulp, promoting downstream suppressive immunity [141]. CCL22, also known as macrophage-derived chemokine (MDC) [142], is a T-cell chemoattractant that binds to and activates CCR4 [143–146]. CCR4 is preferentially expressed by $CD4^+$ and $CD8^+$ lymphocytes particularly Th2/Tc2 subsets, and a subpopulation of $Foxp3^+$ regulatory T cells (Treg), and expression of CCL22 drive Treg recruitment/accumulation and suppression of adaptive immunity [144,147–149]. Of note, inhibition of either CCL22/CCR4 signaling or Treg function was sufficient to block the systemic breakdown of tolerance to self, driven by apoptotic cell injection [141]. Similarly, MMMs and CCL22 were required for DC-dependent regulatory T-cell induction and protection of ischemic injury [150]. Thus, these studies show that splenic macrophages in the marginal zone are critical for the clearance of endogenous material and that they play an important role in apoptotic cell antigen-specific induction of tolerance.

Lymph node

Lymph nodes (LN) are secondary lymphoid organs distributed throughout the body located alongside the lymphatic vasculature that drains peripheral tissues and organs of the body to deliver molecules, antigens, microorganisms, and cells such as lymphocytes and antigen-presenting cells (APCs) from the tissues [151]. These organs are highly compartmentalized by a capsule derived from lymphatic vessels and are divided into four anatomical regions (subcapsular sinus, cortex, paracortex, and medulla) (Fig. 2) characterized by distinct cellular populations (hematopoietic, vascular and mesenchymal cells) and supported by a vascular-stromal compartment that provides structural support, survival factors, regulatory molecules, and guidance cues [152–155]. Although the vast majority of LN cells (approximately 95%) are leukocytes, different populations of non-hematopoietic stromal cells are present, including fibroblastic reticular cells (FRCs), follicular dendritic cells (FDCs), marginal reticular cells (MRCs), blood vessel endothelial cells (BECs), and lymphatic endothelial cells (LECs) that are essential for cell recruitment, compartmentalization, fluid flow, and homeostatic maintenance of immune cells.

The collected lymph, lymph-borne antigens, and dendritic cells enter the LN via several afferent lymphatic vessels, through the planar conduit of the subcapsular sinus (SCS) located immediately beneath the lymph node capsule (Fig. 2). This sinus surrounds the superficial cortex, which contains B-cell follicles, and is adjacent and peripheral to the paracortex (a T-cell-rich region delineated by the fibroblastic reticular cells, where T cells encounter dendritic cells to drive their activation) [156]. Lymphocytes enter the LNs through specialized high endothelial venules (HEV) that have a distinctive appearance, consisting of a plump cuboidal endothelium and various receptors to interact with leukocytes (such as specialized ligands for lymphocytes that support lymphocyte extravasation from the blood) [157–159]. The remaining fluid exits the lymph node through efferent lymphatic vessels, reentering the circulation via the thoracic duct. Finally, the medulla, an antibody-secreting plasma cells-rich region on the LN, contains large blood vessels and interconnected lymphatic sinuses, which serve as an exit zone for lymphocytes. Throughout all of these regions, there exist specialized subsets of macrophages (Fig. 2) that exert unique immunological functions, working as a frontline of immune protection to lymph-borne pathogens. These macrophages have long been categorized according to their anatomical location and on uptake capacity of lymph-borne antigens. Structural studies of lymph nodes have shown the presence of a macrophage population lining the floor of the SCS, overlying follicles, that quickly captures small amounts of lymph-borne antigens but are poorly phagocytic [160–162]. By contrast, the macrophages associated with the medullary sinuses (MSMs) display highly phagocytic activity, trapping large amounts of material, which is evident through their abundant phagolysosomes [161–164]. The medulla

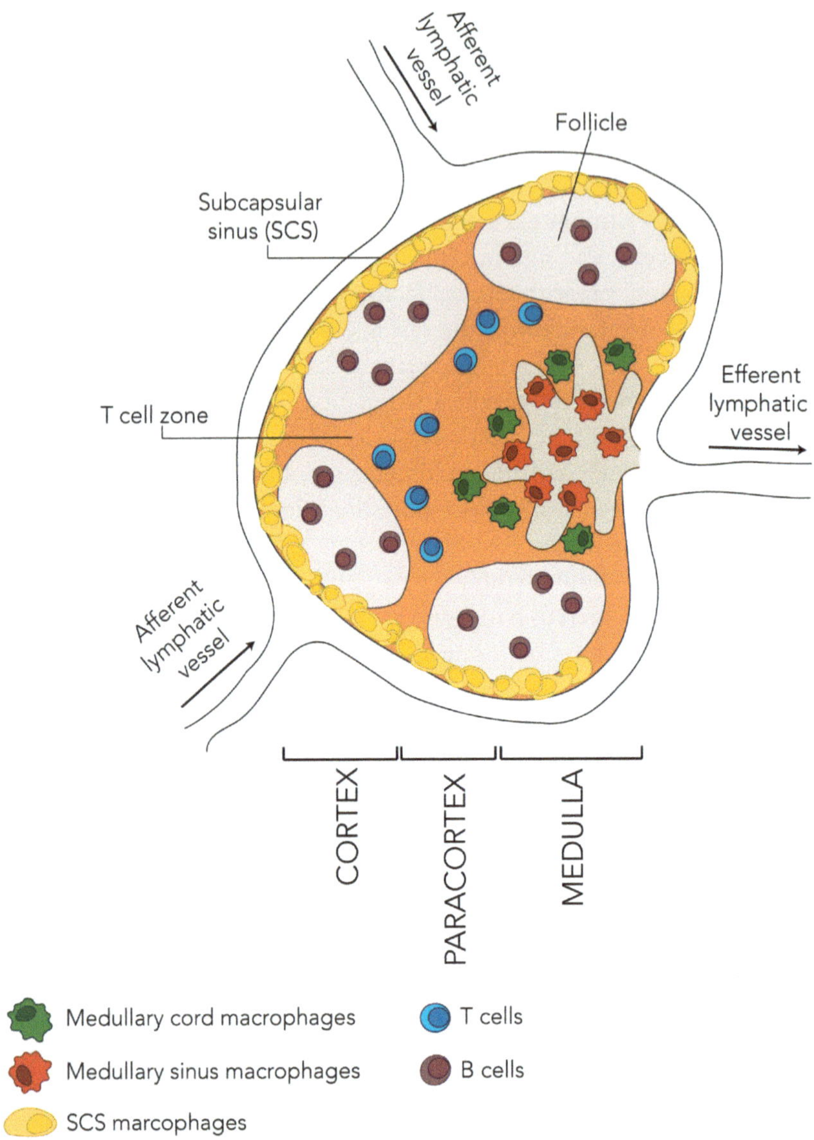

Lima-Junior, Collins & Han. Figure 2

Fig. 2 General structure of the lymph node showing major zones and cellular composition. The lymph node is organized into three main areas: the follicles (B-cell zone), the paracortex (T-cell zone), and the medulla. Lymph arriving from the tissues enters the lymph node by afferent lymphatic vessels and flows into subcapsular sinus (cortex), through the parenchyma (paracortex) to the medulla. After passing through the medulla, lymph then exits via the efferent lymphatic vessel. B lymphocytes are located into follicles in the superficial cortex underneath the subcapsular sinus macrophages (SSMs), where antigens are presented by SSMs to B cells. SSMs line the floor of the subcapsular sinus overlying the B-cell follicles. Medullary sinus macrophages (MSMs) and medullary cords macrophages (MCMs) reside in the medullary sinuses and the medullary cords, respectively.

is also populated by a large number of macrophages (medullar cord macrophages/MCMs) that reside outside the sinuses, situated within the medullary cords [165]. Similar to other lymphoid organs, the lymph node also contains tingible body macrophages, which are responsible for the clearance of apoptotic cells within germinal centers [166].

Subcapsular sinus macrophages

The subcapsular sinus is the main site of entry of pathogens into the lymph node, because it is the site at which afferent lymph drains. The floor of the SCS is lined by the subcapsular sinus macrophages (SSMs) that are unique to LNs, playing an important role in trapping and presenting antigens to B cells. However, compared with MSMs, their ability to internalize and degrade antigens seems to occur at a lower rate [167]. Recent reports have shown that SCS macrophages can also prevent the systemic spread of lymph-borne pathogens and are capable of activating a diverse range of innate and adaptive cells, such as B cells and memory T cells. They are located at the interface between the B-cell follicles and sites of antigen entry and are characterized by the expression of Mac1 (CD11b/CD18), high levels of CD169, and low expression of F4/80 [167]. Unlike DCs, SSMs display limited motility, but have a dendritic morphology presenting a head that projects into the sinus, which facilitate to sample the lymph for antigen, and long stable tails extending across the lymphatic lining cell layer [165,167–169]. This arrangement enables them to efficiently capture and translocate antigen across the floor of the SCS [167] facilitating its recognition by follicular B cells, driving B-cell activation [165,168–170]. Using 2-photon microscopy, elegant studies showed that particulate antigens, lymph-borne viruses, and immune complexes captured by SSC macrophages could be transferred to B cells. Cognate B cells acquired antigens directly from the macrophages via their B-cell receptor [165,168,169]. In the case of immune complexes, SCMs relay them to non-cognate B cells, by mechanisms dependent on the high-affinity complement receptor 2 (CR2 or CD21/CD35) binding, which is posteriorly delivered to follicular dendritic cells (FDCs) in primary follicles [167,169]. Despite the important insights discussed earlier, the contribution of this mode of antigen presentation in promoting humoral immunity or controlling pathogens dissemination remains unclear. SSMs depletion by clodronate liposome treatment induces the impairment of B-cell activation in response to VSV [165], and the transient disruption of the SCS macrophages layer after inflammation promotes a diminished B-cell response to a secondary infection [171]. However, the absence of

SSMs does not impact the germinal center response, and the production of high-affinity antibodies was either not impaired [172,173] or only partially affected [167]. Similarly, the antibody response to VSV was even enhanced [172] and protection to this infection was independent of adaptive immunity [174]. Also, during influenza virus infection, SSMs were shown to be important for limiting the systemic spread of the virus, but not for the generation of a specific humoral response [172]. Thus, further studies are needed to clarify the full extent of the contribution of SSMs to B-cell activation in lymph nodes and other sites.

Subcapsular sinus macrophages are also efficient in capturing pathogens, including viruses, parasites, bacteria, and pathogen-derived antigens, from the lymph. This trapping function of SCS macrophages mediates the first layer of immune defense in the lymph node through the production of antiviral and antibacterial cytokines and peptides to initiate a localized inflammatory response, crucial to prevent the systemic spread of many pathogens [165,175–178]. The ability of SSMs to capture and phagocytose pathogens is achieved through expression of several cell-surface receptors including scavenger receptors, complement receptors, Fc receptors, Toll-like receptors, and CD169, that can bind to sialylated glycoproteins expressed on some viruses [179,180]. Of note, SCS macrophages are very efficient at binding viral particles, but they present poorly endocytic capacity and high ability to retain viral particles on their surface for several hours after capture, which has important implications for the generation of adaptive immune responses and prevention of systemic viral dissemination. Also, these peculiar characteristics can, in part, explain the differential susceptibility to viral infection of SSMs. In contrast to MSMs, SSMs are highly permissive to infection by a range of viruses including vaccinia virus, cytomegalovirus (CMV), vesicular stomatitis virus (VSV), and herpesviruses, as well as the parasite *Toxoplasma gondii*. [173,175,181–184]. This permissive state can contribute to augment lymph node immune responses in several ways. High susceptibility to virus replication may generate high levels of intact viral particles which provoke antibody responses, since viral particles are more effective in promoting B-cell activation than free antigens [185]. Conversely, suppression of VSV replication in $CD169^+$ cells, by depletion of $CD169^+$ cells using diphtheria toxin (DT) treatment of $CD169^+$-DTR mice, or removal of Usp18 (a negative regulator of IFNαR1 signaling), led to a reduced T-cell and antibody response [186]. Interestingly, vaccinia virus induces inflammasome activation (an important molecular platform involved in proteolytic cleavage, maturation, and secretion of pro-inflammatory cytokines IL-1β and interleukin IL-18) in SSMs leading to a robust influx of inflammatory cells and the mobilization ofT cells from the circulation, which maximizes T-cell response [187]. Furthermore, bacterial infection or UV-inactivated influenza virus also results in inflammasome activation as well as the production of several cytokines such as IL-1β, IL-1a, IL-18, and type I interferon that triggers the recruitment and activation of immune cells into the subcapsular niche. Thus, an aberrant virus replication can induce a robust inflammatory response essential for the recruitment and response of effectors cells as well as induction of an antiviral state in immune cells and in other cell types that are more susceptible to detrimental infection, preventing virus from spreading from the LN to the blood circulation or other organs [165,173,188].

Another function of permissive viral replication in SSMs can be to mediate T-cell activation either directly, by pathogen-derived antigens presentation to $CD8^+$ T cells in the context of major histocompatibility complex (MHC) class I molecules, or indirectly, by antigen transfer to nearby DCs. Following vaccinia virus (VV) and vesicular stomatitis virus infection, 85% of virus-infected LN cells were $CD169^+$ macrophages, with the majority of these cells being $F4/80^-$ SSMs [181,184,189]. Antigen-specific CD8 T cells accumulated in the LN peripheral interfollicular region in close proximity to SSMs, but CD8 T-cell interaction with DCs was more frequent and stable. Although the antigen presentation by the macrophages was able to stimulate some T-cell division, it did not fully prime antiviral CD8+T cells resulting in a suboptimal T-cell activation [181,189]. In contrast, with this low capacity to directly priming naive $CD8^+$ T cells, SCS macrophages were shown to play a critical role in the rapid and efficient activation of virus-specific central memory $CD8^+$ T cells [190,191]. The accelerated response of central memory $CD8^+$ T cells to virus infection was coordinated by a cascade of cytokines and chemokines, particularly ligands for CXCR3, which selectively promote central memory $CD8^+$ T-cell recruitment toward infected cells [190,191]. Importantly, disruption of central memory $CD8^+$ T-cell interaction with infected SCS macrophages impaired viral clearance [191]. Collectively, these studies show that SCS macrophages participate in the rapid recruitment of several innate and adaptive immune cells to the inflammatory sites promoting the host defense.

Medullary sinus macrophages and medullary cord macrophages

Medullary sinuses are fluid-containing spaces, surrounded by medullary cords, that are filled with lymphatic fluids and bear high numbers of a macrophage subset known as medullary macrophages. Medullary sinus macrophages are basically identified according to their location inside the medullary sinuses of lymph node and also by their high expression of F4/80 [167,192]. These cells are also characterized by the expression of the MARCO, CD169 (although at lower levels compared with SCMs), SIGN-R1 (CD209b), the mannose receptor (CD206), and the lymphatic vascular endothelial marker LYVE-1 [179,193]. Different from SSMs, MSMs contain large lysosomes, high numbers of heterogeneous cytosolic vesicles, and are highly phagocytic cells responsible for the uptake and clearance of lymph-borne antigens traveling on to the medullary sinuses. Medullary sinus macrophages have been shown to mediate uptake of apoptotic cells. Conversely, TIM-4 expressing F4/80$^+$ medullary macrophages contribute to respiratory tolerance by phagocytosis of activated T cells in lung draining lymph nodes [194]. Another study showed that a population of CD169$^+$ F4/80$^+$ cells that reside in the LN sinus phagocytose apoptotic tumor cells via MFG-E8 and directly cross-present tumor antigens to CTLs, suggesting that uptake of apoptotic cells under inflammatory conditions can promote T-cell immunity [195].

An additional macrophage subset populates the parenchyma between medullary sinuses known as medullary cords [179]. Medullary cords structure generally develops during an immune response when the medulla suffers a structural remodeling with a massive increase of macrophages and B cells. In contrast to their sinus counterparts, medullary cord macrophages (MCMs) express F4/80 but are negative for CD169. In addition, they have distinct morphology, small lysosomes, enzymes contents, and phagocytic activity that are found in medullary sinus macrophages. For instance, after gastrointestinal inflammation, MSMs phagocytosed polymorphonuclear leucocytes and eosinophils and stained strongly for acid phosphatase and weakly for non-specific esterase. In contrast, MCMs stained weakly for acid phosphatase but strongly for non-specific esterase and phagocytosed apoptotic plasma cells [164]. Besides clearance of dying plasma cells, a likely non-phagocytic function of MCMs has been suggested. A study of Mohr et al. showed that during antigen immunization in alum, CD169$^-$ F4/80$^+$ CD11b^{hi} cells were in close association with plasma cells in the medullary region within the medullary cords. Also, the authors found that these cells were the major producers of IL-6 and APRIL, a plasma cell maturation/survival factors, during T-dependent Ab response in the lymph node, suggesting a potential role for MSMs-plasma cells interaction in mediate the quality, magnitude, and duration of antibody responses [196]. Supporting this idea, a live imaging study showed that plasma cells form durable physical interactions with myeloid cells, including MSMs, in the medulla. However, in contrast to the proposed function of medullary macrophages in promoting plasma cell maturation and survival, this study showed that medullary macrophages dampen activation and expansion of plasma cells leading to reduction in antibody responses [197]. Notably, however, in this report adjuvant-free immunization was applied, suggesting that different modes of activation may differentially affect macrophages functions and interactions with plasma cells.

Lymph node macrophages development

The origin and development of lymph node macrophages have long been studied for a better understanding of their characteristics and functions. CSF-1/CSF-1 receptor signaling has been shown to be fundamental for the development of SCS macrophages. Osteopetrotic *op/op* transgenic mice that lack CSF-1 or the administration of an anti-CSF1R antibody (that blocks the CSF1R signaling) provoke a significant reduction of SSMs [23,117,198]. Another key cytokine required for SSMs development is lymphotoxin (LT) $\alpha1\beta2$ [167,198,199]. Although SSMs and MSMs express lymphotoxin receptor LTβR, chimeric mice reconstituted with $Ltbr^{-/-}$ (LTβR-deficient) and wild-type bone marrow after irradiation only show a deficiency in SCS macrophages [167], indicating an intrinsic requirement for LTβR signaling in SSMs or their precursors. Furthermore, treatment with LTβR-Ig, a soluble antagonist of LTβ receptor that blocks its downstream signaling, significantly decreased $CD169^+F4/80^-$ (SSMs) macrophages abundance in the lymph node [167] but not $CD169^+F4/80^+$ (MSMs) cells. Secretion of $LT\alpha1\beta2$ by B cells was essential for development, maintenance, and function of the SSM population [167]. Conversely, μMT mice, which lack mature B cells in the lymph node, showed a marked decrease of SSMs, but increased numbers of $F4/80^+$ MSMs [167,174]. On the order hand, mice that transgenically overexpress $LT\alpha1\beta2$ on B cells [200] have a thicker layer and greater total numbers of SSMs, without affecting B-cell frequencies [167]. These data suggest that $LT\alpha1\beta2$ levels in lymph nodes under homeostatic conditions are limiting for SSM induction or maintenance. The requirements for MSMs and MCMs development are poorly understood. Differently of SSMs, LT blockade or CSF1 deficiency did not interfere in the development of $F4/80^+$ medullary macrophages [23,117,167]. However, CSF-1 receptor-deficient mice show a significant depletion of $F4/80^+$ macrophages, indicating the requirement of CSF-1 receptor activation for F4/80$^+$ macrophage development [27]. Collectively, these findings indicate that while SCS macrophages require CSF-1 receptor and LTβR for their development and the maintenance of their phenotype, MSMs, and MCMs developing is dependent on CSF-1 receptor signaling.

Acknowledgments

The authors thank Xingyao He and Kimberley Tindal for secretarial support. This work was supported by the Division of Intramural Research of the National Institute of Allergy and Infectious Diseases (NIAID), the Office of Dietary Supplements Research Scholar program and the Pew Latin American Fellowship Program in the Biomedical Sciences from the PEW Charitable.

References

[1] Davies LC, Jenkins SJ, Allen JE, Taylor PR. Tissue-resident macrophages. Nat Immunol 2013;14:986–95.
[2] Murray PJ, Wynn TA. Protective and pathogenic functions of macrophage subsets. Nat Rev Immunol 2011;11:723–37.
[3] Wynn TA, Chawla A, Pollard JW. Macrophage biology in development, homeostasis and disease. Nature 2013;496:445–55.
[4] Hashimoto D, Miller J, Merad M. Dendritic cell and macrophage heterogeneity in vivo. Immunity 2011;35:323–35.
[5] Martinez FO, Helming L, Gordon S. Alternative activation of macrophages: an immunologic functional perspective. Annu Rev Immunol 2009;27:451–83.
[6] Gordon S. The macrophage: past, present and future. Eur J Immunol 2007;37(Suppl. 1):S9–17.
[7] Gordon S, Taylor PR. Monocyte and macrophage heterogeneity. Nat Rev Immunol 2005;5:953–64.
[8] Houston Jr EG, Nechanitzky R, Fink PJ. Cutting edge: contact with secondary lymphoid organs drives postthymic T cell maturation. J Immunol 2008;181:5213–7.
[9] Mempel TR, Henrickson SE, Von Andrian UH. T-cell priming by dendritic cells in lymph nodes occurs in three distinct phases. Nature 2004;427:154–9.
[10] Riether C, Schurch CM, Ochsenbein AF. Regulation of hematopoietic and leukemic stem cells by the immune system. Cell Death Differ 2015;22:187–98.
[11] Steiniger BS. Human spleen microanatomy: why mice do not suffice. Immunology 2015;145:334–46.
[12] Bronte V, Pittet MJ. The spleen in local and systemic regulation of immunity. Immunity 2013;39:806–18.
[13] van Krieken JH, te Velde J. Normal histology of the human spleen. Am J Surg Pathol 1988;12:777–85.
[14] Groom AC, Schmidt EE, MacDonald IC. Microcirculatory pathways and blood flow in spleen: new insights from washout kinetics, corrosion casts, and quantitative intravital videomicroscopy. Scanning Microsc 1991;5:159–73 [discussion 173-154].
[15] Drenckhahn D, Wagner J. Stress fibers in the splenic sinus endothelium in situ: molecular structure, relationship to the extracellular matrix, and contractility. J Cell Biol 1986;102:1738–47.
[16] Nolte MA, Hoen EN, van Stijn A, Kraal G, Mebius RE. Isolation of the intact white pulp. Quantitative and qualitative analysis of the cellular composition of the splenic compartments. Eur J Immunol 2000;30:626–34.

[17] Khanna KM, Lefrancois L. Geography and plumbing control the T cell response to infection. Immunol Cell Biol 2008;86:416–22.

[18] Gautier EL, Shay T, Miller J, Greter M, Jakubzick C, Ivanov S, Helft J, Chow A, Elpek KG, Gordonov S, et al. Gene-expression profiles and transcriptional regulatory pathways that underlie the identity and diversity of mouse tissue macrophages. Nat Immunol 2012;13:1118–28.

[19] Kohyama M, Ise W, Edelson BT, Wilker PR, Hildner K, Mejia C, Frazier WA, Murphy TL, Murphy KM. Role for Spi-C in the development of red pulp macrophages and splenic iron homeostasis. Nature 2009;457:318–21.

[20] Hashimoto D, Chow A, Noizat C, Teo P, Beasley MB, Leboeuf M, Becker CD, See P, Price J, Lucas D, et al. Tissue-resident macrophages self-maintain locally throughout adult life with minimal contribution from circulating monocytes. Immunity 2013;38:792–804.

[21] Haldar M, Kohyama M, So AY, Kc W, Wu X, Briseno CG, Satpathy AT, Kretzer NM, Arase H, Rajasekaran NS, et al. Heme-mediated SPI-C induction promotes monocyte differentiation into iron-recycling macrophages. Cell 2014;156:1223–34.

[22] Schulz C, Gomez Perdiguero E, Chorro L, Szabo-Rogers H, Cagnard N, Kierdorf K, Prinz M, Wu B, Jacobsen SE, Pollard JW, et al. A lineage of myeloid cells independent of Myb and hematopoietic stem cells. Science 2012;336:86–90.

[23] Cecchini MG, Dominguez MG, Mocci S, Wetterwald A, Felix R, Fleisch H, Chisholm O, Hofstetter W, Pollard JW, Stanley ER. Role of colony stimulating factor-1 in the establishment and regulation of tissue macrophages during postnatal development of the mouse. Development 1994;120:1357–72.

[24] Takahashi K, Umeda S, Shultz LD, Hayashi S, Nishikawa S. Effects of macrophage colony-stimulating factor (M-CSF) on the development, differentiation, and maturation of marginal metallophilic macrophages and marginal zone macrophages in the spleen of osteopetrosis (op) mutant mice lacking functional M-CSF activity. J Leukoc Biol 1994;55:581–8.

[25] Yamamoto T, Kaizu C, Kawasaki T, Hasegawa G, Umezu H, Ohashi R, Sakurada J, Jiang S, Shultz L, Naito M. Macrophage colony-stimulating factor is indispensable for repopulation and differentiation of Kupffer cells but not for splenic red pulp macrophages in osteopetrotic (op/op) mice after macrophage depletion. Cell Tissue Res 2008;332:245–56.

[26] Rutherford MS, Schook LB. Differential immunocompetence of macrophages derived using macrophage or granulocyte-macrophage colony-stimulating factor. J Leukoc Biol 1992;51:69–76.

[27] MacDonald KP, Palmer JS, Cronau S, Seppanen E, Olver S, Raffelt NC, Kuns R, Pettit AR, Clouston A, Wainwright B, et al. An antibody against the colony-stimulating factor 1 receptor depletes the resident subset of monocytes and tissue- and tumor-associated macrophages but does not inhibit inflammation. Blood 2010;116:3955–63.

[28] Kurotaki D, Kon S, Bae K, Ito K, Matsui Y, Nakayama Y, Kanayama M, Kimura C, Narita Y, Nishimura T, et al. CSF-1-dependent red pulp macrophages regulate CD4 T cell responses. J Immunol 2011;186:2229–37.

[29] Burger P, Hilarius-Stokman P, de Korte D, van den Berg TK, van Bruggen R. CD47 functions as a molecular switch for erythrocyte phagocytosis. Blood 2012;119:5512–21.

[30] Olsson M, Oldenborg PA. CD47 on experimentally senescent murine RBCs inhibits phagocytosis following Fcgamma receptor-mediated but not scavenger receptor-mediated recognition by macrophages. Blood 2008;112:4259–67.

[31] Hemmi H, Idoyaga J, Suda K, Suda N, Kennedy K, Noda M, Aderem A, Steinman RM. A new triggering receptor expressed on myeloid cells (Trem) family member, Trem-like 4, binds to dead cells and is a DNAX activation protein 12-linked marker for subsets of mouse macrophages and dendritic cells. J Immunol 2009;182:1278–86.

[32] Korolnek T, Hamza I. Macrophages and iron trafficking at the birth and death of red cells. Blood 2015;125:2893–7.

[33] Mebius RE, Kraal G. Structure and function of the spleen. Nat Rev Immunol 2005;5:606–16.
[34] White C, Yuan X, Schmidt PJ, Bresciani E, Samuel TK, Campagna D, Hall C, Bishop K, Calicchio ML, Lapierre A, et al. HRG1 is essential for heme transport from the phagolysosome of macrophages during erythrophagocytosis. Cell Metab 2013;17:261–70.
[35] Knutson M, Wessling-Resnick M. Iron metabolism in the reticuloendothelial system. Crit Rev Biochem Mol Biol 2003;38:61–88.
[36] Kovtunovych G, Eckhaus MA, Ghosh MC, Ollivierre-Wilson H, Rouault TA. Dysfunction of the heme recycling system in heme oxygenase 1-deficient mice: effects on macrophage viability and tissue iron distribution. Blood 2010;116:6054–62.
[37] Kristiansen M, Graversen JH, Jacobsen C, Sonne O, Hoffman HJ, Law SK, Moestrup SK. Identification of the haemoglobin scavenger receptor. Nature 2001;409:198–201.
[38] Soe-Lin S, Apte SS, Andriopoulos Jr B, Andrews MC, Schranzhofer M, Kahawita T, Garcia-Santos D, Ponka P. Nramp1 promotes efficient macrophage recycling of iron following erythrophagocytosis in vivo. Proc Natl Acad Sci USA 2009;106:5960–5.
[39] Zhang Z, Zhang F, An P, Guo X, Shen Y, Tao Y, Wu Q, Zhang Y, Yu Y, Ning B, et al. Ferroportin1 deficiency in mouse macrophages impairs iron homeostasis and inflammatory responses. Blood 2011;118:1912–22.
[40] Gruenheid S, Pinner E, Desjardins M, Gros P. Natural resistance to infection with intracellular pathogens: the Nramp1 protein is recruited to the membrane of the phagosome. J Exp Med 1997;185:717–30.
[41] Vidal S, Tremblay ML, Govoni G, Gauthier S, Sebastiani G, Malo D, Skamene E, Olivier M, Jothy S, Gros P. The Ity/Lsh/Bcg locus: natural resistance to infection with intracellular parasites is abrogated by disruption of the Nramp1 gene. J Exp Med 1995;182:655–66.
[42] Fritsche G, Dlaska M, Barton H, Theurl I, Garimorth K, Weiss G. Nramp1 functionality increases inducible nitric oxide synthase transcription via stimulation of IFN regulatory factor 1 expression. J Immunol 2003;171:1994–8.
[43] Enders M, Franken L, Philipp MS, Kessler N, Baumgart AK, Eichler M, Wiertz EJH, Garbi N, Kurts C. Splenic red pulp macrophages cross-prime early effector CTL that provide rapid defense against viral infections. J Immunol 2020;204:87–100.
[44] Akilov OE, Kasuboski RE, Carter CR, McDowell MA. The role of mannose receptor during experimental leishmaniasis. J Leukoc Biol 2007;81:1188–96.
[45] Lee SJ, Zheng NY, Clavijo M, Nussenzweig MC. Normal host defense during systemic candidiasis in mannose receptor-deficient mice. Infect Immun 2003;71:437–45.
[46] Ansel KM, Ngo VN, Hyman PL, Luther SA, Forster R, Sedgwick JD, Browning JL, Lipp M, Cyster JG. A chemokine-driven positive feedback loop organizes lymphoid follicles. Nature 2000;406:309–14.
[47] Gunn MD, Kyuwa S, Tam C, Kakiuchi T, Matsuzawa A, Williams LT, Nakano H. Mice lacking expression of secondary lymphoid organ chemokine have defects in lymphocyte homing and dendritic cell localization. J Exp Med 1999;189:451–60.
[48] Link A, Vogt TK, Favre S, Britschgi MR, Acha-Orbea H, Hinz B, Cyster JG, Luther SA. Fibroblastic reticular cells in lymph nodes regulate the homeostasis of naive T cells. Nat Immunol 2007;8:1255–65.
[49] Luther SA, Tang HL, Hyman PL, Farr AG, Cyster JG. Coexpression of the chemokines ELC and SLC by T zone stromal cells and deletion of the ELC gene in the plt/plt mouse. Proc Natl Acad Sci USA 2000;97:12694–9.
[50] Krautler NJ, Kana V, Kranich J, Tian Y, Perera D, Lemm D, Schwarz P, Armulik A, Browning JL, Tallquist M, et al. Follicular dendritic cells emerge from ubiquitous perivascular precursors. Cell 2012;150:194–206.
[51] Umemoto E, Otani K, Ikeno T, Verjan Garcia N, Hayasaka H, Bai Z, Jang MH, Tanaka T, Nagasawa T, Ueda K, Miyasaka M. Constitutive plasmacytoid dendritic

cell migration to the splenic white pulp is cooperatively regulated by CCR7- and CXCR4-mediated signaling. J Immunol 2012;189:191–9.
[52] Gordon S, Pluddemann A, Martinez Estrada F. Macrophage heterogeneity in tissues: phenotypic diversity and functions. Immunol Rev 2014;262:36–55.
[53] Gordon S, Pluddemann A. Tissue macrophages: heterogeneity and functions. BMC Biol 2017;15:53.
[54] Victora GD, Nussenzweig MC. Germinal centers. Annu Rev Immunol 2012;30:429–57.
[55] MacLennan IC. Germinal centers. Annu Rev Immunol 1994;12:117–39.
[56] Bratton DL, Henson PM. Apoptotic cell recognition: will the real phosphatidylserine receptor(s) please stand up? Curr Biol 2008;18:R76–9.
[57] den Haan JM, Kraal G. Innate immune functions of macrophage subpopulations in the spleen. J Innate Immun 2012;4:437–45.
[58] Khan TN, Wong EB, Soni C, Rahman ZS. Prolonged apoptotic cell accumulation in germinal centers of Mer-deficient mice causes elevated B cell and CD4+ Th cell responses leading to autoantibody production. J Immunol 2013;190:1433–46.
[59] Wong K, Valdez PA, Tan C, Yeh S, Hongo JA, Ouyang W. Phosphatidylserine receptor Tim-4 is essential for the maintenance of the homeostatic state of resident peritoneal macrophages. Proc Natl Acad Sci USA 2010;107:8712–7.
[60] Hanayama R, Tanaka M, Miyasaka K, Aozasa K, Koike M, Uchiyama Y, Nagata S. Autoimmune disease and impaired uptake of apoptotic cells in MFG-E8-deficient mice. Science 2004;304:1147–50.
[61] Kranich J, Krautler NJ, Heinen E, Polymenidou M, Bridel C, Schildknecht A, Huber C, Kosco-Vilbois MH, Zinkernagel R, Miele G, Aguzzi A. Follicular dendritic cells control engulfment of apoptotic bodies by secreting Mfge8. J Exp Med 2008;205:1293–302.
[62] Kraal G. Cells in the marginal zone of the spleen. Int Rev Cytol 1992;132:31–74.
[63] Schmidt EE, MacDonald IC, Groom AC. Comparative aspects of splenic microcirculatory pathways in mammals: the region bordering the white pulp. Scanning Microsc 1993;7:613–28.
[64] Brendolan A, Rosado MM, Carsetti R, Selleri L, Dear TN. Development and function of the mammalian spleen. Bioessays 2007;29:166–77.
[65] van Krieken JH, te Velde J. Immunohistology of the human spleen: an inventory of the localization of lymphocyte subpopulations. Histopathology 1986;10:285–94.
[66] Kraal G, Mebius R. New insights into the cell biology of the marginal zone of the spleen. Int Rev Cytol 2006;250:175–215.
[67] Kraal G, Schornagel K, Streeter PR, Holzmann B, Butcher EC. Expression of the mucosal vascular addressin, MAdCAM-1, on sinus-lining cells in the spleen. Am J Pathol 1995;147:763–71.
[68] Mueller SN, Germain RN. Stromal cell contributions to the homeostasis and functionality of the immune system. Nat Rev Immunol 2009;9:618–29.
[69] Nolte MA, Arens R, Kraus M, van Oers MH, Kraal G, van Lier RA, Mebius RE. B cells are crucial for both development and maintenance of the splenic marginal zone. J Immunol 2004;172:3620–7.
[70] Scapini P, Cassatella MA. Location in the spleen dictates the function of murine neutrophils. J Exp Med 2017;214:1207–9.
[71] You Y, Myers RC, Freeberg L, Foote J, Kearney JF, Justement LB, Carter RH. Marginal zone B cells regulate antigen capture by marginal zone macrophages. J Immunol 2011;186:2172–81.
[72] Attanavanich K, Kearney JF. Marginal zone, but not follicular B cells, are potent activators of naive CD4 T cells. J Immunol 2004;172:803–11.
[73] Kearney JF. B cell subpopulations and secondary lymphoid organ architecture. Semin Immunol 2008;20:1–3.

[74] Lopes-Carvalho T, Foote J, Kearney JF. Marginal zone B cells in lymphocyte activation and regulation. Curr Opin Immunol 2005;17:244–50.
[75] Crowley MT, Reilly CR, Lo D. Influence of lymphocytes on the presence and organization of dendritic cell subsets in the spleen. J Immunol 1999;163:4894–900.
[76] Cupedo T, Lund FE, Ngo VN, Randall TD, Jansen W, Greuter MJ, de Waal-Malefyt R, Kraal G, Cyster JG, Mebius RE. Initiation of cellular organization in lymph nodes is regulated by non-B cell-derived signals and is not dependent on CXC chemokine ligand 13. J Immunol 2004;173:4889–96.
[77] Cupedo T, Vondenhoff MF, Heeregrave EJ, De Weerd AE, Jansen W, Jackson DG, Kraal G, Mebius RE. Presumptive lymph node organizers are differentially represented in developing mesenteric and peripheral nodes. J Immunol 2004;173:2968–75.
[78] Ato M, Nakano H, Kakiuchi T, Kaye PM. Localization of marginal zone macrophages is regulated by C-C chemokine ligands 21/19. J Immunol 2004;173:4815–20.
[79] Yu P, Wang Y, Chin RK, Martinez-Pomares L, Gordon S, Kosco-Vibois MH, Cyster J, Fu YX. B cells control the migration of a subset of dendritic cells into B cell follicles via CXC chemokine ligand 13 in a lymphotoxin-dependent fashion. J Immunol 2002;168:5117–23.
[80] Li H, Fu YX, Wu Q, Zhou Y, Crossman DK, Yang P, Li J, Luo B, Morel LM, Kabarowski JH, et al. Interferon-induced mechanosensing defects impede apoptotic cell clearance in lupus. J Clin Invest 2015;125:2877–90.
[81] Geijtenbeek TB, Groot PC, Nolte MA, van Vliet SJ, Gangaram-Panday ST, van Duijnhoven GC, Kraal G, van Oosterhout AJ, van Kooyk Y. Marginal zone macrophages express a murine homologue of DC-SIGN that captures blood-borne antigens in vivo. Blood 2002;100:2908–16.
[82] Kang YS, Kim JY, Bruening SA, Pack M, Charalambous A, Pritsker A, Moran TM, Loeffler JM, Steinman RM, Park CG. The C-type lectin SIGN-R1 mediates uptake of the capsular polysaccharide of *Streptococcus pneumoniae* in the marginal zone of mouse spleen. Proc Natl Acad Sci USA 2004;101:215–20.
[83] Kang YS, Yamazaki S, Iyoda T, Pack M, Bruening SA, Kim JY, Takahara K, Inaba K, Steinman RM, Park CG. SIGN-R1, a novel C-type lectin expressed by marginal zone macrophages in spleen, mediates uptake of the polysaccharide dextran. Int Immunol 2003;15:177–86.
[84] Elomaa O, Kangas M, Sahlberg C, Tuukkanen J, Sormunen R, Liakka A, Thesleff I, Kraal G, Tryggvason K. Cloning of a novel bacteria-binding receptor structurally related to scavenger receptors and expressed in a subset of macrophages. Cell 1995;80:603–9.
[85] van der Laan LJ, Dopp EA, Haworth R, Pikkarainen T, Kangas M, Elomaa O, Dijkstra CD, Gordon S, Tryggvason K, Kraal G. Regulation and functional involvement of macrophage scavenger receptor MARCO in clearance of bacteria in vivo. J Immunol 1999;162:939–47.
[86] Mebius RE, Nolte MA, Kraal G. Development and function of the splenic marginal zone. Crit Rev Immunol 2004;24:449–64.
[87] Gonzalez NA, Guillen JA, Gallardo G, Diaz M, de la Rosa JV, Hernandez IH, Casanova-Acebes M, Lopez F, Tabraue C, Beceiro S, et al. The nuclear receptor LXRalpha controls the functional specialization of splenic macrophages. Nat Immunol 2013;14:831–9.
[88] Miyake Y, Asano K, Kaise H, Uemura M, Nakayama M, Tanaka M. Critical role of macrophages in the marginal zone in the suppression of immune responses to apoptotic cell-associated antigens. J Clin Invest 2007;117:2268–78.
[89] Grasset EK, Duhlin A, Agardh HE, Ovchinnikova O, Hagglof T, Forsell MN, Paulsson-Berne G, Hansson GK, Ketelhuth DF, Karlsson MC. Sterile inflammation in the spleen during atherosclerosis provides oxidation-specific epitopes that induce a protective B-cell response. Proc Natl Acad Sci USA 2015;112:E2030–8.

[90] Hjelm F, Karlsson MC, Heyman B. A novel B cell-mediated transport of IgE-immune complexes to the follicle of the spleen. J Immunol 2008;180:6604–10.
[91] Karlsson MC, Guinamard R, Bolland S, Sankala M, Steinman RM, Ravetch JV. Macrophages control the retention and trafficking of B lymphocytes in the splenic marginal zone. J Exp Med 2003;198:333–40.
[92] You Y, Zhao H, Wang Y, Carter RH. Cutting edge: primary and secondary effects of CD19 deficiency on cells of the marginal zone. J Immunol 2009;182:7343–7.
[93] Park CG, Takahara K, Umemoto E, Yashima Y, Matsubara K, Matsuda Y, Clausen BE, Inaba K, Steinman RM. Five mouse homologues of the human dendritic cell C-type lectin, DC-SIGN. Int Immunol 2001;13:1283–90.
[94] Galustian C, Park CG, Chai W, Kiso M, Bruening SA, Kang YS, Steinman RM, Feizi T. High and low affinity carbohydrate ligands revealed for murine SIGN-R1 by carbohydrate array and cell binding approaches, and differing specificities for SIGN-R3 and langerin. Int Immunol 2004;16:853–66.
[95] Taylor PR, Brown GD, Herre J, Williams DL, Willment JA, Gordon S. The role of SIGNR1 and the beta-glucan receptor (dectin-1) in the nonopsonic recognition of yeast by specific macrophages. J Immunol 2004;172:1157–62.
[96] Lanoue A, Clatworthy MR, Smith P, Green S, Townsend MJ, Jolin HE, Smith KG, Fallon PG, McKenzie AN. SIGN-R1 contributes to protection against lethal pneumococcal infection in mice. J Exp Med 2004;200:1383–93.
[97] Koppel EA, Ludwig IS, Hernandez MS, Lowary TL, Gadikota RR, Tuzikov AB, Vandenbroucke-Grauls CM, van Kooyk Y, Appelmelk BJ, Geijtenbeek TB. Identification of the mycobacterial carbohydrate structure that binds the C-type lectins DC-SIGN, L-SIGN and SIGNR1. Immunobiology 2004;209:117–27.
[98] Nagaoka K, Takahara K, Tanaka K, Yoshida H, Steinman RM, Saitoh S, Akashi-Takamura S, Miyake K, Kang YS, Park CG, Inaba K. Association of SIGNR1 with TLR4-MD-2 enhances signal transduction by recognition of LPS in gram-negative bacteria. Int Immunol 2005;17:827–36.
[99] Marzi A, Gramberg T, Simmons G, Moller P, Rennekamp AJ, Krumbiegel M, Geier M, Eisemann J, Turza N, Saunier B, et al. DC-SIGN and DC-SIGNR interact with the glycoprotein of Marburg virus and the S protein of severe acute respiratory syndrome coronavirus. J Virol 2004;78:12090–5.
[100] Oehen S, Odermatt B, Karrer U, Hengartner H, Zinkernagel R, Lopez-Macias C. Marginal zone macrophages and immune responses against viruses. J Immunol 2002;169:1453–8.
[101] Gringhuis SI, den Dunnen J, Litjens M, van der Vlist M, Geijtenbeek TB. Carbohydrate-specific signaling through the DC-SIGN signalosome tailors immunity to *Mycobacterium tuberculosis*, HIV-1 and *Helicobacter pylori*. Nat Immunol 2009;10:1081–8.
[102] Gringhuis SI, den Dunnen J, Litjens M, van Het Hof B, van Kooyk Y, Geijtenbeek TB. C-type lectin DC-SIGN modulates toll-like receptor signaling via Raf-1 kinase-dependent acetylation of transcription factor NF-kappaB. Immunity 2007;26:605–16.
[103] Xu H, Bickford JK, Luther E, Carpenito C, Takei F, Springer TA. Characterization of murine intercellular adhesion molecule-2. J Immunol 1996;156:4909–14.
[104] Kang YS, Do Y, Lee HK, Park SH, Cheong C, Lynch RM, Loeffler JM, Steinman RM, Park CG. A dominant complement fixation pathway for pneumococcal polysaccharides initiated by SIGN-R1 interacting with C1q. Cell 2006;125:47–58.
[105] van Rooijen N. Antigen processing and presentation in vivo: the microenvironment as a crucial factor. Immunol Today 1990;11:436–9.
[106] Arredouani M, Yang Z, Ning Y, Qin G, Soininen R, Tryggvason K, Kobzik L. The scavenger receptor MARCO is required for lung defense against pneumococcal pneumonia and inhaled particles. J Exp Med 2004;200:267–72.
[107] Chen Y, Wermeling F, Sundqvist J, Jonsson AB, Tryggvason K, Pikkarainen T, Karlsson MC. A regulatory role for macrophage class A scavenger receptors in TLR4-mediated LPS responses. Eur J Immunol 2010;40:1451–60.

[108] Doyle SE, O'Connell RM, Miranda GA, Vaidya SA, Chow EK, Liu PT, Suzuki S, Suzuki N, Modlin RL, Yeh WC, et al. Toll-like receptors induce a phagocytic gene program through p38. J Exp Med 2004;199:81–90.
[109] Borges da Silva H, Fonseca R, Pereira RM, Cassado Ados A, Alvarez JM, D'Imperio Lima MR. Splenic macrophage subsets and their function during blood-borne infections. Front Immunol 2015;6:480.
[110] Suzuki H, Kurihara Y, Takeya M, Kamada N, Kataoka M, Jishage K, Ueda O, Sakaguchi H, Higashi T, Suzuki T, et al. A role for macrophage scavenger receptors in atherosclerosis and susceptibility to infection. Nature 1997;386:292–6.
[111] Ghosh S, Gregory D, Smith A, Kobzik L. MARCO regulates early inflammatory responses against influenza: a useful macrophage function with adverse outcome. Am J Respir Cell Mol Biol 2011;45:1036–44.
[112] MacLeod DT, Nakatsuji T, Wang Z, di Nardo A, Gallo RL. Vaccinia virus binds to the scavenger receptor MARCO on the surface of keratinocytes. J Invest Dermatol 2015;135:142–50.
[113] MacLeod DT, Nakatsuji T, Yamasaki K, Kobzik L, Gallo RL. HSV-1 exploits the innate immune scavenger receptor MARCO to enhance epithelial adsorption and infection. Nat Commun 2013;4:1963.
[114] Chen Y, Pikkarainen T, Elomaa O, Soininen R, Kodama T, Kraal G, Tryggvason K. Defective microarchitecture of the spleen marginal zone and impaired response to a thymus-independent type 2 antigen in mice lacking scavenger receptors MARCO and SR-A. J Immunol 2005;175:8173–80.
[115] Snook T. Studies on the perifollicular region of the rat's spleen. Anat Rec 1964;148:149–59.
[116] Kraal G, Janse M. Marginal metallophilic cells of the mouse spleen identified by a monoclonal antibody. Immunology 1986;58:665–9.
[117] Witmer-Pack MD, Hughes DA, Schuler G, Lawson L, McWilliam A, Inaba K, Steinman RM, Gordon S. Identification of macrophages and dendritic cells in the osteopetrotic (op/op) mouse. J Cell Sci 1993;104(Pt 4):1021–9.
[118] Jones C, Virji M, Crocker PR. Recognition of sialylated meningococcal lipopolysaccharide by siglecs expressed on myeloid cells leads to enhanced bacterial uptake. Mol Microbiol 2003;49:1213–25.
[119] Heikema AP, Bergman MP, Richards H, Crocker PR, Gilbert M, Samsom JN, van Wamel WJ, Endtz HP, van Belkum A. Characterization of the specific interaction between sialoadhesin and sialylated *Campylobacter jejuni* lipooligosaccharides. Infect Immun 2010;78:3237–46.
[120] Aichele P, Zinke J, Grode L, Schwendener RA, Kaufmann SH, Seiler P. Macrophages of the splenic marginal zone are essential for trapping of blood-borne particulate antigen but dispensable for induction of specific T cell responses. J Immunol 2003;171:1148–55.
[121] Aoshi T, Carrero JA, Konjufca V, Koide Y, Unanue ER, Miller MJ. The cellular niche of *Listeria monocytogenes* infection changes rapidly in the spleen. Eur J Immunol 2009;39:417–25.
[122] Eloranta ML, Alm GV. Splenic marginal metallophilic macrophages and marginal zone macrophages are the major interferon-alpha/beta producers in mice upon intravenous challenge with herpes simplex virus. Scand J Immunol 1999;49:391–4.
[123] Seiler P, Aichele P, Odermatt B, Hengartner H, Zinkernagel RM, Schwendener RA. Crucial role of marginal zone macrophages and marginal zone metallophils in the clearance of lymphocytic choriomeningitis virus infection. Eur J Immunol 1997;27:2626–33.
[124] Backer R, Schwandt T, Greuter M, Oosting M, Jungerkes F, Tuting T, Boon L, O'Toole T, Kraal G, Limmer A, den Haan JM. Effective collaboration between marginal metal-

lophilic macrophages and CD8+ dendritic cells in the generation of cytotoxic T cells. Proc Natl Acad Sci USA 2010;107:216–21.
[125] Gorak PM, Engwerda CR, Kaye PM. Dendritic cells, but not macrophages, produce IL-12 immediately following *Leishmania donovani* infection. Eur J Immunol 1998;28:687–95.
[126] Monteiro VG, Lobato CS, Silva AR, Medina DV, de Oliveira MA, Seabra SH, de Souza W, DaMatta RA. Increased association of Trypanosoma cruzi with sialoadhesin positive mice macrophages. Parasitol Res 2005;97:380–5.
[127] Asano K, Miwa M, Miwa K, Hanayama R, Nagase H, Nagata S, Tanaka M. Masking of phosphatidylserine inhibits apoptotic cell engulfment and induces autoantibody production in mice. J Exp Med 2004;200:459–67.
[128] Cohen PL, Caricchio R, Abraham V, Camenisch TD, Jennette JC, Roubey RA, Earp HS, Matsushima G, Reap EA. Delayed apoptotic cell clearance and lupus-like autoimmunity in mice lacking the c-mer membrane tyrosine kinase. J Exp Med 2002;196:135–40.
[129] Gonzalez NA, Bensinger SJ, Hong C, Beceiro S, Bradley MN, Zelcer N, Deniz J, Ramirez C, Diaz M, Gallardo G, et al. Apoptotic cells promote their own clearance and immune tolerance through activation of the nuclear receptor LXR. Immunity 2009;31:245–58.
[130] Klaas M, Crocker PR. Sialoadhesin in recognition of self and non-self. Semin Immunopathol 2012;34:353–64.
[131] Platt N, Suzuki H, Kurihara Y, Kodama T, Gordon S. Role for the class A macrophage scavenger receptor in the phagocytosis of apoptotic thymocytes in vitro. Proc Natl Acad Sci USA 1996;93:12456–60.
[132] Prabagar MG, Do Y, Ryu S, Park JY, Choi HJ, Choi WS, Yun TJ, Moon J, Choi IS, Ko K, et al. SIGN-R1, a C-type lectin, enhances apoptotic cell clearance through the complement deposition pathway by interacting with C1q in the spleen. Cell Death Differ 2013;20:535–45.
[133] Wermeling F, Chen Y, Pikkarainen T, Scheynius A, Winqvist O, Izui S, Ravetch JV, Tryggvason K, Karlsson MC. Class A scavenger receptors regulate tolerance against apoptotic cells, and autoantibodies against these receptors are predictive of systemic lupus. J Exp Med 2007;204:2259–65.
[134] Freeman GJ, Casasnovas JM, Umetsu DT, DeKruyff RH. TIM genes: a family of cell surface phosphatidylserine receptors that regulate innate and adaptive immunity. Immunol Rev 2010;235:172–89.
[135] Mizui M, Shikina T, Arase H, Suzuki K, Yasui T, Rennert PD, Kumanogoh A, Kikutani H. Bimodal regulation of T cell-mediated immune responses by TIM-4. Int Immunol 2008;20:695–708.
[136] McGaha TL, Chen Y, Ravishankar B, van Rooijen N, Karlsson MC. Marginal zone macrophages suppress innate and adaptive immunity to apoptotic cells in the spleen. Blood 2011;117:5403–12.
[137] Wermeling F, Karlsson MC, McGaha TL. An anatomical view on macrophages in tolerance. Autoimmun Rev 2009;9:49–52.
[138] Iyoda T, Shimoyama S, Liu K, Omatsu Y, Akiyama Y, Maeda Y, Takahara K, Steinman RM, Inaba K. The CD8+ dendritic cell subset selectively endocytoses dying cells in culture and in vivo. J Exp Med 2002;195:1289–302.
[139] Qiu CH, Miyake Y, Kaise H, Kitamura H, Ohara O, Tanaka M. Novel subset of CD8{alpha}+ dendritic cells localized in the marginal zone is responsible for tolerance to cell-associated antigens. J Immunol 2009;182:4127–36.
[140] Ravishankar B, Liu H, Shinde R, Chandler P, Baban B, Tanaka M, Munn DH, Mellor AL, Karlsson MC, McGaha TL. Tolerance to apoptotic cells is regulated by indoleamine 2,3-dioxygenase. Proc Natl Acad Sci USA 2012;109:3909–14.

[141] Ravishankar B, Shinde R, Liu H, Chaudhary K, Bradley J, Lemos HP, Chandler P, Tanaka M, Munn DH, Mellor AL, McGaha TL. Marginal zone CD169+ macrophages coordinate apoptotic cell-driven cellular recruitment and tolerance. Proc Natl Acad Sci USA 2014;111:4215–20.
[142] Godiska R, Chantry D, Raport CJ, Sozzani S, Allavena P, Leviten D, Mantovani A, Gray PW. Human macrophage-derived chemokine (MDC), a novel chemoattractant for monocytes, monocyte-derived dendritic cells, and natural killer cells. J Exp Med 1997;185:1595–604.
[143] Bayry J, Tchilian EZ, Davies MN, Forbes EK, Draper SJ, Kaveri SV, Hill AV, Kazatchkine MD, Beverley PC, Flower DR, Tough DF. In silico identified CCR4 antagonists target regulatory T cells and exert adjuvant activity in vaccination. Proc Natl Acad Sci USA 2008;105:10221–6.
[144] Gobert M, Treilleux I, Bendriss-Vermare N, Bachelot T, Goddard-Leon S, Arfi V, Biota C, Doffin AC, Durand I, Olive D, et al. Regulatory T cells recruited through CCL22/CCR4 are selectively activated in lymphoid infiltrates surrounding primary breast tumors and lead to an adverse clinical outcome. Cancer Res 2009;69:2000–9.
[145] Nishikawa H, Sakaguchi S. Regulatory T cells in tumor immunity. Int J Cancer 2010;127:759–67.
[146] Pere H, Montier Y, Bayry J, Quintin-Colonna F, Merillon N, Dransart E, Badoual C, Gey A, Ravel P, Marcheteau E, et al. A CCR4 antagonist combined with vaccines induces antigen-specific CD8+ T cells and tumor immunity against self antigens. Blood 2011;118:4853–62.
[147] Lee I, Wang L, Wells AD, Dorf ME, Ozkaynak E, Hancock WW. Recruitment of Foxp3+ T regulatory cells mediating allograft tolerance depends on the CCR4 chemokine receptor. J Exp Med 2005;201:1037–44.
[148] Montane J, Bischoff L, Soukhatcheva G, Dai DL, Hardenberg G, Levings MK, Orban PC, Kieffer TJ, Tan R, Verchere CB. Prevention of murine autoimmune diabetes by CCL22-mediated Treg recruitment to the pancreatic islets. J Clin Invest 2011;121:3024–8.
[149] Yu K, Chen Z, Khatri I, Gorczynski RM. CCR4 dependent migration of Foxp3+ Treg cells to skin grafts and draining lymph nodes is implicated in enhanced graft survival in CD200tg recipients. Immunol Lett 2011;141:116–22.
[150] Bajwa A, Huang L, Kurmaeva E, Gigliotti JC, Ye H, Miller J, Rosin DL, Lobo PI, Okusa MD. Sphingosine 1-phosphate receptor 3-deficient dendritic cells modulate splenic responses to ischemia-reperfusion injury. J Am Soc Nephrol 2016;27:1076–90.
[151] Randolph GJ, Angeli V, Swartz MA. Dendritic-cell trafficking to lymph nodes through lymphatic vessels. Nat Rev Immunol 2005;5:617–28.
[152] Chang JE, Turley SJ. Stromal infrastructure of the lymph node and coordination of immunity. Trends Immunol 2015;36:30–9.
[153] Cyster JG, Dang EV, Reboldi A, Yi T. 25-Hydroxycholesterols in innate and adaptive immunity. Nat Rev Immunol 2014;14:731–43.
[154] Fletcher AL, Acton SE, Knoblich K. Lymph node fibroblastic reticular cells in health and disease. Nat Rev Immunol 2015;15:350–61.
[155] Malhotra D, Fletcher AL, Turley SJ. Stromal and hematopoietic cells in secondary lymphoid organs: partners in immunity. Immunol Rev 2013;251:160–76.
[156] Cyster JG. B cell follicles and antigen encounters of the third kind. Nat Immunol 2010;11:989–96.
[157] Baekkevold ES, Yamanaka T, Palframan RT, Carlsen HS, Reinholt FP, von Andrian UH, Brandtzaeg P, Haraldsen G. The CCR7 ligand elc (CCL19) is transcytosed in high endothelial venules and mediates T cell recruitment. J Exp Med 2001;193:1105–12.
[158] Girard JP, Springer TA. High endothelial venules (HEVs): specialized endothelium for lymphocyte migration. Immunol Today 1995;16:449–57.

[159] Stamper Jr HB, Woodruff JJ. Lymphocyte homing into lymph nodes: in vitro demonstration of the selective affinity of recirculating lymphocytes for high-endothelial venules. J Exp Med 1976;144:828–33.
[160] Clark Jr SL. The reticulum of lymph nodes in mice studied with the electron microscope. Am J Anat 1962;110:217–57.
[161] Fossum S. The architecture of rat lymph nodes. IV. Distribution of ferritin and colloidal carbon in the draining lymph nodes after foot-pad injection. Scand J Immunol 1980;12:433–41.
[162] Nossal GJ, Ada GL, Austin CM, Pye J. Antigens in immunity. 8. Localization of 125-I-labelled antigens in the secondary response. Immunology 1965;9:349–57.
[163] Delemarre FG, Kors N, Kraal G, van Rooijen N. Repopulation of macrophages in popliteal lymph nodes of mice after liposome-mediated depletion. J Leukoc Biol 1990;47:251–7.
[164] Steer HW, Foot RA. Changes in the medulla of the parathymic lymph nodes of the rat during acute gastro-intestinal inflammation. J Anat 1987;152:23–36.
[165] Junt T, Moseman EA, Iannacone M, Massberg S, Lang PA, Boes M, Fink K, Henrickson SE, Shayakhmetov DM, Di Paolo NC, et al. Subcapsular sinus macrophages in lymph nodes clear lymph-borne viruses and present them to antiviral B cells. Nature 2007;450:110–4.
[166] Smith JP, Burton GF, Tew JG, Szakal AK. Tingible body macrophages in regulation of germinal center reactions. Dev Immunol 1998;6:285–94.
[167] Phan TG, Green JA, Gray EE, Xu Y, Cyster JG. Immune complex relay by subcapsular sinus macrophages and noncognate B cells drives antibody affinity maturation. Nat Immunol 2009;10:786–93.
[168] Carrasco YR, Batista FD. B cells acquire particulate antigen in a macrophage-rich area at the boundary between the follicle and the subcapsular sinus of the lymph node. Immunity 2007;27:160–71.
[169] Phan TG, Grigorova I, Okada T, Cyster JG. Subcapsular encounter and complement-dependent transport of immune complexes by lymph node B cells. Nat Immunol 2007;8:992–1000.
[170] Szakal AK, Holmes KL, Tew JG. Transport of immune complexes from the subcapsular sinus to lymph node follicles on the surface of nonphagocytic cells, including cells with dendritic morphology. J Immunol 1983;131:1714–27.
[171] Gaya M, Castello A, Montaner B, Rogers N, Reis e Sousa C, Bruckbauer A, Batista FD. Host response. Inflammation-induced disruption of SCS macrophages impairs B cell responses to secondary infection. Science 2015;347:667–72.
[172] Gonzalez SF, Lukacs-Kornek V, Kuligowski MP, Pitcher LA, Degn SE, Kim YA, Cloninger MJ, Martinez-Pomares L, Gordon S, Turley SJ, Carroll MC. Capture of influenza by medullary dendritic cells via SIGN-R1 is essential for humoral immunity in draining lymph nodes. Nat Immunol 2010;11:427–34.
[173] Iannacone M, Moseman EA, Tonti E, Bosurgi L, Junt T, Henrickson SE, Whelan SP, Guidotti LG, von Andrian UH. Subcapsular sinus macrophages prevent CNS invasion on peripheral infection with a neurotropic virus. Nature 2010;465:1079–83.
[174] Moseman EA, Iannacone M, Bosurgi L, Tonti E, Chevrier N, Tumanov A, Fu YX, Hacohen N, von Andrian UH. B cell maintenance of subcapsular sinus macrophages protects against a fatal viral infection independent of adaptive immunity. Immunity 2012;36:415–26.
[175] Chtanova T, Schaeffer M, Han SJ, van Dooren GG, Nollmann M, Herzmark P, Chan SW, Satija H, Camfield K, Aaron H, et al. Dynamics of neutrophil migration in lymph nodes during infection. Immunity 2008;29:487–96.
[176] Frederico B, Chao B, Lawler C, May JS, Stevenson PG. Subcapsular sinus macrophages limit acute gammaherpesvirus dissemination. J Gen Virol 2015;96:2314–27.

[177] Kastenmuller W, Torabi-Parizi P, Subramanian N, Lammermann T, Germain RN. A spatially-organized multicellular innate immune response in lymph nodes limits systemic pathogen spread. Cell 2012;150:1235–48.
[178] Winkelmann ER, Widman DG, Xia J, Johnson AJ, van Rooijen N, Mason PW, Bourne N, Milligan GN. Subcapsular sinus macrophages limit dissemination of West Nile virus particles after inoculation but are not essential for the development of West Nile virus-specific T cell responses. Virology 2014;450–451:278–89.
[179] Gray EE, Cyster JG. Lymph node macrophages. J Innate Immun 2012;4:424–36.
[180] Sewald X, Ladinsky MS, Uchil PD, Beloor J, Pi R, Herrmann C, Motamedi N, Murooka TT, Brehm MA, Greiner DL, et al. Retroviruses use CD169-mediated trans-infection of permissive lymphocytes to establish infection. Science 2015;350:563–7.
[181] Hickman HD, Takeda K, Skon CN, Murray FR, Hensley SE, Loomis J, Barber GN, Bennink JR, Yewdell JW. Direct priming of antiviral CD8+ T cells in the peripheral interfollicular region of lymph nodes. Nat Immunol 2008;9:155–65.
[182] Hsu KM, Pratt JR, Akers WJ, Achilefu SI, Yokoyama WM. Murine cytomegalovirus displays selective infection of cells within hours after systemic administration. J Gen Virol 2009;90:33–43.
[183] Hu JK, Kagari T, Clingan JM, Matloubian M. Expression of chemokine receptor CXCR3 on T cells affects the balance between effector and memory CD8 T-cell generation. Proc Natl Acad Sci USA 2011;108:E118–27.
[184] Norbury CC, Malide D, Gibbs JS, Bennink JR, Yewdell JW. Visualizing priming of virus-specific CD8+ T cells by infected dendritic cells in vivo. Nat Immunol 2002;3:265–71.
[185] Bachmann MF, Rohrer UH, Kundig TM, Burki K, Hengartner H, Zinkernagel RM. The influence of antigen organization on B cell responsiveness. Science 1993;262:1448–51.
[186] Honke N, Shaabani N, Cadeddu G, Sorg UR, Zhang DE, Trilling M, Klingel K, Sauter M, Kandolf R, Gailus N, et al. Enforced viral replication activates adaptive immunity and is essential for the control of a cytopathic virus. Nat Immunol 2011;13:51–7.
[187] Sagoo P, Garcia Z, Breart B, Lemaitre F, Michonneau D, Albert ML, Levy Y, Bousso P. In vivo imaging of inflammasome activation reveals a subcapsular macrophage burst response that mobilizes innate and adaptive immunity. Nat Med 2016;22:64–71.
[188] Farrell HE, Bruce K, Lawler C, Cardin RD, Davis-Poynter NJ, Stevenson PG. Type 1 interferons and NK cells limit murine cytomegalovirus escape from the lymph node subcapsular sinus. PLoS Pathog 2016;12:e1006069.
[189] Hickman HD, Li L, Reynoso GV, Rubin EJ, Skon CN, Mays JW, Gibbs J, Schwartz O, Bennink JR, Yewdell JW. Chemokines control naive CD8+ T cell selection of optimal lymph node antigen presenting cells. J Exp Med 2011;208:2511–24.
[190] Kastenmuller W, Brandes M, Wang Z, Herz J, Egen JG, Germain RN. Peripheral prepositioning and local CXCL9 chemokine-mediated guidance orchestrate rapid memory CD8+ T cell responses in the lymph node. Immunity 2013;38:502–13.
[191] Sung JH, Zhang H, Moseman EA, Alvarez D, Iannacone M, Henrickson SE, de la Torre JC, Groom JR, Luster AD, von Andrian UH. Chemokine guidance of central memory T cells is critical for antiviral recall responses in lymph nodes. Cell 2012;150:1249–63.
[192] Hume DA, Robinson AP, MacPherson GG, Gordon S. The mononuclear phagocyte system of the mouse defined by immunohistochemical localization of antigen F4/80. Relationship between macrophages, Langerhans cells, reticular cells, and dendritic cells in lymphoid and hematopoietic organs. J Exp Med 1983;158:1522–36.
[193] Grigorova IL, Schwab SR, Phan TG, Pham TH, Okada T, Cyster JG. Cortical sinus probing, S1P1-dependent entry and flow-based capture of egressing T cells. Nat Immunol 2009;10:58–65.

[194] Albacker LA, Yu S, Bedoret D, Lee WL, Umetsu SE, Monahan S, Freeman GJ, Umetsu DT, DeKruyff RH. TIM-4, expressed by medullary macrophages, regulates respiratory tolerance by mediating phagocytosis of antigen-specific T cells. Mucosal Immunol 2013;6:580–90.
[195] Asano K, Nabeyama A, Miyake Y, Qiu CH, Kurita A, Tomura M, Kanagawa O, Fujii S, Tanaka M. CD169-positive macrophages dominate antitumor immunity by crosspresenting dead cell-associated antigens. Immunity 2011;34:85–95.
[196] Mohr E, Serre K, Manz RA, Cunningham AF, Khan M, Hardie DL, Bird R, MacLennan IC. Dendritic cells and monocyte/macrophages that create the IL-6/APRIL-rich lymph node microenvironments where plasmablasts mature. J Immunol 2009;182:2113–23.
[197] Fooksman DR, Nussenzweig MC, Dustin ML. Myeloid cells limit production of antibody-secreting cells after immunization in the lymph node. J Immunol 2014;192:1004–12.
[198] Tumanov AV, Grivennikov SI, Shakhov AN, Rybtsov SA, Koroleva EP, Takeda J, Nedospasov SA, Kuprash DV. Dissecting the role of lymphotoxin in lymphoid organs by conditional targeting. Immunol Rev 2003;195:106–16.
[199] Rennert PD, Browning JL, Mebius R, Mackay F, Hochman PS. Surface lymphotoxin alpha/beta complex is required for the development of peripheral lymphoid organs. J Exp Med 1996;184:1999–2006.
[200] Cascalho M, Ma A, Lee S, Masat L, Wabl M. A quasi-monoclonal mouse. Science 1996;272:1649–52.

CHAPTER 5

Kidney—Inflammation and remodeling

Stefanie Steiger, Julia Lichtnekert, and Hans-Joachim Anders
LMU Hospital, Department of Medicine IV, Division of Nephrology, Munich, Germany

Macrophages are a heterogeneous cell population that displays an important part of the mononuclear phagocytic system [1,2]. Macrophages were first recognized for their ability to clear infectious microorganisms [3] but subsequent studies reported that macrophages are multifunctional cells with distinct biological functions depending on the local tissue microenvironment [4–8]. Macrophages take a sentinel role in pathogen clearance and repair of injured tissues [9–11] but can also promote tissue damage. Macrophages can be found in a variety of tissues including the liver, heart, kidney, lung, spleen, and brain. In the kidney, under healthy physiological conditions, the macrophage compartment is comprised of long-lived tissue-resident macrophages as well as macrophages that originate from circulating monocytes and bone marrow precursors [12,13].

A pathogenic role for macrophages has been proposed in many types of kidney diseases [14–18]. Animal models elucidated the contribution of macrophages and their soluble mediators to kidney injury, necroinflammation, and repair [19–25]. Some of the latest developments in this area of research have established the role of macrophage plasticity during the progression from kidney inflammation to tubular atrophy and interstitial fibrosis, an important step in the development of kidney failure [16,26–28]. In this book chapter, we provide an update on (a) The most recent data on resident and infiltrating macrophages in the kidney, (b) How tissue microenvironments determine macrophage phenotype to assist and orchestrate all phases of acute and chronic kidney injury, inflammation and repair, and (c) Which therapeutic avenues can target macrophage-dependent kidney remodeling and failure.

Macrophages in the Human Body
https://doi.org/10.1016/B978-0-12-821385-8.00003-2

Macrophage heterogeneity and plasticity

Kidney-resident macrophages originate either from embryonic yolk sac precursors [29] or from the bone marrow [30]. In peripheral organs, macrophages, which can self-maintain and repopulate locally [31] contribute to important physiological functions, such as matrix turnover and phagocytosis of debris (Fig. 1) [31,32]. For example, macrophages are required for the clearance of nuclei that get expelled from erythroblasts during their maturation toward erythrocytes in the bone marrow [33].

Kidney injury triggered either by pathogens, ischemia, toxins, crystal deposits or trauma induces the recruitment of circulating monocytes that differentiate into two major macrophage phenotypes, which can be broadly categorized as pro-inflammatory (M1) macrophages and anti-inflammatory (M2) macrophages. The local tissue microenvironment plays an important role in the differentiation process into the different macrophage phenotypes. It is important to note that the paradigm of M1 and M2 macrophages represents a major oversimplification of macrophage phenotypes in the kidney and does not describe fixed subpopulations, because functional data are mainly based on in vitro studies and in vivo animal disease models. In fact, innovative techniques with extensive analysis of monocyte/macrophage transcriptomes implied more information on tissue macrophage origin, and common and unique activation pathways, e.g., lineage tracing [34], and single-cell RNA-sequencing (scRNA-seq). So far, scRNA-seq has revealed a large complexity of macrophage populations at a steady stage and in the development and progression of diseases such as cancer, atherosclerosis, and kidney [30,35–37]. It is possible that more insights from such analyses at a single-cell level will lead to new approaches for categorizing macrophages according to their phenotype and function.

The current understanding of macrophage biology during kidney disease is as follows: In response to kidney injury, inflammatory responses are needed to avoid, for example, pathogen entry to control infections [38]. Because pathogens release pathogen-associated molecular patterns (PAMPs) and damaged tissue cells release damage-associated molecular patterns (DAMPs). These PAMPs and DAMPs are recognized by toll-like receptors (TLRs) and other pattern recognition receptors on immune and non-immune cells in the tissue, which triggers the secretion of pro-inflammatory cytokines and chemokines [39–42]. Furthermore, pro-inflammatory macrophages release matrix metalloproteinases (MMPs) and hyaluronidase to digest extracellular matrix (ECM) proteins and thereby reduce the viscosity

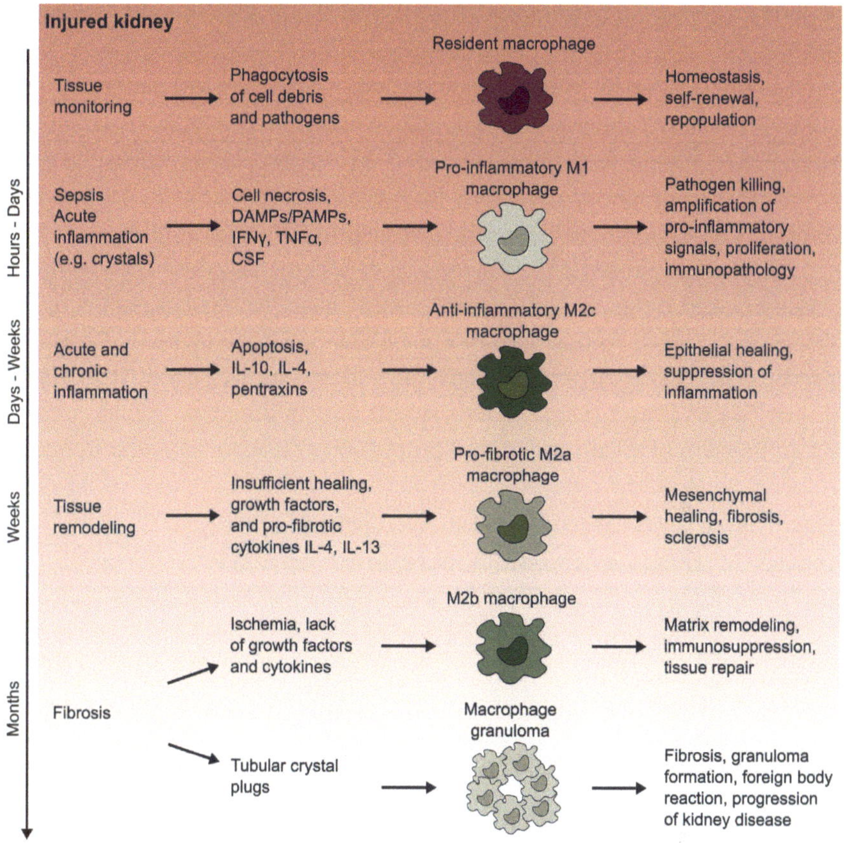

Fig. 1 Kidney environment and macrophage phenotypes in health and disease. In health, tissue-resident macrophages (F4/80^{high}) maintain homeostasis in the kidney. Upon kidney injury, tissue-resident macrophages coordinate monocyte recruitment to the kidney, where they differentiate into pro-inflammatory M1 macrophages and amplify inflammation in the days after acute injury. This differentiation is followed by a change in the microenvironment and regulation of negative regulators, whereby macrophages predominantly differentiate into anti-inflammatory M2 macrophages throughout the later stages of tissue repair. During the tissue repair phase, M2b macrophages and resident tissue macrophages are mainly responsible for immunosuppression, matrix remodeling, and wound healing once the underlying injury has resolved. By contrast, persistent kidney inflammation triggers M2a and M2b macrophage polarization and the formation of macrophage granulomas, which results in increased transforming growth factor (TGF)β1 production and kidney fibrosis. CSF, colony-stimulating factor; interferon (IFN)γ. *(Modified from Lech M, et al. Tissues use resident dendritic cells and macrophages to maintain homeostasis and to regain homeostasis upon tissue injury: the immunoregulatory role of changing tissue environments. Mediat Inflamm 2012;2012:951390; Tang PM, Nikolic-Paterson DJ, Lan HY. Macrophages: versatile players in renal inflammation and fibrosis. Nat Rev Nephrol 2019;15(3):144–158.)*

of the ECM. This process, together with an enhanced vascular permeability promotes the recruitment of circulating monocytes and subsequent PAMP- and DAMP-mediated inflammation [43–48]. Such an inflammatory microenvironment will drive the polarization toward a pro-inflammatory M1 macrophage phenotype. These activated M1 macrophages release a variety of pro-inflammatory mediators in a vicious cycle with tubular necrosis (Table 1; Fig. 1), a process referred to as necroinflammation [49].

The phenotypic switch of macrophages is determined by the microenvironment and programmed upregulation of negative regulators during the course of acute and chronic kidney disease [54–56]. Monocytes that infiltrate the damaged tissue can differentiate into alternative activated M2 macrophages [57]. M2 macrophages have a wide range of functions, which include the suppression of inflammation and promotion of tissue repair after ischemic

Table 1 Characteristics of macrophage subsets upon kidney injury.

Subsets	Surface marker/gene expression	Secretion	Refs.
Tissue-resident	Surface: F4/80^{high}, CD11b^{+}, MHCII^{+}, CD11c^{int}, CD81, CD74 Genes: *Apoe, C1q*		[30,50,51]
M1	Surface: F4/80^{low}, CD206^{-}, CD68^{+}, Ly6C^{int}, CCR2^{+}, CSF1R^{+}, CD11c^{+}, CX3CR1^{+}, MHCII^{+}, CD86^{+}, CD11b^{+} Genes: *pStat1, Irf5, Tnfaip3, Socs3, Csf2 and 3*	TNFα, IL-1β, IL-6, ROS, IL-12, MMPs, MCP1, iNOS, CCL2, CCL5, IL-27, CCL1, CXCL10, CXCL9	[52,53]
M2a	Surface: F4/80^{high}, CD206^{+}, CD68^{+}, CD163^{+}, MHCII^{+}, CD11b^{+}, Dectin-1, Scavenger receptor Genes: *Arg1, Irf4, Mrc1*	IL-10, IL-12, IL-4, IL-13, VEGF, FGF2, TGFβ1	[8,52,53]
M2b	Surface: F4/80^{high}, CD206^{+}, CD68^{+}, CD163^{+}, CD36, MHCII^{+}, CD86^{+}, CD11b^{+} Genes: *Irf4, Mrc1, Irak-m*	IL-10, TGFβ1, FGF, TNFα, IL-1β, IL-6, CCL1	[52–54]
M2c	Surface: F4/80^{high}, CD206^{+}, CD68^{+}, CD163^{+}, CD11b^{+}, B7-H4, CD150 Genes: *Irf4, Mrc1, Arg1, Timp1*	IL-10, MMPs, CCL16, CCL18, TGFβ1	[8,52,53]

kidney injury [58–60], and can be broadly classified into three M2 subpopulations (that is, M2a-M2c) based on their phenotype and function (Fig. 1). The polarization into M2 macrophages is driven by a wide range of stimuli: IL-4 and IL-13 induce the differentiation of pro-fibrotic M2a macrophages [61,62], immune complexes, and toll-like receptor (TLR) and/or IL-1 receptor ligands induce M2b macrophages [63,64], and IL-10, transforming growth factor (TGF)β, and glucocorticoids trigger M2c macrophage polarization [65]. While M2a macrophages induce an anti-inflammatory immune response and promote wound healing and tissue fibrosis, M2b macrophages participate in immune regulation and contribute to Th2-like activation, and M2c anti-inflammatory macrophages contribute to immunosuppression, matrix remodeling, and tissue repair [66–69]. However, studies on resident macrophages as well as M1 and M2 macrophage subsets during kidney homeostasis, injury, and repair are hampered by the lack of cell-specific markers that can clearly distinguish these subsets. Known macrophage markers are listed in Table 1 but one should mention here that expression of these markers changes along with changes in the local microenvironment; therefore, it is still challenging to dissect specific macrophage subsets during the different stages of disease in mice and humans (Box 1).

It is important to mention that macrophage biology clearly overlaps with that of dendritic cells in terms of their cellular origin and morphology,

BOX 1 Differences between mouse and human macrophage polarization

Differences between gene and surface expression in rodent and human polarized macrophages have been previously discussed, along with similarities and differences between the species in terms of immunity in general [202], e.g. using transcriptome-based network analysis [203]. In essence, although there are major apparent differences between species, especially regarding amino acid metabolizing enzymes, technical issues are at the root of the controversies. For example, monocytes are generally used as the source of macrophages in humans, whereas bone marrow-derived monocytes or peritoneal cavity macrophages are often used in rodents. The key experiment of a side-by-side comparison of monocyte-derived mouse and human macrophages polarized by different endogenous mediator (M1 and M2a,b,c) has yet to be reported. These issues are openly acknowledged in the field, and the use of standardized single-cell methods and rigorous bioinformatics should yield the information necessary to compare polarization between species and different disease contexts.

function (phagocytosis and antigen presentation), and cell surface marker expression. Therefore, macrophages and dendritic cells in the kidney are somewhat referred to as kidney mononuclear phagocytes [70]. However, the role of dendritic cells in kidney disease is beyond the scope of this chapter and has been reviewed elsewhere [71,72].

Monocyte recruitment and monocyte-derived macrophage diversity

Monocytes in the mouse blood are defined according to their expression of the surface marker Ly6C and chemokine receptors, e.g., CCR2, Cx3CR1 [73]. Upon kidney injury, locally secreted chemokines, such as CCL2, CCL3, CCL4, CCL5, and CXCL10 by resident macrophages and tubular epithelial cells initiate the recruitment of Ly6C+ monocytes from the bloodstream to the site of tissue inflammation. Monocyte recruitment strongly depends on chemotactic mechanisms [74]. For example, CCL2/CCR2 signaling is critical for monocyte recruitment to the site of injury. Recruitment of circulating monocytes into the kidney was significantly reduced 24 h after IRI in CCR2 knockout mice, resulting in less functional acute tissue injury [75,76]. Other chemokines including CX_3CL1 [76,77], CXCL16 [78], CCL5 [79], macrophage migration inhibitory factor (MIF) [80], CCL1 [81], and osteopontin [82] contribute to macrophage accumulation and kidney inflammation and damage.

The functional role of chemokines in kidney disease was also confirmed by several acute and chronic animal studies using neutralizing antibodies, chemokine antagonists, and targeted disruption of genes encoding chemokines and their receptors. For example, the blockade of the chemokine CCL2 has renoprotective effects by suppressing monocyte recruitment and M1 macrophage proliferation in IRI [75]. Blocking CCL2, CCR1, and/or CXCL12 suppresses the progression of murine proliferative lupus nephritis [83–85], type 2 diabetes [86], and obstructive nephropathy [87]. These findings point toward a role for chemokines and their receptors in leukocyte-mediated acute and progressive forms of kidney damage and fibrosis.

Following transmigration into the inflamed tissue, monocytes can differentiate into mononuclear phagocytes, including dendritic cells and macrophages. Most kidney-infiltrating macrophages observed in models of acute kidney injury (AKI) are derived from circulating classical $Ly6C^{high}$ monocytes [88]. Infiltrating $Ly6C^{high}$ macrophages within the kidney are morphologically immature and monocyte-like, whereas the $Ly6C^{low}$ population (derived from $Ly6C^{high}$ monocytes) is more mature and exerts pro-fibrotic functions [89,90]. $Ly6C^{int}$ macrophages have wound healing characteristics and are present in high numbers during the kidney repair phase. This phenotype was confirmed by inhibiting macrophage colony-stimulating factor 1 receptor (CSF1R) to suppress macrophage proliferation at different stages after IRI in mice [91]. There exists also a $Ly6C^-$ population of circulating monocytes characterized by prolonged crawling along the endothelium; these cells can phagocytose microparticles and soluble material from the bloodstream and respond to microbial infection and tissue injury [92]. These patrolling monocytes have been identified in the kidney and can present antigen to CD4 + T cells within the glomerular capillaries [93].

Macrophage plasticity

Monocytes and macrophages can polarize into particular phenotypes upon exposure to specific stimuli, but whether the same cell can then undergo a further phenotype switch secondary to changes in the microenvironment is currently unclear. This possibility is particularly relevant in conditions in which a clear M1-to-M2 phenotypic switch occurs, such as during kidney ischemia-reperfusion injury (IRI), calcium oxalate crystal-induced kidney injury, and unilateral ureteric obstruction (UUO), whereby pro-inflammatory M1 macrophages are mainly present after the acute injury phase but M2 macrophages predominantly exist during the subsequent phase of kidney repair (Fig. 1). This suggests that different macrophage phenotypes coexist in the kidney in the different phases of injury and repair [59,60,69]. Whether M1 macrophages switch their phenotype to become M2 macrophages or whether these are separate cell populations remains unknown. A recent study addressed this question, wherein macrophages were differentiated into an M2 phenotype in vitro before being adoptively transferred into recipient mice with Adriamycin nephropathy. While polarized M2 splenic macrophages maintained their phenotype in the injured kidney and were protected from Adriamycin nephropathy, M2 bone marrow-derived macrophages underwent a phenotype switch and failed to provide protection [94], indicating that bone marrow-derived macrophages have considerable plasticity.

Macrophages in acute kidney injury

Acute kidney disease is characterized by an abrupt loss of kidney function arising from different events, such as sepsis/septic shock, IRI, calcium oxalate crystal-induced kidney injury, nephrotoxicity, hypoxia, and UUO. In the different forms of AKI, kidney-resident $F4/80^{high}$ macrophages accumulate either in the glomerular or tubulointerstitial compartments and respond to injury and inflammation. In vivo four phases of regulated inflammation after

AKI have been described: (1) recruitment of monocytes to the injured tissue, (2) differentiation and activation of M1 macrophages, (3) M1-to-M2 macrophage phenotype switching and (4) re-establishment of tissue homeostasis or persistence to chronic non-resolving inflammation [95].

Once infiltrated, monocytes differentiate into-inflammatory M1 macrophages and contribute to the release of pro-inflammatory cytokines and chemokines that initiate local inflammation (Fig. 2) [96,97]. The degree of M1 macrophage accumulation correlates with the severity of kidney function, glomerular hypercellularity, segmental lesions, and crescent formation

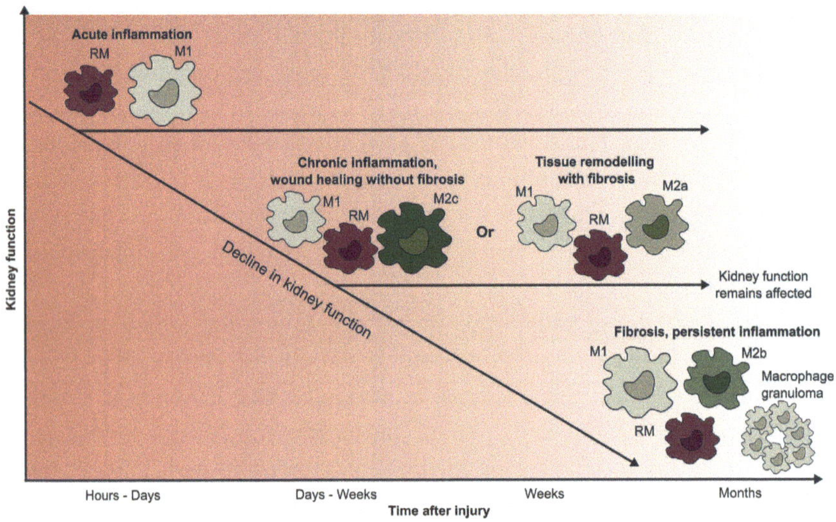

Fig. 2 Macrophage subsets during acute and chronic kidney disease. Kidney function declines over time, depending on the type of injury, the persistence and severity of damage, and the reparative ability of the kidney. In acute kidney injury, the pro-inflammatory microenvironment and resident macrophages (RM) drive the differentiation of infiltrating monocytes into pro-inflammatory M1 macrophages associated with the release of pro-inflammatory mediators and local proliferation. If the acute inflammation resolves, tissue homeostasis will be reached with the help of resident macrophages and kidney function remains. In chronic kidney disease, infiltrating M1 macrophages and RM release pro-inflammatory mediators, which promote inflammation. If the injury resolves, kidney function and kidney tissue ameliorate, depending on the damage severity and duration. M1 macrophages switch to M2c or M2a phenotypes, leading to wound healing that may involve tissue fibrosis. However, if there is no injury resolution, M1 macrophages persist at injury sites and there is a decrease in the number of M2 macrophages, which could also subsequently undergo an M1 phenotype switch and form macrophage granulomas. The persistent release of pro-fibrotic and inflammatory factors promotes interstitial fibrosis, leading to kidney failure. *(Modified from Tang PM, Nikolic-Paterson DJ, Lan HY. Macrophages: versatile players in renal inflammation and fibrosis. Nat Rev Nephrol 2019;15(3):144–158.)*

in glomerulonephritis [98], allograft rejection in models of kidney transplantation [99,100], and crystalline-induced kidney injury. Studies using macrophage depletion techniques (such as clodronate liposomes or administration of diphtheria toxin to transgenic mice) have demonstrated that macrophages can induce kidney injury during inflammation in animal models of crescentic glomerulonephritis, acute renal allograft rejection, and IRI [101–105]. However, one limitation of such techniques is that they deplete different macrophage populations. For example, diphtheria toxin-induced depletion of $CD11c^+$ cells (that are macrophages and dendritic cells) resulted in more depletion of $F4/80^+$ macrophages than the use of clodronate liposomes, which did not deplete $CD206^+$("M2") macrophages. Macrophage depletion with clodronate liposomes but not $CD11c^+$ cell depletion was protective in IRI-induced AKI [103]. Only macrophage phenotype-specific fate mapping approaches will overcome this issue, which has recently been demonstrated for dendritic cells using *Clec9a*$^{cre/cre}$ *Rosa26*^{YFP} mice in AKI [50].

M1 macrophage proliferation

Macrophages in the injured kidney are almost entirely derived from circulating monocytes that arise from the bone marrow. The local proliferation of these M1 macrophages is a feature of inflammatory kidney injury [12,106]. Evidence comes from studies showing that administration of a neutralizing CSF1R antibody that inhibits macrophage proliferation in the kidney without affecting blood monocyte counts resulted in decreased proliferation of M1 macrophages, and therefore less inflammation in animal models of renal allograft rejection, UUO, and diabetic kidney disease [107–109]. In addition, inhibition of CSF1R signaling suppresses monocyte generation in the bone marrow, which dramatically reduces the accumulation of kidney M1 macrophages and protects from kidney injury in crescentic glomerulonephritis and acute renal allograft rejection [20,110].

M1 macrophage activation and function

A range of endogenous factors can induce M1 macrophage activation and polarization [52]. Macrophages express innate pattern recognition receptors like TLRs, C-type lectin receptors, RIG- and NOD-like receptors that recognize PAMPs such as LPS and bacterial (CpG) DNA (TLR4 and TLR9 ligands, respectively), which induce M1 macrophage responses in vitro [111,112]. Administration of LPS or CpG in models of kidney disease increases the M1 macrophage-mediated response and exacerbates kidney injury [113–115]. The release of endogenous DAMPs by damaged or necrotic cells during, e.g., crystal-induced AKI [116,117] or IRI

[50,118], can also induce M1 macrophage polarization. Examples of these DAMPs include histones [118,119], uromodulin [120,121], high mobility group protein B1 (HMGB1) [122], as well as advanced glycation end-products and C-reactive protein [123]. These PAMPs and DAMPs trigger M1 macrophage polarization and activation via NF-κB [43,124,125] and Janus kinase-signal transducer and activator of transcription (JAK/STAT) [126,127] signaling pathways, and NLRP3 inflammasome activation [128]. Subsequently, M1 macrophages release a variety of inflammatory mediators including IL-1β, TNFα, IL-6, and IFN-γ, chemokines, reactive oxygen species, nitric oxide, and promote Th1 and Th17 responses [129]. Besides endogenous regulators, intrinsic transcriptional factors also contribute to M1 polarization. Examples include the transcription factors interferon regulatory factor 5 (IRF5) [130,131], Krüppel-like factor 4 and 6 (KLF4 and 6) [132,133] and JunB [134]. However, the precise contribution of multiple endogenous factors to M1 polarization is not well understood because the underlying mechanisms might vary greatly dependent on the inflammatory context.

Pro-inflammatory M1 macrophages contribute to the initiation of kidney injury and sustain inflammatory processes that drive IRI [54,104]. Ongoing injury leads to the persistent release of DAMPs and M1 macrophage polarization. Macrophage depletion directly after kidney injury, at a time when M1 macrophages are present in high numbers, was beneficial on the outcome in immunologic, toxin-, ischemia-, and hypertension-mediated models of kidney [101,103,135]. Genetic inhibition of NF-κB in macrophages reduced the production of IL-12, TNF, and NO signaling in M1 macrophages, which attenuated interstitial fibrosis, less crescent formation, and improved albuminuria in a nephrotoxic nephritis model [43]. In addition, the release of pro-inflammatory mediators by activated macrophages can further cause endothelial dysfunction and kidney cell death [136,137], while at the same time, the immune system sets anti-inflammatory mechanisms into place. For example, a study with multi-trauma patients showed that elevated DAMP levels correlated with high IL-10 and decreased pro-inflammatory cytokines, potentially linking the release of DAMPs to the frequently observed immunosuppression after trauma [138].

Anti-inflammatory macrophages

Overshooting organ inflammation holds the risk of acute organ dysfunction; therefore, numerous anti-inflammatory mechanisms are required to keep the balance, a process that allows the resolution of inflammation. Resolution of inflammation is initiated by a shift in the tissue from a pro- into an anti-inflammatory environment, a process associated with the clearance of apoptotic neutrophils by macrophages leading to the release of IL-10 and TGFβ [139–141]. Serum amyloid-P, also known as pentraxin 2, can opsonize apoptotic cells, which supports an M1-to-M2 phenotype change [142]. In addition, M2c macrophages primed with IL-10 and TGFβ reduce tissue inflammation and protect mice from kidney injury [143–147]. The fact that M2c macrophages themselves produce large amounts of IL-10 illustrates how macrophages can amplify their environment by secreting similar cytokines [148]. These cells are needed to enforce the resolution of inflammation and wound healing.

To enhance the regeneration process, M2 macrophages acquire an environment of growth factor-producing cells that actively drive epithelial repair. For example, postischemic AKI is driven by factors released by dying tubular epithelial cells and by phagocytosis of apoptotic and necrotic cells [59,60,149]. In vivo experiments, in which macrophages are depleted during the healing phase of kidney injury showed a delay in recovery [150], suggesting that M2 macrophages contribute to the recovery from injury. The deactivation of pro-inflammatory M1 macrophages involves the transcription factor IRF4 [56], which competes with IRF5, a non-redundant factor of TLR and IL-1R signaling [151–154]. IFR4 deficiency does not allow this phenotype switch [152], hence, persistent activation of macrophages contributes to ongoing immunopathology [155]. In addition, IRAK-M deficiency in mice is sufficient to turn AKI recovery into progressive CKD, suggesting that IRAK-M is required to suppress persistent M1 macrophage-related kidney inflammation [54]. Adoptive transfer of IFNγ-activated bone marrow-derived M1 macrophages can switch from an M1-to-M2 macrophage phenotype during the repair phase. Adoptive transfer of IL-4-polarized M2 macrophages but not IFNγ-polarized M1 macrophages promoted kidney tubular proliferation, which is indicative of kidney repair [61]. Further animal studies in which kidney injury resolves have identified a phenotypic change from M1-to-M2 macrophage during recovery and this confers a protective and reparative role, like in rhabdomyolysis-induced AKI [156], IRI [54,58,150], UUO [157], and nephrotoxic serum nephritis [158]. Thus, M1 macrophages are important players in the pathogenesis of AKI by contributing to the vicious cycle of cell death—release of DAMPs—TLR signaling—M1 polarization, while sufficient tissue repair fosters a phenotype switch from M1-to-M2 macrophages that is crucial for the recovery from AKI and tissue homeostasis.

Macrophages in chronic kidney disease

It is clear that resident macrophages are important for kidney homeostasis, but their role in different stages of kidney disease is rather speculative, e.g., in wound healing. As mentioned above, pro-inflammatory M1 macrophages are predominantly present and drive necroinflammation during the acute phase of kidney injury [159]. This scenario changes in the chronic phase of the disease, where infiltrating monocytes differentiate toward a M2 macrophage phenotype, although pro-inflammatory M1 macrophages remain a feature of chronic inflammation (Fig. 2). The progression of chronic kidney disease (CKD) to kidney failure is linked to an increased deposition of collagen and other extracellular matrix proteins, which involves the loss of normal kidney architecture [160,161]. Regarding kidney pathology, fibrosis of the tubulointerstitial compartment is the strongest predictor of disease progression not only in animal models but also in human kidney diseases [162,163].

Pro-fibrotic macrophages

In the face of ongoing damage and insufficient repair, sustained M2 macrophage infiltration, as a result, of microenvironmental changes with constant production of several wound healing growth factors, such as TGFβ1, collagen, and fibronectin [16,148,164], and what initially begins as a reparative mechanism may subsequently become harmful. Martinez et al. demonstrated that when ECM clearance is deficient due to defects in the engulfment of dying cells, functional markers of renal injuries, such as serum creatinine, blood urea nitrogen, and proteinuria, progressively increase [165]. In fact, persistent wound healing could be pathological, resulting in irreversible fibrosis and progressive kidney tissue destruction [164]. In vitro studies have shown that IL-4 and IL-13 induce STAT6 signaling, which drives an M2a macrophages phenotype that releases fibronectin and other ECM molecules, and expresses mannose and scavenger receptors, including FIZZ and YM-1 [166]. Animal models of kidney disease have revealed the presence of M2 macrophages during the fibrotic phase [158,167–170]. For example, depletion of macrophages protects against progressive interstitial collagen deposition in a severe IRI model of AKI progression to CKD [65]. Similarly, macrophage depletion from day 4 in a model of UUO significantly reduced fibrosis, whereas adoptive transfer of M2 macrophages promoted the accumulation of α smooth muscle actin (αSMA) + cells [171]. The Wnt ligand Wnt3a enhances M2 macrophage polarization induced by

IL-4 or TGFβ1; conditional depletion of *Wnt3a* in myeloid cells attenuated M2 macrophage polarization and reduced interstitial fibrosis [172]. M2 macrophages also promote glomerulosclerosis in which deletion via CSF1R inhibition of the M2 macrophage infiltrate over days 14–35 significantly reduced both glomerulosclerosis and interstitial fibrosis [158]. In a model of nephrocalcinosis-related CKD, neutralizing TGFβ1-inhibited interstitial fibrosis by reducing the number of pro-fibrotic M2 macrophages and slowing the decline of glomerular filtration rate (GFR) [173]. Thus, the in vivo data suggest that the presence of M2 macrophages promotes interstitial fibrosis and disease progression. The M2 macrophage scavenger receptor CD36 regulates oxidative fibrogenic signaling and promotes kidney fibrosis, although it is unclear whether these effects are related to insufficient clearance of apoptotic cells [174]. A number of other molecules secreted by macrophages, including IL-1, MMP2, MMP9, MMP12, galectin 3, chemokine receptors 1, 2 and 5, and platelet-derived growth factor, have also been implicated in promoting fibroblast proliferation and interstitial fibrosis [23,81,175–181]. In addition, conditional deletion of the mineralocorticoid receptor in myeloid cells reduced interstitial fibrosis due to direct effects on macrophage-mediated tissue injury [182,183].

Macrophage-to-myofibroblast transition

Myofibroblasts are a subset of activated fibroblasts that express αSMA and primarily produce collagen during tissue fibrosis. Studies suggest that macrophages derived from the bone marrow can differentiate into αSMA+ myofibroblasts within injured kidneys [16,26–28,184], a process termed macrophage-to-myofibroblast transition (MMT) driven by TGFβ1-Smad3 signaling [185]. Experimental evidence indicates that bone marrow-derived fibroblasts make a substantial contribution to the development and progression of tissue scarring [186,187] and that myofibroblasts within fibrotic lesions originate from the bone marrow in chimeric mice [186,188–191].

A biopsy study involving patients with different forms of kidney diseases identified CD68+αSMA+ cells in patients with active fibrotic lesions but not in kidney biopsies with acute inflammation without fibrosis or with inactive fibrosis [27,28,192]. The majority of CD68+αSMA+ cells co-expressed CD206, suggesting that macrophages that undergo MMT are predominantly M2 [28]. In line with human biopsy studies, cells that co-express markers of macrophages (F4/80) and myofibroblasts (αSMA) have been identified in mouse models of UUO and chronic renal allograft rejection [26–28]. Adoptive transfer of GFP+F4/80+ macrophages isolated from the bone marrow into irradiated UUO mice gave rise to GFP+ αSMA+ myofibrocytes within the obstructed kidney and increased interstitial fibrosis [26].

A recent study showed that bone marrow-derived hematopoietic cells contribute to about 38%–50% of the deposition of collagen I in fibrosis models of reversible UUO and adenine-induced nephropathy using cell type-specific and bone marrow chimeric *col1a1*-deficient mice [193]. The effect of fibrosis on kidney function was dependent on the type of kidney damage, e.g., in the reversible UUO model, collagen production was required to preserve kidney function, while it was detrimental to the kidney function in the more chronic model of adenine-induced nephropathy [193]. Fibrocytes seem to predominate in more chronic types of fibrosis associated with persistent inflammation. Currently, it is difficult to distinguish bone marrow-derived fibrocytes from MMT-derived myofibroblasts. Studies suggest that immature CD34+CD11b+CD45+ myeloid precursor bone marrow cells differentiate into CD45+type I collagen+ fibrocytes, which can be found in the blood and spleen [194–196]. Fibrocytes are known to promote angiogenesis and wound healing, but they may also contribute to collagen deposition in chronic fibrotic diseases [195]. Whether fibrocytes maintain the expression of macrophage lineage markers outside the bone marrow is unclear, although one report describes CD11b expression by splenic and kidney fibroblasts [194]. In addition, the lack of specific fibrocyte markers makes it difficult to isolate this population for detailed characterization and lineage tracing studies. Fibrocytes and MMT-derived myofibroblasts share a common feature in that they originate from bone marrow myeloid progenitors that infiltrate the kidney and contribute substantially to the myofibroblast population and promote renal fibrosis [185].

Macrophage granulomas

Granulomas are comprised of macrophages and other immune cells. The evolutionary conservation of the granulomatous response in animals and humans suggests that this reaction is protective to destroy or sequester particles, e.g., bacteria, but can also be pathological in response to persistent particulate stimuli—infectious or noninfectious—that individual macrophages cannot eradicate, a process known as foreign body reaction [197]. For example, silica particles and asbestos fibers cannot be cleared by phagocytes from the lung and hence trigger granuloma formation, which promotes remodeling and fibrosis [198,199]. In the kidney, crystal plugs within tubules may cause fibrosis as a secondary response to nephron loss, e.g., in chronic calcium oxalate crystal-induced nephropathy [19,173]. Granuloma formation can also occur upon "extratubulation" of large intratubular crystal plugs into the interstitial compartment during 2,8-dihydroxyadenine crystal nephropathy [200]. Further evidence suggests that pre-existing fibrosis with tubular atrophy is required for the formation of uric acid crystal granulomas by M1 macrophages in a model of chronic urate nephropathy with granulomatous nephritis [201] (Figs. 1 and 2). Thus, interstitial macrophage granulomas drive the progression of CKD.

Macrophages during different kidney disease stages

The immune system maintains body homeostasis during physiological fluctuations and following tissue injury, in which macrophages play an important role. Thus, distinct macrophage subsets can coexist in kidney tissue and certain subsets can exist at different disease stages, from the beginning of kidney injury to the recovery phase, to counter-regulate and balance immune responses (Fig. 3) [53]. For example, in acute kidney injury, M1 macrophages promote inflammation, while M2 macrophages exert anti-inflammatory properties. This in turn leads to a short-lasting inflammatory phase and drives wound healing with less fibrosis due to the self-renewal of resident macrophages (Fig. 3) [204]. By contrast, inflammation persists upon repetitive or ongoing ischemia or toxin exposure. This means that in CKD due to progressive and persistent injury and inflammation, M1 macrophages remain at sites of tissue injury and consequently reduce the number of anti-inflammatory M2 macrophages but increase the number of pro-fibrotic and fibrolytic M2 macrophages in the kidney (Fig. 3). Clinical studies showed that the accumulation of M2 macrophages expressing CD206 and/or CD163 correlates with interstitial fibrosis in human kidney disease, e.g., in diabetic kidney disease [205]. In adults with newly diagnosed immunoglobulin A (IgA) nephropathy, interstitial CD163+ and CD204+ macrophages are present in areas of active fibrosis and produce connective tissue growth factor; the degree of mesangial matrix expansion in biopsy samples from these patients correlates with the number of CD163+ macrophages [206]. A retrospective study of kidney allograft biopsy samples found that CD163+ macrophages frequently localize to areas of interstitial fibrosis, characterized by the deposition of type I collagen and the accumulation of myofibroblasts that express αSMA [184]. These clinical data point toward a pro-fibrotic role for a subset of M2 macrophages that contributes to interstitial fibrosis. Rather than using immunohistochemistry, the application of

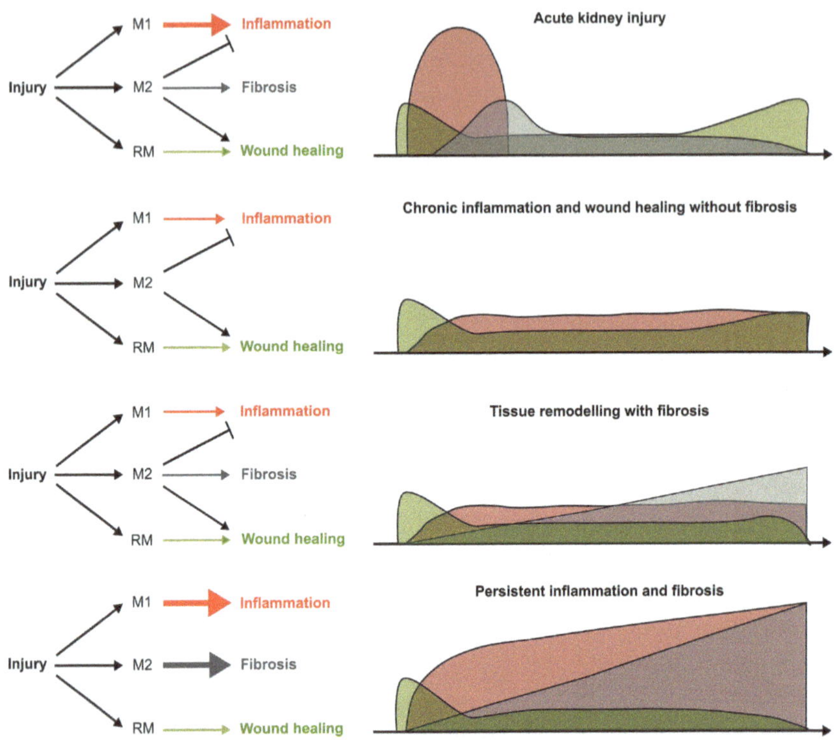

Fig. 3 The role of different macrophage subsets in the clinical context. Pro-inflammatory M1 macrophages promote tissue inflammation and immunopathology based on their role in host defense against intracellular pathogens. Extracellular pathogens are mostly attached by humoral factors such as complement, but when they persist, M2 macrophages provide means of host defense that involve anti-inflammatory, pro-generative, and pro-fibrotic elements. The balance of inflammation and fibrosis varies over time and is different depending on the disease states and often operates in parallel. For this reason, tissue biopsies often become difficult to read and display a mixture of all these elements. The figure provides examples of common disease entities to illustrate how changing tissue environments involve resident macrophages (RM) as well as M1 and M2 macrophage-mediated pathology either in a sequential manner, in an intermittent manner, or in a parallel manner, which largely depends on the associated underlying disease processes and cofactors. The sequential pattern shown at the top was the one that dominated during the evolution of wound healing from the stage of the first multicellular organisms, for example, healing of mechanical trauma in nonsterile environments. All other patterns that physicians often get to see in pathology textbooks and their clinics originate from that and represent maladaptive variants of this underlying danger response program that was otherwise extremely successful during evolution.

scRNA-seq. of human kidney biopsies might further enable us to define macrophage phenotypes across different disease states, characterize macrophage populations during the progressive versus resolution phase, identify candidate macrophage pathways, and how the macrophage phenotype is modified by immunosuppressive therapy [207]. In addition, comprehensive flow cytometry on urinary macrophage polarization, e.g., sCD163, could serve as a biomarker of macrophage polarization for diagnostic purposes during different disease stages [208].

Together, injuries change the homeostatic tissue toward DAMP- and/or PAMP-rich environments, which activate resident and infiltrating macrophages. They produce additional immunostimulatory mediators that aggravate local inflammation, a process that is evolutionally conserved to control invading pathogens. This danger response program is often associated with significant immunopathology, especially in sterile inflammation. Finally, this causes unnecessary tissue injury and becomes a maladaptive disease pathomechanism, which provides the rationale for immunosuppressive, cell-specific, and anti-inflammatory therapies [32].

Therapeutic prospects of macrophage function

Targeting macrophages during acute and chronic kidney injury must be context-dependent because what is a "good" or a "bad" phenotype depends on the phase of the disease and the disease context. On the one hand, inhibition of pro-inflammatory and pro-fibrotic macrophages that cause acute and progressive kidney disease is needed (Fig. 4) but on the other hand, it is important to preserve the ability of macrophages to contribute to tissue repair and to somewhat homeostasis, e.g., by regulating the immune response and defend against infection.

As mentioned previously, chemokines play an important role in the recruitment of circulating monocytes to the site of injury. Targeting, for example, CCL2 or its receptor CCR2 is protective in models of kidney inflammation and fibrosis [80,82,89,92,209], findings that led to a successful phase III clinical trial of a CCR2 antagonist in diabetic kidney disease [210]. Tubular cell production of CXCL16 promotes the recruitment of CXCR6-expressing monocytes and/or macrophages from the bone marrow, which results in the accumulation of CD45+type I collagen+ cells and increased fibrosis in the UUO model [211,212]. Thus, targeting

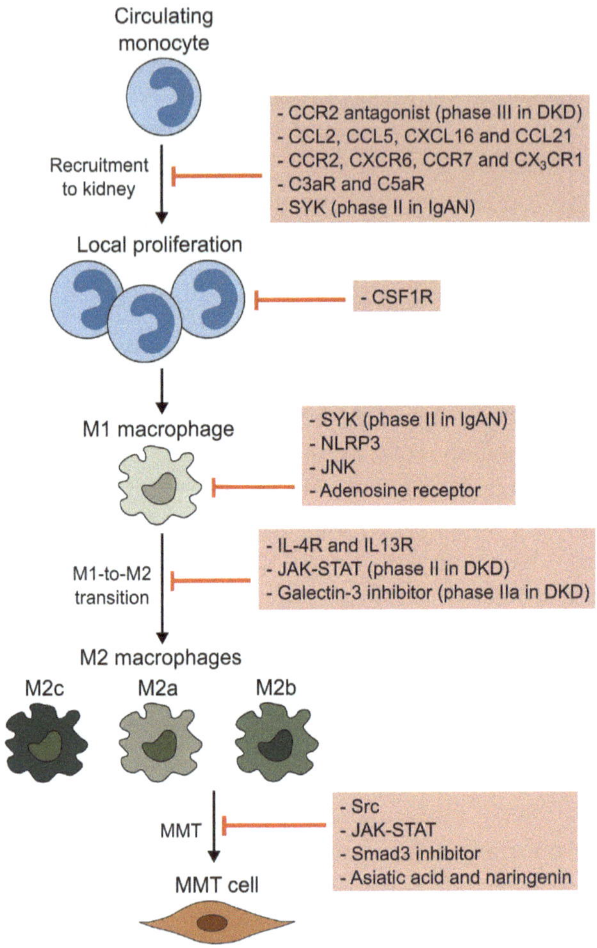

Fig. 4 Potential targets in macrophage-mediated acute and chronic kidney injury. Monocytes can differentiate into M1 and M2 macrophages depending on the environment in response to kidney injury. These differentiation processes of monocyte recruitment and macrophage polarization can be targeted by specific inhibitors to suppress inflammation and improve kidney function, e.g., inhibition of chemokines and other inflammatory signaling pathways and surface receptors. *MMT*, macrophage-to-myofibroblast transition; *DKD*, diabetic kidney disease; *IgAN*, IgA nephropathy. *(Modified from Tang PM, Nikolic-Paterson DJ, Lan HY. Macrophages: versatile players in renal inflammation and fibrosis. Nat Rev Nephrol 2019;15(3):144–158.)*

CXCL16 or its receptor CXCR6 might provide a more selective approach for preventing the accumulation of bone marrow-derived myofibroblasts in CKD. Similar results were obtained by targeting CCL21 and its receptor CCR7 in the UUO model [213] as well as inhibiting integrins, FcRs, complement receptors C3bR and C5aR via spleen tyrosine kinase (SYK) signaling [16,214–216], to selectively reduce macrophage recruitment and activation. In addition, the non-selective SYK inhibitor fostamatinib, which is also a potent janus kinase (JAK)2 inhibitor, is effective in models of glomerulonephritis and lupus nephritis [217,218] and is now in a phase II trial in patients with IgA nephropathy [219]. Blocking CSF1R can reduce macrophage accumulation and kidney injury without affecting the number of blood monocytes [107–109]. However, CSF1R inhibitors have yet to be tested in patients with kidney disease.

Targeting M1 macrophage polarization using c-jun n-terminal kinase (JNK) inhibitors was effective in models of renal IRI [220], UUO-induced fibrosis [221], and glomerulonephritis [222], even with delayed treatment [223]. A role for the kinase Src in interstitial fibrosis has been reported [224], and Src inhibitors were protective in animal models [225–228]. However, current Src inhibitors lack specificity, a major limitation in applying these inhibitors to CKD patients. Moreover, NLRP3 inflammasome inhibition using 1,3-butanediol treatment induced a shift of infiltrating pro-inflammatory and pro-fibrotic macrophages toward an anti-inflammatory phenotype, and prevented interstitial fibrosis in a model of nephrocalcinosis-related CKD [19]. In a model of hyperuricemia with crystalluria-induced CKD, adenosine A_2 receptor inhibition using adenosine suppressed M1 macrophages and attenuated granulomatous nephritis and the progressive decline of kidney function but not inhibiting the JAK/STAT inflammatory pathway with the JAK1/3 inhibitor tofacitinib [201]. Further evidence for targeting adenosine receptors [229], such as A_{2A} and A_{2B}, has been reported in models of diabetic kidney disease [230–232], glomerulonephritis [233], UUO [234,235], and Adriamycin nephropathy [236] that showed renoprotective effects. Although tofacitinib therapy was not beneficial in mice with chronic urate nephropathy [201], a phase II study of baracitinib, a JAK1/2 inhibitor, reduced albuminuria in patients with diabetic kidney disease [237]. Galectin-3 expression and secretion by macrophages promote interstitial fibrosis via activating fibroblasts in the UUO model [177], and serum galectin-3 levels correlate with progressive kidney disease in type 2 diabetes [238]. A galectin-3 antagonist (GCS-100) is currently in phase II trials in diabetic kidney disease.

The last checkpoint is to target the MMT process, specifically TGFβ1-Smad3 signaling because excessive signaling can drive kidney fibrosis, whereas too little signaling could affect immune regulation. This balance has been achieved in animal studies that used Smad3 inhibitors, such as synthetic inhibitor SIS3 and compounds derived from Chinese medicine (asiatic acid and naringenin) [188,225,239] or a neutralizing TGFβ1 antibody

[173], to fine-tune TGFβ1-Smad3 signaling and inhibit interstitial fibrosis. These strategies may hold promise for future clinical studies. However, further studies are needed to define specific macrophage phenotypes across different disease stages, define which macrophage populations predict disease progression versus resolution, identify candidate pathways by which macrophages cause renal damage, determine how macrophages are modified by immunosuppressive therapy, and provide a detailed validation of how well experimental disease models translate to the clinic [185].

Conclusions

Over the last decade, research on macrophages in the kidney has grown rapidly and improved our understanding on the role of macrophages and macrophage polarization under steady-state conditions and during the different phases of acute and chronic kidney injury. Nevertheless, it is important to note that the different macrophage phenotypes most likely coexist in the kidney, which makes it difficult to define the functional contribution of each distinct macrophage subset and also due to the overlapping and changing cell surface marker and gene expression patterns with dendritic cells in physiological and inflammatory conditions. The M1-M2 macrophage concept was proposed 20 years ago and still remains a simplified view on processes including plasticity, heterogeneity, activation, and polarization of macrophages, which are primarily based on in vitro findings. In vivo, the processes are more complex in the tissue, wherein macrophage dynamics and reprogramming depend on the local microenvironment, activation and inflammatory cell status, crosstalk with other kidney cells, and origin of macrophages. However, investigation on the origin and development of resident kidney macrophages and the recruited monocyte-derived macrophages during kidney disease has provided new insights as summarized in this chapter, but their contribution to polarization in the pathogenesis is still not well understood. New approaches using improved cell separation, single-cell, and deep sequencing technologies, and lineage tracing are needed to compare macrophage subsets including resident, anti- and pro-inflammatory, pro-fibrotic, fibrolytic, and granuloma macrophages during the resting state (homeostasis), as well as the different inflammatory and remodeling phase during kidney injury. These approaches will hopefully contribute to a detailed classification of macrophage subsets according to their function (uniform macrophage nomenclature) as well as states of mononuclear phagocytes based on unique discriminatory parameters. Translating the findings from kidney injury models in rodents to kidney disease patients is a challenging task for researchers but necessary to create new opportunities for macrophage-based therapeutic interventions. Current clinical trials targeting mechanisms involved in macrophage recruitment and activation, together with the multitude of preclinical studies, hold promise that it might be possible to protect the kidneys from macrophage-dependent inflammation and fibrosis.

References

[1] Viehmann SF, et al. The multifaceted role of the renal mononuclear phagocyte system. Cell Immunol 2018;330:97–104.
[2] MacPherson GG, Murphy Jr MJ, Morris B. The traffic of mononuclear phagocytes through renal allografts in sheep. Transplantation 1977;24(1):16–28.
[3] Cline MJ, Warner NL, Metcalf D. Identification of the bone marrow colony mononuclear phagocyte as a macrophage. Blood 1972;39(3):327–30.
[4] Muller S, et al. Single-cell profiling of human gliomas reveals macrophage ontogeny as a basis for regional differences in macrophage activation in the tumor microenvironment. Genome Biol 2017;18(1):234.
[5] Lavin Y, et al. Tissue-resident macrophage enhancer landscapes are shaped by the local microenvironment. Cell 2014;159(6):1312–26.
[6] Gabius HJ, Vehmeyer K. Effect of microenvironment and cell-line type on carbohydrate-binding proteins of macrophage-like cells. Biochem Cell Biol 1988;66(11):1169–76.
[7] Robertson MJ, et al. Retinal microenvironment controls resident and infiltrating macrophage function during uveoretinitis. Invest Ophthalmol Vis Sci 2002;43(7):2250–7.
[8] Weidenbusch M, Anders HJ. Tissue microenvironments define and get reinforced by macrophage phenotypes in homeostasis or during inflammation, repair and fibrosis. J Innate Immun 2012;4(5–6):463–77.
[9] Knighton DR, Fiegel VD. Macrophage-derived growth factors in wound healing: regulation of growth factor production by the oxygen microenvironment. Am Rev Respir Dis 1989;140(4):1108–11.
[10] Das A, et al. Monocyte and macrophage plasticity in tissue repair and regeneration. Am J Pathol 2015;185(10):2596–606.
[11] Rosenberger CM, Finlay BB. Phagocyte sabotage: disruption of macrophage signalling by bacterial pathogens. Nat Rev Mol Cell Biol 2003;4(5):385–96.
[12] Isbel NM, et al. Tubules are the major site of M-CSF production in experimental kidney disease: correlation with local macrophage proliferation. Kidney Int 2001;60(2):614–25.
[13] Lever JM, et al. Parabiosis reveals leukocyte dynamics in the kidney. Lab Investig 2018;98(3):391–402.
[14] Gendelman HE, Skillman DR, Meltzer MS. Interferon alpha (IFN)-macrophage interactions in human immunodeficiency virus (HIV) infection: role of IFN in the tempo and progression of HIV disease. Int Rev Immunol 1992;8(1):43–54.
[15] Li H, et al. Cigarette smoke extract-treated mast cells promote alveolar macrophage infiltration and polarization in experimental chronic obstructive pulmonary disease. Inhal Toxicol 2015;27(14):822–31.
[16] Nikolic-Paterson DJ, Wang S, Lan HY. Macrophages promote renal fibrosis through direct and indirect mechanisms. Kidney Int Suppl 2014;4(1):34–8.
[17] Sabelli M, et al. Human macrophage ferroportin biology and the basis for the ferroportin disease. Hepatology 2017;65(5):1512–25.
[18] Brune B, Weigert A, Dehne N. Macrophage polarization in the tumor microenvironment. Redox Biol 2015;5:419.

[19] Anders HJ, et al. The macrophage phenotype and inflammasome component NLRP3 contributes to nephrocalcinosis-related chronic kidney disease independent from IL-1-mediated tissue injury. Kidney Int 2018;93(3):656–69.
[20] Han Y, et al. C-fms blockade reverses glomerular macrophage infiltration and halts development of crescentic anti-GBM glomerulonephritis in the rat. Lab Investig 2011;91(7):978–91.
[21] Ma FY, et al. TGF-beta1-activated kinase-1 regulates inflammation and fibrosis in the obstructed kidney. Am J Physiol Ren Physiol 2011;300(6):F1410–21.
[22] Ma R, et al. Intrarenal macrophage infiltration induced by T cells is associated with podocyte injury in lupus nephritis patients. Lupus 2016;25(14):1577–86.
[23] Tan TK, et al. Macrophage matrix metalloproteinase-9 mediates epithelial-mesenchymal transition in vitro in murine renal tubular cells. Am J Pathol 2010;176(3):1256–70.
[24] Wise AF, et al. Human mesenchymal stem cells alter macrophage phenotype and promote regeneration via homing to the kidney following ischemia-reperfusion injury. Am J Physiol Ren Physiol 2014;306(10):F1222–35.
[25] Zhang XL, et al. Vitamin D prevents podocyte injury via regulation of macrophage M1/M2 phenotype in diabetic nephropathy rats. Endocrinology 2014;155(12):4939–50.
[26] Wang S, et al. TGF-beta/Smad3 signalling regulates the transition of bone marrow-derived macrophages into myofibroblasts during tissue fibrosis. Oncotarget 2016;7(8):8809–22.
[27] Wang YY, et al. Macrophage-to-myofibroblast transition contributes to interstitial fibrosis in chronic renal allograft injury. J Am Soc Nephrol 2017;28(7):2053–67.
[28] Meng XM, et al. Inflammatory macrophages can transdifferentiate into myofibroblasts during renal fibrosis. Cell Death Dis 2016;7(12):e2495.
[29] Stremmel C, et al. Yolk sac macrophage progenitors traffic to the embryo during defined stages of development. Nat Commun 2018;9(1):75.
[30] Zimmerman KA, et al. Single-cell RNA sequencing identifies candidate renal resident macrophage gene expression signatures across species. J Am Soc Nephrol 2019;30(5):767–81.
[31] Hashimoto D, et al. Tissue-resident macrophages self-maintain locally throughout adult life with minimal contribution from circulating monocytes. Immunity 2013;38(4):792–804.
[32] Lech M, et al. Tissues use resident dendritic cells and macrophages to maintain homeostasis and to regain homeostasis upon tissue injury: the immunoregulatory role of changing tissue environments. Mediat Inflamm 2012;2012:951390.
[33] Yoshida H, et al. Phosphatidylserine-dependent engulfment by macrophages of nuclei from erythroid precursor cells. Nature 2005;437(7059):754–8.
[34] Liu F, et al. Distinct fate, dynamics and niches of renal macrophages of bone marrow or embryonic origins. Nat Commun 2020;11(1):2280.
[35] Kiss M, et al. Myeloid cell heterogeneity in cancer: not a single cell alike. Cell Immunol 2018;330:188–201.
[36] Cochain C, et al. Single-cell RNA-seq reveals the transcriptional landscape and heterogeneity of aortic macrophages in murine atherosclerosis. Circ Res 2018;122(12):1661–74.
[37] Kim CC, Nakamura MC, Hsieh CL. Brain trauma elicits non-canonical macrophage activation states. J Neuroinflammation 2016;13(1):117.
[38] Medzhitov R. Origin and physiological roles of inflammation. Nature 2008;454(7203):428–35.
[39] Kalish SV, et al. Macrophages reprogrammed in vitro towards the M1 phenotype and activated with LPS extend lifespan of mice with ehrlich ascites carcinoma. Med Sci Monit Basic Res 2015;21:226–34.

[40] Murphy BS, et al. Azithromycin alters macrophage phenotype. J Antimicrob Chemother 2008;61(3):554–60.
[41] Ishizuka EK, et al. Role of interplay between IL-4 and IFN-gamma in the in regulating M1 macrophage polarization induced by Nattectin. Int Immunopharmacol 2012;14(4):513–22.
[42] Venturin GL, et al. M1 polarization and the effect of PGE2 on TNF-alpha production by lymph node cells from dogs with visceral leishmaniasis. Parasite Immunol 2016;38(11):698–704.
[43] Wilson HM, et al. Inhibition of macrophage nuclear factor-kappaB leads to a dominant anti-inflammatory phenotype that attenuates glomerular inflammation in vivo. Am J Pathol 2005;167(1):27–37.
[44] Hennemann B, et al. Effect of granulocyte-macrophage colony-stimulating factor treatment on phenotype, cytokine release and cytotoxicity of circulating blood monocytes and monocyte-derived macrophages. Br J Haematol 1998;102(5):1197–203.
[45] Karuppagounder V, et al. Curcumin alleviates renal dysfunction and suppresses inflammation by shifting from M1 to M2 macrophage polarization in daunorubicin induced nephrotoxicity in rats. Cytokine 2016;84:1–9.
[46] Lv LL, et al. The pattern recognition receptor, mincle, is essential for maintaining the M1 macrophage phenotype in acute renal inflammation. Kidney Int 2017;91(3):587–602.
[47] Iacopino AM, et al. Phenytoin and cyclosporine A specifically regulate macrophage phenotype and expression of platelet-derived growth factor and interleukin-1 in vitro and in vivo: possible molecular mechanism of drug-induced gingival hyperplasia. J Periodontol 1997;68(1):73–83.
[48] Onore CE, et al. Inflammatory macrophage phenotype in BTBR T+tf/J mice. Front Neurosci 2013;7:158.
[49] Mulay SR, Linkermann A, Anders HJ. Necroinflammation in kidney disease. J Am Soc Nephrol 2016;27(1):27–39.
[50] Salei N, et al. The kidney contains ontogenetically distinct dendritic cell and macrophage subtypes throughout development that differ in their inflammatory properties. J Am Soc Nephrol 2020;31(2):257–78.
[51] Puranik AS, et al. Kidney-resident macrophages promote a proangiogenic environment in the normal and chronically ischemic mouse kidney. Sci Rep 2018;8(1):13948.
[52] Murray PJ. Macrophage polarization. Annu Rev Physiol 2017;79:541–66.
[53] Cao Q, Harris DC, Wang Y. Macrophages in kidney injury, inflammation, and fibrosis. Physiology (Bethesda) 2015;30(3):183–94.
[54] Lech M, et al. Macrophage phenotype controls long-term AKI outcomes—kidney regeneration versus atrophy. J Am Soc Nephrol 2014;25(2):292–304.
[55] Gunthner R, Anders HJ. Interferon-regulatory factors determine macrophage phenotype polarization. Mediat Inflamm 2013;2013:731023.
[56] Lorenz G, et al. IFN regulatory factor 4 controls post-ischemic inflammation and prevents chronic kidney disease. Front Immunol 2019;10:2162.
[57] Gordon S, Pluddemann A. Tissue macrophages: heterogeneity and functions. BMC Biol 2017;15(1):53.
[58] Vinuesa E, et al. Macrophage involvement in the kidney repair phase after ischaemia/reperfusion injury. J Pathol 2008;214(1):104–13.
[59] Huen SC, Cantley LG. Macrophage-mediated injury and repair after ischemic kidney injury. Pediatr Nephrol 2015;30(2):199–209.
[60] Lee S, et al. Distinct macrophage phenotypes contribute to kidney injury and repair. J Am Soc Nephrol 2011;22(2):317–26.
[61] Zhang MZ, et al. IL-4/IL-13-mediated polarization of renal macrophages/dendritic cells to an M2a phenotype is essential for recovery from acute kidney injury. Kidney Int 2017;91(2):375–86.

[62] Cassol E, et al. M1 and M2a polarization of human monocyte-derived macrophages inhibits HIV-1 replication by distinct mechanisms. J Immunol 2009;182(10):6237–46.
[63] Zhao X, et al. PI3K/Akt signaling pathway modulates influenza virus induced mouse alveolar macrophage polarization to M1/M2b. PLoS One 2014;9(8):e104506.
[64] Lisi L, et al. Proinflammatory-activated glioma cells induce a switch in microglial polarization and activation status, from a predominant M2b phenotype to a mixture of M1 and M2a/B polarized cells. ASN Neuro 2014;6(3):171–83.
[65] Kim MG, et al. The role of M2 macrophages in the progression of chronic kidney disease following acute kidney injury. PLoS One 2015;10(12):e0143961.
[66] Spiller KL, et al. The role of macrophage phenotype in vascularization of tissue engineering scaffolds. Biomaterials 2014;35(15):4477–88.
[67] Lu J, et al. Discrete functions of M2a and M2c macrophage subsets determine their relative efficacy in treating chronic kidney disease. Kidney Int 2013;84(4):745–55.
[68] Tang L, et al. M2A and M2C macrophage subsets ameliorate inflammation and fibroproliferation in acute lung injury through interleukin 10 pathway. Shock 2017;48(1):119–29.
[69] Chaves LD, et al. Contrasting effects of systemic monocyte/macrophage and CD4+ T cell depletion in a reversible ureteral obstruction mouse model of chronic kidney disease. Clin Dev Immunol 2013;2013:836989.
[70] Rogers NM, et al. Dendritic cells and macrophages in the kidney: a spectrum of good and evil. Nat Rev Nephrol 2014;10(11):625–43.
[71] Kurts C, Ginhoux F, Panzer U. Kidney dendritic cells: fundamental biology and functional roles in health and disease. Nat Rev Nephrol 2020;16(7):391–407.
[72] Schraml BU, Reis e Sousa C. Defining dendritic cells. Curr Opin Immunol 2015;32:13–20.
[73] Ginhoux F, Jung S. Monocytes and macrophages: developmental pathways and tissue homeostasis. Nat Rev Immunol 2014;14(6):392–404.
[74] Anders HJ, Vielhauer V, Schlondorff D. Chemokines and chemokine receptors are involved in the resolution or progression of renal disease. Kidney Int 2003;63(2):401–15.
[75] Furuichi K, et al. CCR2 signaling contributes to ischemia-reperfusion injury in kidney. J Am Soc Nephrol 2003;14(10):2503–15.
[76] Li L, et al. The chemokine receptors CCR2 and CX3CR1 mediate monocyte/macrophage trafficking in kidney ischemia-reperfusion injury. Kidney Int 2008;74(12):1526–37.
[77] Peng X, et al. CX3CL1-CX3CR1 interaction increases the population of Ly6C(−) CX3CR1(hi) macrophages contributing to unilateral ureteral obstruction-induced fibrosis. J Immunol 2015;195(6):2797–805.
[78] Ma Z, et al. CXCL16 regulates renal injury and fibrosis in experimental renal artery stenosis. Am J Physiol Heart Circ Physiol 2016;311(3):H815–21.
[79] Lloyd CM, et al. RANTES and monocyte chemoattractant protein-1 (MCP-1) play an important role in the inflammatory phase of crescentic nephritis, but only MCP-1 is involved in crescent formation and interstitial fibrosis. J Exp Med 1997;185(7):1371–80.
[80] Lan HY, et al. Expression of macrophage migration inhibitory factor in human glomerulonephritis. Kidney Int 2000;57(2):499–509.
[81] Anders HJ, et al. A chemokine receptor CCR-1 antagonist reduces renal fibrosis after unilateral ureter ligation. J Clin Invest 2002;109(2):251–9.
[82] Yu XQ, et al. A functional role for osteopontin in experimental crescentic glomerulonephritis in the rat. Proc Assoc Am Physicians 1998;110(1):50–64.
[83] Devarapu SK, et al. Dual blockade of the pro-inflammatory chemokine CCL2 and the homeostatic chemokine CXCL12 is as effective as high dose cyclophosphamide in murine proliferative lupus nephritis. Clin Immunol 2016;169:139–47.
[84] Kulkarni O, et al. Spiegelmer inhibition of CCL2/MCP-1 ameliorates lupus nephritis in MRL-(Fas)lpr mice. J Am Soc Nephrol 2007;18(8):2350–8.

[85] Kulkarni O, et al. Anti-Ccl2 Spiegelmer permits 75% dose reduction of cyclophosphamide to control diffuse proliferative lupus nephritis and pneumonitis in MRL-Fas(lpr) mice. J Pharmacol Exp Ther 2009;328(2):371–7.
[86] Sayyed SG, et al. Podocytes produce homeostatic chemokine stromal cell-derived factor-1/CXCL12, which contributes to glomerulosclerosis, podocyte loss and albuminuria in a mouse model of type 2 diabetes. Diabetologia 2009;52(11):2445–54.
[87] Vielhauer V, et al. CCR1 blockade reduces interstitial inflammation and fibrosis in mice with glomerulosclerosis and nephrotic syndrome. Kidney Int 2004;66(6):2264–78.
[88] Menezes S, et al. The heterogeneity of Ly6C(hi) monocytes controls their differentiation into iNOS(+) macrophages or monocyte-derived dendritic cells. Immunity 2016;45(6):1205–18.
[89] Lin SL, et al. Bone marrow Ly6Chigh monocytes are selectively recruited to injured kidney and differentiate into functionally distinct populations. J Immunol 2009;183(10):6733–43.
[90] Rodero MP, et al. Reduced Il17a expression distinguishes a Ly6c(lo)MHCII(hi) macrophage population promoting wound healing. J Invest Dermatol 2013;133(3):783–92.
[91] Clements M, et al. Differential Ly6C expression after renal ischemia-reperfusion identifies unique macrophage populations. J Am Soc Nephrol 2016;27(1):159–70.
[92] Carlin LM, et al. Nr4a1-dependent Ly6C(low) monocytes monitor endothelial cells and orchestrate their disposal. Cell 2013;153(2):362–75.
[93] Westhorpe CLV, et al. Effector CD4(+) T cells recognize intravascular antigen presented by patrolling monocytes. Nat Commun 2018;9(1):747.
[94] Cao Q, et al. Failed renoprotection by alternatively activated bone marrow macrophages is due to a proliferation-dependent phenotype switch in vivo. Kidney Int 2014;85(4):794–806.
[95] Murray PJ, Wynn TA. Protective and pathogenic functions of macrophage subsets. Nat Rev Immunol 2011;11(11):723–37.
[96] Isbel NM, et al. Local macrophage proliferation correlates with increased renal M-CSF expression in human glomerulonephritis. Nephrol Dial Transplant 2001;16(8):1638–47.
[97] Yang N, et al. Local macrophage proliferation in human glomerulonephritis. Kidney Int 1998;54(1):143–51.
[98] Nolasco FE, et al. Intraglomerular T cells and monocytes in nephritis: study with monoclonal antibodies. Kidney Int 1987;31(5):1160–6.
[99] Brasen JH, et al. Macrophage density in early surveillance biopsies predicts future renal transplant function. Kidney Int 2017;92(2):479–89.
[100] Matturri L, et al. Renal allograft rejection: immunohistochemistry of inflammatory cellular subsets and vascular lesions. Basic Appl Histochem 1986;30(2):267–77.
[101] D'Souza MJ, et al. Macrophage depletion by albumin microencapsulated clodronate: attenuation of cytokine release in macrophage-dependent glomerulonephritis. Drug Dev Ind Pharm 1999;25(5):591–6.
[102] Duffield JS, et al. Conditional ablation of macrophages halts progression of crescentic glomerulonephritis. Am J Pathol 2005;167(5):1207–19.
[103] Ferenbach DA, et al. Macrophage/monocyte depletion by clodronate, but not diphtheria toxin, improves renal ischemia/reperfusion injury in mice. Kidney Int 2012;82(8):928–33.
[104] Jo SK, et al. Macrophages contribute to the initiation of ischaemic acute renal failure in rats. Nephrol Dial Transplant 2006;21(5):1231–9.
[105] Jose MD, et al. Macrophages act as effectors of tissue damage in acute renal allograft rejection. Transplantation 2003;76(7):1015–22.
[106] Lan HY, et al. Local macrophage proliferation in the progression of glomerular and tubulointerstitial injury in rat anti-GBM glomerulonephritis. Kidney Int 1995;48(3):753–60.

[107] Jose MD, et al. Blockade of macrophage colony-stimulating factor reduces macrophage proliferation and accumulation in renal allograft rejection. Am J Transplant 2003;3(3):294–300.
[108] Le Meur Y, et al. Macrophage accumulation at a site of renal inflammation is dependent on the M-CSF/c-fms pathway. J Leukoc Biol 2002;72(3):530–7.
[109] Lim AK, et al. Antibody blockade of c-fms suppresses the progression of inflammation and injury in early diabetic nephropathy in obese db/db mice. Diabetologia 2009;52(8):1669–79.
[110] Ma FY, et al. Macrophages contribute to cellular but not humoral mechanisms of acute rejection in rat renal allografts. Transplantation 2013;96(11):949–57.
[111] Ikezumi Y, et al. Macrophage-mediated renal injury is dependent on signaling via the JNK pathway. J Am Soc Nephrol 2004;15(7):1775–84.
[112] Schroder K, Sweet MJ, Hume DA. Signal integration between IFNgamma and TLR signalling pathways in macrophages. Immunobiology 2006;211(6–8):511–24.
[113] Anders HJ, et al. Activation of toll-like receptor-9 induces progression of renal disease in MRL-Fas(lpr) mice. FASEB J 2004;18(3):534–6.
[114] Ryu M, et al. Bacterial CpG-DNA accelerates Alport glomerulosclerosis by inducing an M1 macrophage phenotype and tumor necrosis factor-alpha-mediated podocyte loss. Kidney Int 2011;79(2):189–98.
[115] Tomosugi NI, et al. Modulation of antibody-mediated glomerular injury in vivo by bacterial lipopolysaccharide, tumor necrosis factor, and IL-1. J Immunol 1989;142(9):3083–90.
[116] Mulay SR, et al. Hyperoxaluria requires TNF receptors to initiate crystal adhesion and kidney stone disease. J Am Soc Nephrol 2017;28(3):761–8.
[117] Mulay SR, et al. Mitochondria permeability transition versus necroptosis in oxalate-induced AKI. J Am Soc Nephrol 2019;30(10):1857–69.
[118] Nakazawa D, et al. Histones and neutrophil extracellular traps enhance tubular necrosis and remote organ injury in ischemic AKI. J Am Soc Nephrol 2017;28(6):1753–68.
[119] Allam R, et al. Histones from dying renal cells aggravate kidney injury via TLR2 and TLR4. J Am Soc Nephrol 2012;23(8):1375–88.
[120] Bullen AL, et al. The SPRINT trial suggests that markers of tubule cell function in the urine associate with risk of subsequent acute kidney injury while injury markers elevate after the injury. Kidney Int 2019;96(2):470–9.
[121] Darisipudi MN, et al. Uromodulin triggers IL-1beta-dependent innate immunity via the NLRP3 inflammasome. J Am Soc Nephrol 2012;23(11):1783–9.
[122] Wu H, et al. HMGB1 contributes to kidney ischemia reperfusion injury. J Am Soc Nephrol 2010;21(11):1878–90.
[123] Trial J, Potempa LA, Entman ML. The role of C-reactive protein in innate and acquired inflammation: new perspectives. Inflamm Cell Signal 2016;3(2):e1409.
[124] Glass CK, Natoli G. Molecular control of activation and priming in macrophages. Nat Immunol 2016;17(1):26–33.
[125] Tomita N, et al. In vivo administration of a nuclear transcription factor-kappaB decoy suppresses experimental crescentic glomerulonephritis. J Am Soc Nephrol 2000;11(7):1244–52.
[126] Arnold CE, et al. A critical role for suppressor of cytokine signalling 3 in promoting M1 macrophage activation and function in vitro and in vivo. Immunology 2014;141(1):96–110.
[127] Rutschman R, et al. Cutting edge: Stat6-dependent substrate depletion regulates nitric oxide production. J Immunol 2001;166(4):2173–7.
[128] Mulay SR, et al. Calcium oxalate crystals induce renal inflammation by NLRP3-mediated IL-1beta secretion. J Clin Invest 2013;123(1):236–46.

[129] Lichtnekert J, et al. Changes in macrophage phenotype as the immune response evolves. Curr Opin Pharmacol 2013;13(4):555–64.
[130] Krausgruber T, et al. IRF5 promotes inflammatory macrophage polarization and TH1-TH17 responses. Nat Immunol 2011;12(3):231–8.
[131] Dalmas E, et al. Irf5 deficiency in macrophages promotes beneficial adipose tissue expansion and insulin sensitivity during obesity. Nat Med 2015;21(6):610–8.
[132] Date D, et al. Kruppel-like transcription factor 6 regulates inflammatory macrophage polarization. J Biol Chem 2014;289(15):10318–29.
[133] Liao X, et al. Kruppel-like factor 4 regulates macrophage polarization. J Clin Invest 2011;121(7):2736–49.
[134] Fontana MF, et al. JUNB is a key transcriptional modulator of macrophage activation. J Immunol 2015;194(1):177–86.
[135] Wang Y, et al. Partial depletion of macrophages by ED7 reduces renal injury in adriamycin nephropathy. Nephrology (Carlton) 2005;10(5):470–7.
[136] Ma S, et al. Perirenal fat promotes renal arterial endothelial dysfunction in obese swine through tumor necrosis factor-alpha. J Urol 2016;195(4 Pt 1):1152–9.
[137] Yang N, et al. Local macrophage and myofibroblast proliferation in progressive renal injury in the rat remnant kidney. Nephrol Dial Transplant 1998;13(8):1967–74.
[138] Timmermans K, et al. Plasma levels of danger-associated molecular patterns are associated with immune suppression in trauma patients. Intensive Care Med 2016;42(4):551–61.
[139] Savill J. Apoptosis in resolution of inflammation. J Leukoc Biol 1997;61(4):375–80.
[140] Savill J, Gregory C, Haslett C. Cell biology. Eat me or die. Science 2003;302(5650):1516–7.
[141] Lucas M, et al. Apoptotic cells and innate immune stimuli combine to regulate macrophage cytokine secretion. J Immunol 2003;171(5):2610–5.
[142] Duffield JS. Macrophages and immunologic inflammation of the kidney. Semin Nephrol 2010;30(3):234–54.
[143] Barczyk K, et al. Glucocorticoids promote survival of anti-inflammatory macrophages via stimulation of adenosine receptor A3. Blood 2010;116(3):446–55.
[144] Zheng D, et al. Transfused macrophages ameliorate pancreatic and renal injury in murine diabetes mellitus. Nephron Exp Nephrol 2011;118(4):e87–99.
[145] Cao Q, et al. IL-10/TGF-beta-modified macrophages induce regulatory T cells and protect against adriamycin nephrosis. J Am Soc Nephrol 2010;21(6):933–42.
[146] Kluth DC, et al. Macrophages transfected with adenovirus to express IL-4 reduce inflammation in experimental glomerulonephritis. J Immunol 2001;166(7):4728–36.
[147] Wang Y, et al. Ex vivo programmed macrophages ameliorate experimental chronic inflammatory renal disease. Kidney Int 2007;72(3):290–9.
[148] Mosser DM, Edwards JP. Exploring the full spectrum of macrophage activation. Nat Rev Immunol 2008;8(12):958–69.
[149] Filardy AA, et al. Proinflammatory clearance of apoptotic neutrophils induces an IL-12(low)IL-10(high) regulatory phenotype in macrophages. J Immunol 2010;185(4):2044–50.
[150] Zhang MZ, et al. CSF-1 signaling mediates recovery from acute kidney injury. J Clin Invest 2012;122(12):4519–32.
[151] Negishi H, et al. Negative regulation of Toll-like-receptor signaling by IRF-4. Proc Natl Acad Sci USA 2005;102(44):15989–94.
[152] Satoh T, et al. The Jmjd3-Irf4 axis regulates M2 macrophage polarization and host responses against helminth infection. Nat Immunol 2010;11(10):936–44.
[153] El Chartouni C, Schwarzfischer L, Rehli M. Interleukin-4 induced interferon regulatory factor (Irf) 4 participates in the regulation of alternative macrophage priming. Immunobiology 2010;215(9–10):821–5.

[154] Ishii M, et al. Epigenetic regulation of the alternatively activated macrophage phenotype. Blood 2009;114(15):3244–54.
[155] Lassen S, et al. Ischemia reperfusion induces IFN regulatory factor 4 in renal dendritic cells, which suppresses postischemic inflammation and prevents acute renal failure. J Immunol 2010;185(3):1976–83.
[156] Belliere J, et al. Specific macrophage subtypes influence the progression of rhabdomyolysis-induced kidney injury. J Am Soc Nephrol 2015;26(6):1363–77.
[157] Lin L, Hu K. Tissue-type plasminogen activator modulates macrophage M2 to M1 phenotypic change through annexin A2-mediated NF-kappaB pathway. Oncotarget 2017;8(50):88094–103.
[158] Han Y, et al. Role of macrophages in the fibrotic phase of rat crescentic glomerulonephritis. Am J Physiol Ren Physiol 2013;304(8):F1043–53.
[159] Anders HJ, et al. One concept does not fit all: the immune system in different forms of acute kidney injury. Nephrol Dial Transplant 2020;36:29–38.
[160] Meng XM, Nikolic-Paterson DJ, Lan HY. TGF-beta: the master regulator of fibrosis. Nat Rev Nephrol 2016;12(6):325–38.
[161] Rockey DC, Bell PD, Hill JA. Fibrosis—a common pathway to organ injury and failure. N Engl J Med 2015;372(12):1138–49.
[162] Danilewicz M, Wagrowska-Danielwicz M. Morphometric and immunohistochemical insight into focal segmental glomerulosclerosis in obese and non-obese patients. Nefrologia 2009;29(1):35–41.
[163] Ikezumi Y, et al. The sialoadhesin (CD169) expressing a macrophage subset in human proliferative glomerulonephritis. Nephrol Dial Transplant 2005;20(12):2704–13.
[164] Ricardo SD, van Goor H, Eddy AA. Macrophage diversity in renal injury and repair. J Clin Invest 2008;118(11):3522–30.
[165] Martinez J, et al. Noncanonical autophagy inhibits the autoinflammatory, lupus-like response to dying cells. Nature 2016;533(7601):115–9.
[166] Mantovani A, et al. The chemokine system in diverse forms of macrophage activation and polarization. Trends Immunol 2004;25(12):677–86.
[167] Ikezumi Y, et al. Contrasting effects of steroids and mizoribine on macrophage activation and glomerular lesions in rat thy-1 mesangial proliferative glomerulonephritis. Am J Nephrol 2010;31(3):273–82.
[168] Braga TT, et al. MyD88 signaling pathway is involved in renal fibrosis by favoring a TH2 immune response and activating alternative M2 macrophages. Mol Med 2012;18:1231–9.
[169] Yamate J, et al. Participation of different macrophage populations and myofibroblastic cells in chronically developed renal interstitial fibrosis after cisplatin-induced renal injury in rats. Vet Pathol 2002;39(3):322–33.
[170] Kushiyama T, et al. Alteration in the phenotype of macrophages in the repair of renal interstitial fibrosis in mice. Nephrology (Carlton) 2011;16(5):522–35.
[171] Shen B, et al. Macrophages regulate renal fibrosis through modulating TGFbeta superfamily signaling. Inflammation 2014;37(6):2076–84.
[172] Feng Y, et al. Wnt/beta-catenin-promoted macrophage alternative activation contributes to kidney fibrosis. J Am Soc Nephrol 2018;29(1):182–93.
[173] Steiger S, et al. Anti-transforming growth factor beta IgG elicits a dual effect on calcium oxalate crystallization and progressive nephrocalcinosis-related chronic kidney disease. Front Immunol 2018;9:619.
[174] Pennathur S, et al. The macrophage phagocytic receptor CD36 promotes fibrogenic pathways on removal of apoptotic cells during chronic kidney injury. Am J Pathol 2015;185(8):2232–45.
[175] Du X, et al. Involvement of matrix metalloproteinase-2 in the development of renal interstitial fibrosis in mouse obstructive nephropathy. Lab Investig 2012;92(8):1149–60.

[176] Eitner F, et al. PDGF-C is a proinflammatory cytokine that mediates renal interstitial fibrosis. J Am Soc Nephrol 2008;19(2):281–9.
[177] Henderson NC, et al. Galectin-3 expression and secretion links macrophages to the promotion of renal fibrosis. Am J Pathol 2008;172(2):288–98.
[178] Jones LK, et al. IL-1RI deficiency ameliorates early experimental renal interstitial fibrosis. Nephrol Dial Transplant 2009;24(10):3024–32.
[179] Niu H, et al. Matrix metalloproteinase 12 modulates high-fat-diet induced glomerular fibrogenesis and inflammation in a mouse model of obesity. Sci Rep 2016;6:20171.
[180] Vielhauer V, et al. Obstructive nephropathy in the mouse: progressive fibrosis correlates with tubulointerstitial chemokine expression and accumulation of CC chemokine receptor 2- and 5-positive leukocytes. J Am Soc Nephrol 2001;12(6):1173–87.
[181] Eis V, et al. Chemokine receptor CCR1 but not CCR5 mediates leukocyte recruitment and subsequent renal fibrosis after unilateral ureteral obstruction. J Am Soc Nephrol 2004;15(2):337–47.
[182] Huang LL, et al. Myeloid mineralocorticoid receptor activation contributes to progressive kidney disease. J Am Soc Nephrol 2014;25(10):2231–40.
[183] Barrera-Chimal J, et al. The myeloid mineralocorticoid receptor controls inflammatory and fibrotic responses after renal injury via macrophage interleukin-4 receptor signaling. Kidney Int 2018;93(6):1344–55.
[184] Ikezumi Y, et al. Alternatively activated macrophages in the pathogenesis of chronic kidney allograft injury. Pediatr Nephrol 2015;30(6):1007–17.
[185] Tang PM, Nikolic-Paterson DJ, Lan HY. Macrophages: versatile players in renal inflammation and fibrosis. Nat Rev Nephrol 2019;15(3):144–58.
[186] Broekema M, et al. Bone marrow-derived myofibroblasts contribute to the renal interstitial myofibroblast population and produce procollagen I after ischemia/reperfusion in rats. J Am Soc Nephrol 2007;18(1):165–75.
[187] Haudek SB, et al. Bone marrow-derived fibroblast precursors mediate ischemic cardiomyopathy in mice. Proc Natl Acad Sci USA 2006;103(48):18284–9.
[188] LeBleu VS, et al. Origin and function of myofibroblasts in kidney fibrosis. Nat Med 2013;19(8):1047–53.
[189] Jang HS, et al. Recruitment and subsequent proliferation of bone marrow-derived cells in the postischemic kidney are important to the progression of fibrosis. Am J Physiol Ren Physiol 2014;306(12):F1451–61.
[190] Jang HS, et al. Bone marrow-derived cells play a major role in kidney fibrosis via proliferation and differentiation in the infiltrated site. Biochim Biophys Acta 2013;1832(6):817–25.
[191] Li J, et al. The contribution of bone marrow-derived cells to the development of renal interstitial fibrosis. Stem Cells 2007;25(3):697–706.
[192] Grimm PC, et al. Neointimal and tubulointerstitial infiltration by recipient mesenchymal cells in chronic renal-allograft rejection. N Engl J Med 2001;345(2):93–7.
[193] Buchtler S, et al. Cellular origin and functional relevance of collagen I production in the kidney. J Am Soc Nephrol 2018;29(7):1859–73.
[194] Reich B, et al. Fibrocytes develop outside the kidney but contribute to renal fibrosis in a mouse model. Kidney Int 2013;84(1):78–89.
[195] Herzog EL, Bucala R. Fibrocytes in health and disease. Exp Hematol 2010;38(7):548–56.
[196] Bucala R, et al. Circulating fibrocytes define a new leukocyte subpopulation that mediates tissue repair. Mol Med 1994;1(1):71–81.
[197] Pagan AJ, Ramakrishnan L. The formation and function of granulomas. Annu Rev Immunol 2018;36:639–65.
[198] Liu G, Cheresh P, Kamp DW. Molecular basis of asbestos-induced lung disease. Annu Rev Pathol 2013;8:161–87.
[199] Mulay SR, et al. A guide to crystal-related and nano- or microparticle-related tissue responses. FEBS J 2020;287(5):818–32.

[200] Klinkhammer BM, et al. Cellular and molecular mechanisms of kidney injury in 2,8-dihydroxyadenine nephropathy. J Am Soc Nephrol 2020;31(4):799–816.
[201] Sellmayr M, et al. Only hyperuricemia with crystalluria, but not asymptomatic hyperuricemia, drives progression of chronic kidney disease. J Am Soc Nephrol 2020;31:2773–92.
[202] Murray PJ, Wynn TA. Obstacles and opportunities for understanding macrophage polarization. J Leukoc Biol 2011;89(4):557–63.
[203] Xue J, et al. Transcriptome-based network analysis reveals a spectrum model of human macrophage activation. Immunity 2014;40(2):274–88.
[204] Swaminathan S, Griffin MD. First responders: understanding monocyte-lineage traffic in the acutely injured kidney. Kidney Int 2008;74(12):1509–11.
[205] Klessens CQF. Macrophages in diabetic nephropathy in patients with type 2 diabetes. Nephrol Dial Transplant 2017;32(8):1322–9. https://doi.org/10.1093/ndt/gfw260.
[206] Ikezumi Y, et al. Identification of alternatively activated macrophages in new-onset paediatric and adult immunoglobulin A nephropathy: potential role in mesangial matrix expansion. Histopathology 2011;58(2):198–210.
[207] Wu H, et al. Single-cell transcriptomics of a human kidney allograft biopsy specimen defines a diverse inflammatory response. J Am Soc Nephrol 2018;29(8):2069–80.
[208] Sun PP, et al. Urine macrophages reflect kidney macrophage content during acute tubular interstitial and glomerular injury. Clin Immunol 2019;205:65–74.
[209] Tesch GH. Diabetic nephropathy—is this an immune disorder? Clin Sci (Lond) 2017;131(16):2183–99.
[210] de Zeeuw D, et al. The effect of CCR2 inhibitor CCX140-B on residual albuminuria in patients with type 2 diabetes and nephropathy: a randomised trial. Lancet Diabetes Endocrinol 2015;3(9):687–96.
[211] Xia Y, Entman ML, Wang Y. CCR2 regulates the uptake of bone marrow-derived fibroblasts in renal fibrosis. PLoS One 2013;8(10):e77493.
[212] Chen G, et al. CXCL16 recruits bone marrow-derived fibroblast precursors in renal fibrosis. J Am Soc Nephrol 2011;22(10):1876–86.
[213] Sakai N, et al. Secondary lymphoid tissue chemokine (SLC/CCL21)/CCR7 signaling regulates fibrocytes in renal fibrosis. Proc Natl Acad Sci USA 2006;103(38):14098–103.
[214] Ryan J, et al. Myeloid cell-mediated renal injury in rapidly progressive glomerulonephritis depends upon spleen tyrosine kinase. J Pathol 2016;238(1):10–20.
[215] Ryan J, et al. Spleen tyrosine kinase signaling promotes myeloid cell recruitment and kidney damage after renal ischemia/reperfusion injury. Am J Pathol 2016;186(8):2032–42.
[216] Han Y, et al. An inhibitor of spleen tyrosine kinase suppresses experimental crescentic glomerulonephritis. Int J Immunopathol Pharmacol 2018;32. 2058738418783404.
[217] Bahjat FR, et al. An orally bioavailable spleen tyrosine kinase inhibitor delays disease progression and prolongs survival in murine lupus. Arthritis Rheum 2008;58(5):1433–44.
[218] McAdoo SP, et al. Spleen tyrosine kinase inhibition attenuates autoantibody production and reverses experimental autoimmune GN. J Am Soc Nephrol 2014;25(10):2291–302.
[219] Ma TK, McAdoo SP, Tam FW. Spleen tyrosine kinase: a crucial player and potential therapeutic target in renal disease. Nephron 2016;133(4):261–9.
[220] Kanellis J, et al. JNK signalling in human and experimental renal ischaemia/reperfusion injury. Nephrol Dial Transplant 2010;25(9):2898–908.
[221] Ma FY, et al. A pathogenic role for c-Jun amino-terminal kinase signaling in renal fibrosis and tubular cell apoptosis. J Am Soc Nephrol 2007;18(2):472–84.
[222] Flanc RS, et al. A pathogenic role for JNK signaling in experimental anti-GBM glomerulonephritis. Kidney Int 2007;72(6):698–708.
[223] Ma FY, et al. Blockade of the c-Jun amino terminal kinase prevents crescent formation and halts established anti-GBM glomerulonephritis in the rat. Lab Investig 2009;89(4):470–84.

[224] Anguita E, Villalobo A. Ca(2+) signaling and Src-kinases-controlled cellular functions. Arch Biochem Biophys 2018;650:59–74.
[225] Tang PM, et al. The proto-oncogene tyrosine protein kinase Src is essential for macrophage-myofibroblast transition during renal scarring. Kidney Int 2018;93(1):173–87.
[226] Skhirtladze C, et al. Src kinases in systemic sclerosis: central roles in fibroblast activation and in skin fibrosis. Arthritis Rheum 2008;58(5):1475–84.
[227] Wang J, Zhuang S. Src family kinases in chronic kidney disease. Am J Physiol Ren Physiol 2017;313(3):F721–8.
[228] Yan Y, et al. Src inhibition blocks renal interstitial fibroblast activation and ameliorates renal fibrosis. Kidney Int 2016;89(1):68–81.
[229] Roberts VS, et al. The role of adenosine receptors A2A and A2B signaling in renal fibrosis. Kidney Int 2014;86(4):685–92.
[230] Awad AS, et al. Adenosine A2A receptor activation attenuates inflammation and injury in diabetic nephropathy. Am J Physiol Ren Physiol 2006;290(4):F828–37.
[231] Tak E, et al. CD73-dependent generation of adenosine and endothelial Adora2b signaling attenuate diabetic nephropathy. J Am Soc Nephrol 2014;25(3):547–63.
[232] Cardenas A, et al. Adenosine A(2B) receptor-mediated VEGF induction promotes diabetic glomerulopathy. Lab Investig 2013;93(1):135–44.
[233] Garcia GE, et al. Adenosine A(2A) receptor activation prevents progressive kidney fibrosis in a model of immune-associated chronic inflammation. Kidney Int 2011;80(4):378–88.
[234] Xiao H, et al. Adenosine A2A receptor: a target for regulating renal interstitial fibrosis in obstructive nephropathy. PLoS One 2013;8(4):e60173.
[235] Xiao H, et al. The effects of adenosine A2A receptor knockout on renal interstitial fibrosis in a mouse model of unilateral ureteral obstruction. Acta Histochem 2013;115(4):315–9.
[236] Wang YM, et al. Regulatory T cells participate in CD39-mediated protection from renal injury. Eur J Immunol 2012;42(9):2441–51.
[237] Tuttle KR, et al. JAK1/JAK2 inhibition by baricitinib in diabetic kidney disease: results from a phase 2 randomized controlled clinical trial. Nephrol Dial Transplant 2018;33(11):1950–9.
[238] Tan KCB, et al. Galectin-3 is independently associated with progression of nephropathy in type 2 diabetes mellitus. Diabetologia 2018;61(5):1212–9.
[239] Meng XM, et al. Treatment of renal fibrosis by rebalancing TGF-beta/Smad signaling with the combination of asiatic acid and naringenin. Oncotarget 2015;6(35):36984–97.

CHAPTER 6

Dermal macrophages in health and disease

Ana Carolina Guerta Salina[a], Nathan Klopfenstein[b,c], Jacqueline Pinon[b], and C. Henrique Serezani[b,c,d,e]

[a]Department of Cellular, Molecular Biology and Pathogenic Bioagents, School of Medicine of Ribeirão Preto, University of São Paulo, Ribeirão Preto, São Paulo, Brazil
[b]Department of Medicine, Division of Infectious Diseases, Vanderbilt University Medical Center, Nashville, TN, United States
[c]Department of Pathology, Microbiology, and Immunology, Vanderbilt University Medical Center, Nashville, TN, United States
[d]Vanderbilt Institute of Infection, Immunology, and Inflammation, Vanderbilt University Medical Center, Nashville, TN, United States
[e]Vanderbilt Center for Immunobiology, Vanderbilt University Medical Center, Nashville, TN, United States

Introduction

The skin and its appendages are the most extensive and fast-growing organs in the body. Skin plays pivotal roles varying from the maintenance of the body homeostasis to protect against infections. It is a multilayer barrier that interacts with the outside environment and acts as the first line of host defense [1,2]. The skin is continuously exposed to mechanical, chemical, and microbial stressors requiring an orchestrated response between immunological and non-immunological cells to maintain tissue homeostasis [2]. This tissue is divided into two main anatomical parts, epidermis and dermis, and each layer is constituted by different immune and structural cells [2,3].

The epidermis is the outermost water-impermeable layer of the skin comprised of stratified epithelium, hair follicles, and appendages. It consists mostly of dead and live keratinocytes, Langerhans cells (LCs), CD8+ resident memory T (TRM) cells, dendritic epidermal γδ T cells (DETCs—mice only), melanocytes, and hair follicles [2]. The dermis, in addition to the abundant presence of extracellular matrix (ECM) - mainly collagen and elastin fibers, has an upper region or the papillary dermis that is separated from the epidermis by the basement membrane. The deeper layer (reticular dermis) is rich in the vasculature and comprises several different resident innate immune cells, including dermal macrophages (DMs), dermal dendritic cells (DCs), mast cells, innate lymphoid cells (ILCs), natural killer cells, γδ T cells, and adaptive CD4+ and CD8+ T cells [4] which interact with non-resident cells to keep the skin integrity.

Macrophages in the Human Body
https://doi.org/10.1016/B978-0-12-821385-8.00008-1

Dermal macrophages play a pivotal role in responding to pathogens, inflammation, tissue remodeling, and maintaining tissue homeostasis. The dermal tissue-resident macrophage population is developmental imprinted and originated either from the embryogenic phase of fetal development or circulating blood monocytes. However, during life, the embryogenic population is gradually replaced by monocytes so that, in adulthood, this monocyte-derived pool of macrophages represents the majority of cells present in the tissue [5].

Although DM studies have focused mainly on addressing the pro- or anti-inflammatory actions during wound healing, infection, chemical irritants, and autoimmune diseases, less is known about their role during skin homeostasis. However, recent work suggests that DMs are key in controlling different aspects of skin homeostasis and integrity, such as salt levels, tissue expansion, stress responses, and hair follicle regeneration [6]. Since macrophages are often plastic cells and able to adapt to the systemic and localized cues, studies aiming to investigate a role for DMs in different conditions are likely to generate new therapeutic opportunities to treat inflammatory skin diseases, improve host defense, and maintenance of the skin integrity and homeostasis.

In this review, we discuss (1) the ontology of DMs, (2) their main effector functions during homeostasis and tissue injury, (3) the potential role of DMs in skin morbidities associated with metabolic diseases.

Dermal macrophage characterization and phenotype

Because DMs show high similarity with DCs and the two different origins (fetal vs. blood—see below), isolating and fully characterizing these cells is challenging. Mouse DMs can be differentiated from dermal monocyte-derived DCs, based on their high phagocytic activity, low expression of CCR2, high levels of CD64 (the high-affinity IgG receptor FcγRI), and the receptor tyrosine kinase MERTK [7–9]. DMs cannot migrate to draining lymph nodes in mice and consequently show an inadequate capacity to present antigen to T cells compared with differentiated dermal DCs. DMs also express gene sets that further suggest roles in scavenging cell-matrix and other self-macromolecules, as well as antimicrobial effector functions.

Studies have been shown that the expression of $CD11b^{+}CD24^{lo}$cells in the dermis is a heterogeneous cell subset composed of both DCs and macrophages [10,11]. Furthermore, classic macrophages markers such as CX_3CR_1 (the fractalkine chemokine receptor), F4/80, CD68, and lysozyme

M should not be used alone to distinguish macrophages from the other $CD11b^+$ cells in the dermis [7]. To fully characterize DMs, one should use the classic macrophage markers mentioned above, along with CD64, which allows the separation DCs from DMs. DMs show high expression of these molecules while they are absent in DCs [7–9]. Tamoutounour and colleagues demonstrated that in the skin of adult animals, DCs are $CCR2^+$, while DMs are $CCR2^-$ [7]. Furthermore, Kolter and colleagues showed that blood monocytes downregulated the expression of CX_3CR_1 and Ly6C after skin infiltration [12]. Human dermal macrophages are large $CD1a^-$ $CD14^+$ $FXIIIa^+$ cells, autofluorescent, and granular filled with cytoplasmic vacuoles. Corroborating these findings, a recent study divided DMs into two distinct cellular subsets through the expression of CX_3CR_1. Resident DMs derived from embryonic phases shows high expression of $CX_3CR_{1,}$ and monocyte-derived DMs do not [13]. Moreover, the activation profile of these two DM populations is different: $CX_3CR_1{}^+$ are predominantly alternatively activated, whereas $CX_3CR_1{}^-$cells exhibit both classical and alternative states of activation [13]. The importance of the macrophage activation profiles for wound healing will be discussed below.

To better characterize DM populations and imply functional characteristics, Lee et al. identified four distinct populations designated P1–P4. P1 was categorized predominantly as inflammatory monocytes, P2 was characterized as monocyte-derived DCs (moDCs; P2), and P3 was identified as dermal DCs ($CD11b^+Ly6C^-CD64^-$). Both P2 and P3 populations showed phagocyte morphology with a larger cytoplasm. Finally, P4 cells were identified as DMs. This cell population showed characteristics of mature macrophages with abundant cytoplasmic vacuoles and melanin granules, high expression of mannose receptor, CD206, plus $CD64^+CCR2^{low}MHCII^-$ expression [14]. Curiously, Tamoutounour et al. found five distinct populations. The first four (P1-P4) were the same populations characterized by Lee and colleagues; however, they determined a P5 population with high expression of MHCII. The author suggested that P4 might mature and acquire a P5 phenotype [7].

DM activation profile

Macrophages are versatile cells acting both in the context of pathologies and homeostasis. Because of this wide range of effector functions, these cells show different activation profiles with metabolic alterations depending on the local microenvironment [15].

Although macrophages display in vivo heterogeneity in their activation profile, they are didactically separated into two cellular subsets. M1 macrophages (classical activation state) display pro-inflammatory characteristics and are mainly involved in eradicating invading pathogens and promoting inflammatory responses [16]. Conversely, M2 macrophages (alternative activation sate) correlated with the expression and secretion of anti-inflammatory molecules, such as those involved in dead cell scavenging, angiogenesis, and connective tissue remodeling, and resolution of inflammation [16].

In the resting state, the P1-P4 DMs mentioned above might play specific roles in maintaining homeostasis, inflammation, and infection control. Lee and colleagues reported that *Leishmania major* infects and survives preferentially in P4 DMs that express high levels of CD206 [14]. Although the activation profile of these different DM populations in resting state is not clear, studies have shown that the entry of pathogens into the dermis followed by their recognition via innate immunity receptors such as toll-like receptors (TLRs), NOD-like receptors (NLRs), and (RIG1)-like helicase receptors (RLRs), triggers intracellular responses that culminate in the M1 profile activation in macrophages and release of inflammatory cytokines, TNF-α and IL-1β among others [17,18]. Also, M1 DMs are linked to skin lesions associated with pathogens [19] or in sterile inflammatory diseases, such as lupus. Conditions [20,21].

During the inflammatory phase of wound healing, classically activated macrophages are quite prevalent due to recognizing invasive agents and/or damage-associated molecular pattern molecules (DAMPs) released by tissue cells [22]. These cells are responsible for the induction of the inflammatory response and release of pro-inflammatory mediators such as cytokines, chemokines, and reactive oxygen and nitrogen intermediates that contribute to pathogen killing and inflammation resolution [23]. However, over time, DMs are prone to acquire a pro-resolution phenotype (M2) that favors wound healing [24]. The switch from the inflammatory phenotype to an anti-inflammatory macrophage phenotype is necessary for efficient wound healing [24]. In the inflammation phase of wound healing, there is an intense migration of neutrophils that die from apoptosis in subsequent stages that must be removed from this microenvironment. Studies have shown that the phagocytosis of apoptotic-neutrophil drives macrophages to a pro-resolution phenotype [25].

Moreover, anti-inflammatory mediators released during the clearance of these apoptotic cells can sustain the alternative macrophage profile by bolstering the tissue repair processes [26]. The persistence of M1 macrophages for an extended period or the absence/depletion of the M2 macrophage population results in impaired tissue regeneration and wound closure [27–29].

Dermal macrophage ontogeny

Since Metchnikoff first described macrophages, these cells have been widely studied and characterized [30]. At the beginning of the 1970s, Van Furth and colleagues believed that all macrophages were part of a mononuclear phagocytic system with a common precursor derived from the bone marrow (BM) [31]. They suggested that promonocytes present in BM differentiated into monocytes, and once in the blood, these cells could remain circulating as monocytes or enter tissues to become tissue-specific macrophages [31].

With the advent of new cell tracing techniques, our understanding of macrophage ontogeny and differentiation has grown dramatically. Tissue-resident macrophages are present throughout the body in all organ systems, connective tissue, and serous cavities [31]. Although the most diverse tissue-resident macrophages display common markers, it is essential to note that between tissues, resident macrophages display differences in morphology and expression of surface markers and activation profile, effector functions, and embryonic origins.

Defining a macrophage's ontogeny can be extremely challenging because mammalian embryonic hematopoiesis occurs in multiple successive stages [32]. Although mice and humans exhibit many biological differences, these species share similar embryonic hematopoietic systems [33]. Studies have demonstrated that macrophages can arise from embryos dependent or independent of the generation of hematopoietic stem cells (HSCs). However, the exact origin of dermal resident macrophage remains to be fully determined.

Initial studies in mice suggested that macrophages develop within 7 days of embryonic development (E7.0) from the posterior plate mesoderm in the blood islands of the extra-embryonic yolk sac (YS) [34,35]. However, experiments in adult animals using a fate-mapping model that traced Runt-related transcription factor 1 ($RUNX1^+$) expression in a tamoxifen-inducible $Runx1^{-Mer-cre-Mer}$ mouse [36] revealed that only microglia remained labeled in these mice. In contrast, other resident macrophage populations are not labeled, which suggests that YS-derivate macrophages are replaced by a different cell precursor [37]. The following studies described two waves for erythro-myeloid precursor (EMP) development derived from the YS hemogenic endothelium [38]. The first wave emerges from the YS around E7.5, generating the "early" EMPs [39]. These cells do not express the transcription factor required to expand and differentiate all hematopoietic cell lineages, c-Myb [40], and are the microglia precursors in the brain [37]. On the other hand, between E8.0 and E8.5, the second pool of EMPs derived from YS hemogenic endothelium arises, producing the "late" c-Myb^{+} EMPs [38,41]. Most of these cells migrate to the fetal liver (FL) [42] where they will generate granulocytes, monocytes, and macrophages [43]. Hoeffel and colleagues proposed that the late EMPs represent the main precursors for tissue-resident macrophages through a monocytic intermediate in the fetal liver [41]. Studies have been shown that fetal liver monocytes (FLM) emerge in the FL around E12.5 and colonize the tissues 2 days later (E14.5) through blood dispersion [41,44,45].

Although previous studies have shown a differential EMPs' participation ("early" or "late") in the origin of tissue-resident macrophage pools, the contribution of HSC precursors to the establishment of these cell populations remains unclear. Using a fate-mapping model in a Kit$^{Mer\text{-}cre\text{-}Mer}$ mouse, Sheng and colleagues demonstrated that hematopoietic stem cells (HSCs) contribute to the origin of all tissue-resident macrophage, except microglia and partially LC, which are YS derived [46]. Between E8.5 - E10.5 of embryogenesis, the hematopoietic progenitors start generating immature HSCs in the para-aortic splanchnopleura region, giving rise to fetal HSCs and fetal BM [47]. It is clear that there is a pool of DMs originating in the prenatal phase through all this. However, this subset of cells accounts for only 20% of dermal macrophages [7] and remains stable until around 6 weeks post-birth [12]. Experiments using animals deficient in CCR2, a critical molecule related to the emigration of circulating Ly6C^{hi}blood monocytes from BM [48], revealed a dramatic decrease in the number of DMs when compared to WT animals [7]. Furthermore, these data were confirmed in bone marrow transplant mouse experiments. Here, mice co-expressing CD45.1 and CD45.2 were irradiated and reconstituted with BM cells isolated from wild-type (WT) CD45.1^{+} B6 mice and from CD45.2^{+} B6 deficient in CCR2. After 8 weeks, 85% of dermal macrophages were CD45.1^{+}, demonstrating that the blood monocytes are essential to establishing the dermal macrophages pool [7].

Although at birth, all tissues are populated by fetal macrophages, these cells are partially or almost entirely replaced by blood monocytes with time in some tissues [5]. Dermal tissue-resident macrophages show a continuous turnover in adulthood via the migration and proliferation of Ly6C^{Hi} circulating blood monocytes independently of microbiota stimuli [7]. Although

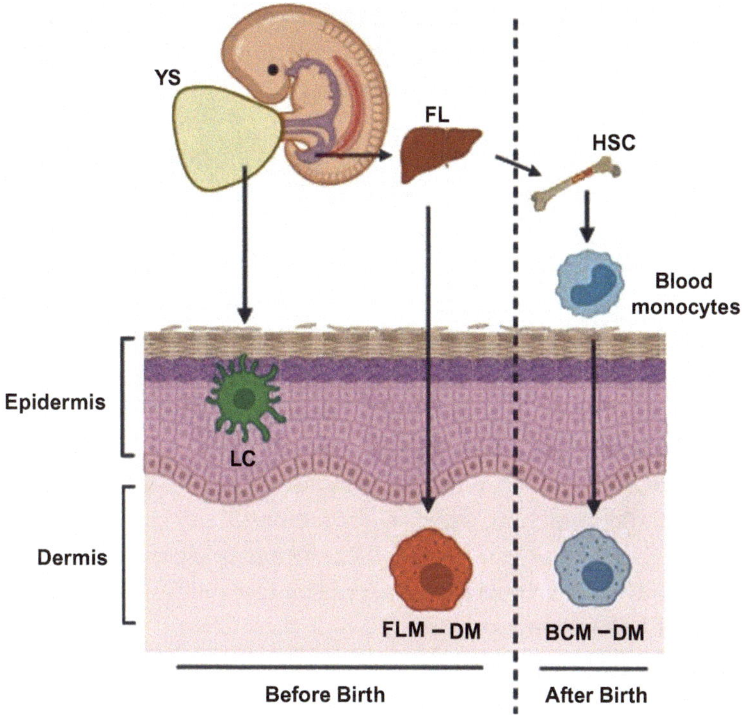

Fig. 1 DMs ontology. In adults, macrophages from different sources can be found in the skin. The first macrophage precursor originates in the YS at E7.5 days of embryonic development and gives rise to LCs in the skin epidermis layer. On the other hand, the macrophages that populate the dermis have two distinct origins. In smaller numbers are the DMs derived from monocytes of the fetal liver (FLM-DM) that migrate to the skin tissues around the 14th day of embryonic development from the fetal liver, and in larger numbers are the DMs derived from blood circulating monocytes (BCM-DM) that continuously migrate to the skin tissue to maintain the DM homeostatic renewal of this cellular population after the birth. *(Figure created with BioRender.com.)*

the skin microbiota is not the determining factor for migration to tissues, studies suggest that it controls cell turnover speed [49]. DMs have an estimated half-life of 4–6 weeks and fail to self-maintain, needing a constant replacement from circulating monocytes [7,50] (Fig. 1).

Dermal macrophage functions

Homeostasis

DMs are essential for maintaining homeostasis of the skin in both healthy and pathological conditions. At a steady-state, tissue-resident macrophages are abundant hematopoietic cells in the skin and contribute to homeostatic tissue renewal and resident microbiota control [51]. Unlike monocyte-derived DCs or cDCs also present in the dermis, DMs cannot migrate to the lymph nodes and present antigens efficiently [7]. These cells are specialized in phagocytosis and elimination of both pathogens and dead cells.

Recently, a new role for DMs has been unveiled. DMs are critical cells in the development of hair follicles. Although the first observation of the role of DMs in skin hair follicle formation was in the 1970s [52], recent studies have dissected the molecular programs by which DMs influence hair development. During homeostatic states, DMs cluster around the hair follicle and ingest and degrade collagen fibers in the basal lamina. Also, DMs enhance the expression of intercellular adhesion molecule (ICAM)-1 in a manner dependent on the hair cycle [53] and ICAM-1 deficient mice showed premature hair follicle regression [53].

Moreover, DMs enhance the proliferation of basal epidermal cells, hair follicle cells, and stem cells, which raises the possibility that DMs also promote the early stages of hair development. Importantly, macrophage depletion with clodronate promotes premature hair loss and is associated with the activation of hair follicle stem cells (HFSC) via activation of the B-catenin/Wnt signaling pathways [54].

Another important role for DMs in maintaining homeostasis is the elimination of dead and senescent cells. In the dermis, the accumulation of senescent cells is involved with skin aging and cancer. Also, it has been proposed that senescent cells maintain the pro-inflammatory tone, contributing to inadequate wound healing [55]. Clearing these cells is a promising approach to treat age-related diseases and improve physical function. However, although the mechanisms underlying the elimination of senescent cells in the skin remains elusive, a role for DMs in this clearance has been suggested [55].

DMs have an extensive apparatus of receptor and intracellular programs essential for removing dead cells and cellular debris. These cells express scavenger receptors involved in recognizing lipids expressed in the outer layer of the cellular membrane in dying cells such as MerTK, CD36, and Stabilin-1 [7]. The subsequent engulfment and degradation of apoptotic bodies by macrophages promotes anti-inflammatory cytokines, such as IL-10 and TGF-β and prostaglandin E2 (PGE_2) [25], crucial to the creation of a pro-resolution microenvironment that favors homeostasis after inflammation. In the steady-state, DMs show high *Il-10* expression [7] and an M2 phenotype [56]. Furthermore, eliminating dead cells is both a homeostatic condition and required to improve wound healing in healthy individuals and during preexisting conditions (as noted below).

DM and infections

The skin is continuously exposed to microbes that are either part of the skin microbiota or pathogens once a skin breach occurs. Invading microbes are recognized by keratinocytes and then by resident immune cells, such as LCs and DMs. Once the pathogens reach the dermis, DMs and other resident immune cells, such as mast cells, NK, and αβ T cells work together, surveilling the dermis for pathogens [2,4]. The entry of a small number of microorganisms into the dermis induces a quick and vigorous immune response triggering inflammation, pathogen clearance, and later promoting the return to homeostasis. However, skin infections caused by a breach in the skin barrier or extensive tissue invasion results in a more robust immune response to control infection [57].

During skin infection, DMs and structural cells express pattern recognition receptors (PRRs) that recognize invading microorganisms via recognition of pathogen-associated molecular patterns (PAMPs). The PAMP-PRR interaction results in the production of antimicrobial effectors, cytokines, chemokines, and lipid mediators, promoting monocyte and neutrophil recruitment to the site of infection and pathogen clearance [18,58].

The skin is commonly infected with virus, fungus, protozoan parasites, and bacteria. The most common infectious agents in the skin are *Corynebacterium* spp, and *Streptococcus pyogenees* (group A β-hemolytic streptococci), and *Staphylococcus aureus* [59,60]. Since these pathogens can cause invasive infections, such as osteomyelitis and sepsis, understanding the mechanism underlying microbial clearance in the skin is of high relevance.

DMs are required for the initial production of chemoattractants during infection and therefore are essential for neutrophil and monocyte recruitment, suggesting that DMs are crucial for fast and coordinated defense against pathogens in the skin [61,62]. Resident DMs are required to produce CXCL2, CXCL1, IL-1β, and the bioactive lipid leukotriene B_4 (LTB_4) during *S. aureus* infection [63,64]. Using two independent methods of DM depletion (MMDTR—challenged with diphtheria toxin) and liposomal clodronate revealed a delay in neutrophils migration to the skin, resulting in an uncontrolled infection, low abscess formation, and IL-1β accumulation [62,63]. Moreover, we have shown that DM-mediated LTB_4 production is essential for abscess formation during methicillin-resistant *S. aureus* (MRSA) skin infection [63]. Since there are four different resident DM populations (P1–P4-characterized by different surface markers), it is reasonable to speculate that such DMs might play distinct roles in skin host defense. Interestingly, Lee et al., have shown that Leishmania major preferentially infects resident P4 DMs.

An emerging concept in the field of host defense is the acquisition of innate immune memory by macrophages that have been previously exposed to microbial ligands or pathogens. Innate memory is heavily dependent on epigenetic modifications that allow faster transcription of genes involved in the inflammatory response and antimicrobial effector function. Therefore, it has been proposed as a tool to improve host defense [65]. The role of innate memory in DMs and skin host defense is emerging. Chan and colleagues revealed that DMs are the key to protecting *S. aureus* skin infection in a model of recurrent infection. The authors observed that DMs shape both early innate and adaptive cytokine production, ultimately driving Th17 activation and increased M1 macrophage accumulation in the skin. Interestingly, the innate memory and protection are localized to the infection site and did not confer protection during disseminated infection [66]. Recently, Feuerstein et al. showed that DMs could acquire a memory signature in situ, leading to a stronger STAT1 and CXCL9 production, leading to abundant neutrophil and monocyte recruitment and increased bacterial killing. However, as the infection progresses, BM-derived cells' increased macrophage turnover leads to an eventual loss of innate memory [67]. Studies are needed to identify whether different DM populations also develop innate memory and whether such events are also seen in various infection models.

Dermal macrophages and wound healing

Wound healing is a complex, highly coordinated, and dynamic process with sequential and overlapping phases such as inflammation and coagulation, resolution of the inflammatory response, proliferative or tissue formation (also known as tissue neovascularization remodeling), with dermal macrophages playing an essential role in each phase [3].

After an injury, platelets are quickly recruited to the site of injury, followed by the production of inflammatory mediators released by injured tissues, that coordinate the infection control, clearance of apoptotic and necrotic cells, and the beginning of the wound healing process [68,69]. When the second phase of healing starts, an intense proliferation of structural cells occurs, along with the removal of dead cells, the formation of new blood vessels, and epithelial closure. The last phase is characterized by the generation of newly formed tissues that are remodeled to improve their integrity [69]. A myriad of cells participates in these processes, including platelets, mast cells, neutrophils, lymphocytes, keratinocytes, fibroblasts, endothelial cells, monocytes, and macrophages. Many studies have demonstrated that

DMs are indispensable for wound healing, angiogenesis, collagen deposition, and closure of skin wounds in both sterile and infected skin wounds [28,70,71].

The wound healing issue represents a great scenario to illustrate the M1 and M2 macrophage dichotomy in vivo. During the acute phase of the inflammatory phase of wound healing, M1 macrophages are quite prevalent due to recognizing invasive agents and/or DAMPs released by tissue cells [22]. Local activation of M1 DMs and keratinocytes create an inflammatory environment characterized by the presence of the cytokines IL-1β, TNF-α, IL-6 [23,71], chemokines CXCL1, CXCL2, CCL2, and CXCL8 [72], lipid mediators LTB_4, CysLTs, and sphingosine-1-phosphate [73,74] that culminate in the generation of reactive oxygen species that cause tissue injury and induce the recruitment of blood neutrophils and monocytes that contribute to wound sterilization [75–77]. Whereas macrophages are present and participate in every wound healing stage, neutrophil migration is restricted to the early stages of the inflammatory response [78]. After the first wave of neutrophil recruitment, monocytes arrive at the inflammation site and differentiate into tissue macrophages.

Klar and colleagues noticed that although M1 macrophages are predominant in the early stages of wound healing, over the process of healing, these cells acquire a resolution phenotype (M2 macrophages) [24]. The switch from the inflammatory phenotype to an anti-inflammatory macrophage phenotype is necessary for efficient wound healing [24]. As previously discussed, in the inflammation phase of wound healing, there is an intense migration of neutrophils that die from apoptosis in subsequent stages that must be removed from this microenvironment. Studies have shown that the phagocytosis of apoptotic-neutrophil drives macrophage to a pro-resolution phenotype [25]. Anti-inflammatory mediators released during the clearance of these apoptotic cells can sustain the alternative macrophage profile by increasing the tissue repair processes [26]. The persistence of M1 macrophages for an extended period or the absence/depletion of the M2 macrophage population results in impaired tissue regeneration decreased proliferation of granulation tissue, reduced vascularization, and wound closure [27–29].

DMs support the neovascularization in different ways: (1) formation of new lymphatic vessels by forming non-endothelial vessels (vascular mimicry) and (2) production of high quantities of vascular endothelial growth factor (VEGF) that is required for blood vessel sprouting. DMs along with fibroblasts, keratinocytes, and endothelial cells release proangiogenic factors such as VEGF, fibroblast growth factors (FGF), platelet-derived growth factor

(PDGF), and transforming growth factor (TGF-β) that results in the neovascularization of the wound site [3,28,79,80]. Furthermore, DM secretion of anti-inflammatory mediators (from macrophages alternately activated) modulates the phenotypic state of dermal fibroblasts. $CD206^{+}CD301b^{+}$ M2-DMs increase fibroblast proliferation and collagen deposition by inducing the transformation of fibroblasts to alpha-smooth muscle actin (α-SMA) depositing myofibroblasts [81,82].

During the next stage of repair, new tissue formation is triggered by the recruitment and accumulation of immune and non-immune cells in the lesion area. Fibroblasts and macrophages produce an extracellular matrix that facilitates adhesion, migration, and proliferation of new cells, promoting tissue formation as seen by the abundance of [83]. Furthermore, while pro-fibrotic DMs drive fibroblast activation and extracellular matrix deposition, it remains to be entirely determined whether other newly recruited monocytes/derived macrophage and resident DMs influence skin repair differently. This process can take weeks to months, and although skin remodeling is the final determination of healing response, it is the least studied of all stages of wound healing.

At the final stage of repair, keratinocytes are stimulated to increase the epidermal barrier restoration, closure of the wound, and promotion of tissue regeneration. In this phase, macrophages and fibroblasts produce type I collagen that is more organized and replaces the tissue previously built in the initial stages of the wound healing, forming a low cellularity scar [84,85]. During the maturation tissue phase, macrophage depletion was not associated with significant differences in wound repair or scar formation [28].

DM and comorbidities associated with diabetes

Diabetes is a metabolic disorder associated with a deficiency in insulin secretion (type 1) or inefficient insulin recognition (type 2), resulting in hyperglycemia, which causes a state of sterile inflammation, characterized by a low-grade inflammatory response [86]. Diabetes is associated with complications in chronic nonhealing wounds and increased susceptibility to systemic and localized infections, particularly in the skin [87].

Skin and soft tissue infections occur more frequently in patients with diabetes mellitus than in the general population. *S. aureus* and MRSA is the main bacteria associated with these infections [88]. In a series of studies from our lab and others, we noted that the skin of infected diabetic mice is accompanied by increased and sustained production of inflammatory mediators, aberrant neutrophil recruitment to the site of infection, and exaggerated collagen production [89,90]. When we depleted skin resident macrophages in the skin of diabetic mice, we observed an inflammatory response similar to the one observed in infected non-diabetic mice [89]. These data incriminated skin resident DMs in poor skin host defense in diabetic mice.

Another interesting and potentially translational observation was the unbalanced production of the bioactive lipid mediator eicosanoids during diabetic skin infection. While we and others have shown that skin infection is accompanied by increased production of both leukotrienes (LTs) and prostaglandins (PGs), we found that MRSA skin infection in hyperglycemic mice leads to excessive levels of LTB_4 both at days 1 and 9 post-infection and lower levels of PGE_2 after 48 h of infection [89,91]. Blocking LTB_4 actions restore neutrophil migration, cytokine, and chemokine production, and wound healing. However, PGE_2 effects were related to a deficient migration of LCs to the draining lymph nodes and Th17 generation. Notably, when we employed a topical ointment containing a PGE_2 analog termed misoprostol (an FDA approved drug used to treat NSAID-induced stomach ulcers) and restoring PGE_2 levels in the site of infection enhanced LC migration to the lymph nodes and increased Th17 cell numbers that induced wound healing and host defense in diabetic mice [91]. Our lab is currently studying the mechanisms underlying deficient PGE_2 production and the PGE_2 receptors involved in misoprostol effects in skin host defense.

Several studies have also demonstrated that both exaggerated accumulation and state of activation of macrophages dictate the levels of wound healing. M1 macrophages are essential in the early stages of the inflammatory response; however, the presence of these cells in the subsequent stages of healing due to the chronic inflammatory state induced by diabetes contributes to a delay in wound closure. Several studies have demonstrated that DM shows a predominantly M1 activation phenotype in wounds from type II diabetic patients and mice with high *Nos2*, *Stat 1,* and *Il12* expression and increased TNF-α, IL-1β, IL-6, and nitrite levels [92–95] (Fig. 2). Furthermore, the prolonged presence of inflammatory macrophages in the wound of hyperglycemic mice is also associated with the accumulation of proteases, reduced levels of growth factors, delayed wound closure, and decreased angiogenesis, and impaired matrix deposition. Results similar to those found in diabetic mice were detected in wound biopsies from ulcers from people with diabetes, with increased M1 macrophage accumulation and decreased M2 macrophages [96–98]. Inhibition of pro-inflammatory

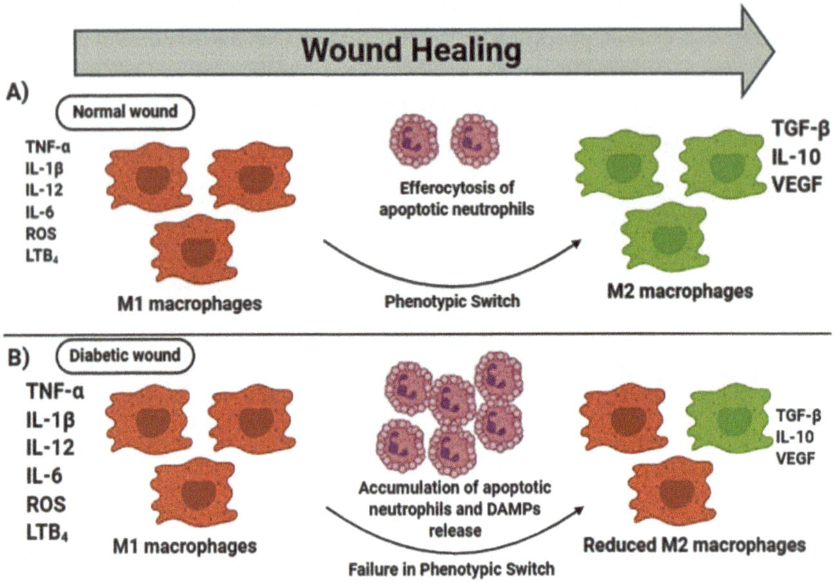

Fig. 2 Macrophage dysfunction in diabetic wounds. (A) Upon injury, DMs under normal conditions acquire an M1 phenotype. These macrophages have a high phagocytic capacity that promotes the clearance of the injured site. After this phase, DMs suffer a phenotype switch to the M2 profile that supports the inflammation and tissue formation resolution. (B) In diabetic wounds, hyperglycemia induces an inflammatory state with increased production of LTB_4, IL-12, IL-6, TNF-α, IL-1β, and reactive oxygen species (ROS) that decreases efferocytosis receptors expression, impairing the clearance of apoptotic neutrophils, generating necrotic cells and promoting the DAMP release. This inflammatory environment favors the maintenance of an M1 phenotype, reducing the release of proliferative mediators such as TGF-β, IL-10, and VEGF, impairing wound healing. *(Figure created with BioRender.com.)*

cytokines, such as IL-1β, using either a blocking antibody or IL-1R knockout mice results in improved wound healing in hyperglycemic mice via the induction of a reparative macrophage phenotype [97]. Moreover, in the absence of the 5-Lipoxygenase (5-LO) enzyme, which produces LTB_4, diabetic wounds showed an increase in the number of M2 macrophages [94]. Mirza and colleagues demonstrated that in the wound of diabetic mice, there is an impaired transition from M1 to M2 activation phenotypes resulting in an environment that impedes healing [99]. A longitudinal study developed an M1/M2 score based on pro-and anti-inflammatory cytokines production in human chronic wounds as a predictive model of healing. Patients with chronic diabetic wounds showed a higher score in the later phases of healing, suggesting a prolonged hyper inflammation that impairs the wound healing process [100]. Moreover, DMs from diabetic wounds in mice show a decrease in VEGF production and reduced VEGFR1 signaling contributing to impaired angiogenesis and vessel maturation in these animals [79,101,102]. Thus, macrophages activation plays a decisive role in diabetic wound closure in both mice and humans.

Another event contributing to the persistence of an inflammatory state in diabetic patients' wounds is a decrease in apoptotic-neutrophil uptake. Hyperglycemia downregulated the expression of two efferocytosis receptors, CD36 and SCARB-1 (Class B scavenger type I receptors), in macrophages [103] resulting in an accumulation of apoptotic cells and promoting DAMPs release in this microenvironment. Type 2 diabetic patients show wound macrophages with impaired efferocytosis, elevated TNF-α, and IL-6 secretion, and impaired wound healing [104] (Fig. 2).

The increased inflammatory response can probably be explained in part by epigenetic modifications induced by hyperglycemia in both resident and BM-derived cells. Gallagher et al. showed that in the wounds of non-diabetic mice, inflammatory $Ly6C^{Hi}$ macrophages ($Gr1^{Hi}$ $CCR2^{+}CX_3CR_1{}^{Lo}$) accumulate in the wounded tissue in the first 24–48 h post-injury, followed by epigenetic reprogramming to the generative anti-inflammatory $Ly6C^{Lo}$ ($Gr1^{Lo}$ $CCR2^{-}CX_3CR_1{}^{Hi}$) monocytes/macrophages [98]. In the wounds of obese and hyperglycemic mice, the authors observed a second influx of $Ly6C^{Hi}$ macrophages in the later phases of healing (day 3–4 posts injury) followed by an accumulation of macrophages when compared to non-diabetic wounds [99,105,106]. Kimball et al. using the adoptive transfer of *tdTomato*-labeled $CD11b^{+}Ly6C^{+}$ cells noticed that the phenotype transition $Ly6C^{Hi}$ to $Ly6C^{Lo}$ is delayed in diabetic wounds leading to persistent inflammation [105]. Moreover, diabetic mice showed a prolonged infiltration of blood $Ly6C^{+}$ cells that might potentialize the observed defect in the phenotype transition [105]. In further studies, the same group of scientists showed that wound DMs in obese and diabetic mice exhibit increased methylation in the histone residues histone H3 lysine-4 (H3K4) and histone H3 lysine-27 (H3K27) methylation that represses M2 genes. These effects are dependent on demethylase Jmjd3, which drives the overexpression of inflammatory IL-12 promoting wound inflammation [98,107].

Determining phenotype and effector functions of macrophages under normal and hyperglycemic conditions is essential to understanding molecular and cellular mechanisms that govern impaired wound healing in diabetic patients and guide new studies that favor diabetic wound care improvement.

Final considerations and future directions

The present review summarizes the currently available evidence, supporting multiple roles of macrophages and Langerhans cells in maintaining cutaneous tissue homeostasis. Furthermore, we describe mechanisms by which macrophages act as a double-edged sword, depending on the timing of the tissue injury and the health status. We review how macrophages are required for skin homeostasis, wound healing, and skin host defense. We also covered important events in skin injury in both murine models and people with preexisting conditions, such as hyperglycemia and aging. The studies included in this review, which show different experimental approaches for the ablation of myeloid cells in the skin are limited. More specific genetic models are required to clearly define the role of macrophages in the skin compared to the role of the rest of the myeloid-derived cells. The variety of "omic" approaches including single-cell RNAseq, ATAC-Seq coupled with metabolomics promises to break new grounds of investigation regarding the role of DMs in both homeostasis and disease. Understanding the array of skin resident functions versus monocyte-derived macrophages during skin homeostasis and disease is critical for the effective design of therapeutic strategies that aim to restore skin homeostasis after injury.

Acknowledgments

This work was supported by NIH grants R01HL124159-01 and RAI149207A (to CHS), T32AI138932 (to NK) and FAPESP grants 2018/01622-9, 2017/04786-0 and 2020/10133-1 (to ACGS).

References

[1] Belkaid Y, Tamoutounour S. The influence of skin microorganisms on cutaneous immunity. Nat Rev Immunol 2016;16(6):353–66.
[2] Kabashima K, et al. The immunological anatomy of the skin. Nat Rev Immunol 2019;19(1):19–30.
[3] Cañedo-Dorantes L, Cañedo-Ayala M. Skin acute wound healing: a comprehensive review. Int J Inflamm 2019;2019, 3706315.
[4] Tong PL, et al. The skin immune atlas: three-dimensional analysis of cutaneous leukocyte subsets by multiphoton microscopy. J Investig Dermatol 2015;135(1):84–93.
[5] Wolf AA, et al. The ontogeny of monocyte subsets. Front Immunol 2019;10:1642.
[6] Yanez DA, et al. The role of macrophages in skin homeostasis. Pflugers Arch - Eur J Physiol 2017;469(3–4):455–63.
[7] Tamoutounour S, et al. Origins and functional specialization of macrophages and of conventional and monocyte-derived dendritic cells in mouse skin. Immunity 2013;39(5):925–38.

[8] Tamoutounour S, et al. CD 64 distinguishes macrophages from dendritic cells in the gut and reveals the T h1-inducing role of mesenteric lymph node macrophages during colitis. Eur J Immunol 2012;42(12):3150–66.
[9] Gautier EL, et al. Gene-expression profiles and transcriptional regulatory pathways that underlie the identity and diversity of mouse tissue macrophages. Nat Immunol 2012;13(11):1118–28.
[10] Haniffa M, et al. Human tissues contain CD141hi cross-presenting dendritic cells with functional homology to mouse CD103+ nonlymphoid dendritic cells. Immunity 2012;37(1):60–73.
[11] Henri S, et al. CD207+ CD103+ dermal dendritic cells cross-present keratinocyte-derived antigens irrespective of the presence of Langerhans cells. J Exp Med 2010;207(1):189–206.
[12] Kolter J, et al. A subset of skin macrophages contributes to the surveillance and regeneration of local nerves. Immunity 2019;50(6):1482–1497. e7.
[13] Burgess M, et al. Cx3CR1 expression identifies distinct macrophage populations that contribute differentially to inflammation and repair. ImmunoHorizons 2019;3(7):262–73.
[14] Lee SH, et al. Mannose receptor high, M2 dermal macrophages mediate nonhealing Leishmania major infection in a Th1 immune environment. J Exp Med 2018;215(1):357–75.
[15] Shapouri-Moghaddam A, et al. Macrophage plasticity, polarization, and function in health and disease. J Cell Physiol 2018;233(9):6425–40.
[16] Martinez FO, et al. Macrophage activation and polarization. Front Biosci 2008;13:453.
[17] Kumar V. Macrophages: the potent immunoregulatory innate immune cells. In: Macrophage at the crossroads of innate and adaptive immunity. IntechOpen; 2019.
[18] Miller LS. Toll-like receptors in skin. Adv Dermatol 2008;24:71.
[19] Hanke ML, et al. Targeting macrophage activation for the prevention and treatment of Staphylococcus aureus biofilm infections. J Immunol 2013;190(5):2159–68.
[20] Ma W-T, et al. The role of monocytes and macrophages in autoimmune diseases: a comprehensive review. Front Immunol 2019;10:1140.
[21] Li F, et al. Macrophage polarization modulates development of systemic lupus erythematosus. Cell Physiol Biochem 2015;37(4):1279–88.
[22] Koh TJ, DiPietro LA. Inflammation and wound healing: the role of the macrophage. Expert Rev Mol Med 2011;13:e23.
[23] Barrientos S, et al. Growth factors and cytokines in wound healing. Wound Repair Regen 2008;16(5):585–601.
[24] Klar AS, et al. Characterization of M1 and M2 polarization of macrophages in vascularized human dermo-epidermal skin substitutes in vivo. Pediatr Surg Int 2018;34(2):129–35.
[25] Fadok VA, et al. Macrophages that have ingested apoptotic cells in vitro inhibit proinflammatory cytokine production through autocrine/paracrine mechanisms involving TGF-beta, PGE2, and PAF. J Clin Invest 1998;101(4):890–8.
[26] Arienti S, et al. Regulation of apoptotic cell clearance during resolution of inflammation. Front Pharmacol 2019;10:891.
[27] Ambarus CA, et al. Systematic validation of specific phenotypic markers for in vitro polarized human macrophages. J Immunol Methods 2012;375(1–2):196–206.
[28] Lucas T, et al. Differential roles of macrophages in diverse phases of skin repair. J Immunol 2010;184(7):3964–77.
[29] Sindrilaru A, et al. An unrestrained proinflammatory M1 macrophage population induced by iron impairs wound healing in humans and mice. J Clin Invest 2011;121(3):985–97.

[30] Metchnikoff E. Leçons sur la pathologie comparée de l'inflammation: faites à l'Institut Pasteur en avril et mai 1891. G. Masson; 1892.
[31] van Furth R, et al. The mononuclear phagocyte system: a new classification of macrophages, monocytes, and their precursor cells. Bull World Health Organ 1972;46(6):845.
[32] Ginhoux F, Guilliams M. Tissue-resident macrophage ontogeny and homeostasis. Immunity 2016;44(3):439–49.
[33] Tavian M, Peault B. Embryonic development of the human hematopoietic system. Int J Dev Biol 2003;49(2-3):243–50.
[34] Palis J, et al. Development of erythroid and myeloid progenitors in the yolk sac and embryo proper of the mouse. Development 1999;126(22):5073–84.
[35] Tober J, et al. The megakaryocyte lineage originates from hemangioblast precursors and is an integral component both of primitive and of definitive hematopoiesis. Blood 2007;109(4):1433–41.
[36] Samokhvalov IM, Samokhvalova NI, Nishikawa S-I. Cell tracing shows the contribution of the yolk sac to adult haematopoiesis. Nature 2007;446(7139):1056–61.
[37] Ginhoux F, et al. Fate mapping analysis reveals that adult microglia derive from primitive macrophages. Science 2010;330(6005):841–5.
[38] Frame JM, McGrath KE, Palis J. Erythro-myeloid progenitors: "definitive" hematopoiesis in the conceptus prior to the emergence of hematopoietic stem cells. Blood Cell Mol Dis 2013;51(4):220–5.
[39] Perdiguero EG, et al. Tissue-resident macrophages originate from yolk-sac-derived erythro-myeloid progenitors. Nature 2015;518(7540):547–51.
[40] Ramsay RG, Gonda TJ. MYB function in normal and cancer cells. Nat Rev Cancer 2008;8(7):523–34.
[41] Hoeffel G, et al. C-Myb+ erythro-myeloid progenitor-derived fetal monocytes give rise to adult tissue-resident macrophages. Immunity 2015;42(4):665–78.
[42] Palis J, Yoder MC. Yolk-sac hematopoiesis: the first blood cells of mouse and man. Exp Hematol 2001;29(8):927–36.
[43] Bertrand JY, et al. Three pathways to mature macrophages in the early mouse yolk sac. Blood 2005;106(9):3004–11.
[44] Naito M, et al. Development, differentiation, and phenotypic heterogeneity of murine tissue macrophages. J Leukoc Biol 1996;59(2):133–8.
[45] Hoeffel G, et al. Adult Langerhans cells derive predominantly from embryonic fetal liver monocytes with a minor contribution of yolk sac-derived macrophages. J Exp Med 2012;209(6):1167–81.
[46] Sheng J, Ruedl C, Karjalainen K. Most tissue-resident macrophages except microglia are derived from fetal hematopoietic stem cells. Immunity 2015;43(2):382–93.
[47] Ginhoux F, et al. New insights into the multidimensional concept of macrophage ontogeny, activation and function. Nat Immunol 2016;17(1):34–40.
[48] Boring L, et al. Impaired monocyte migration and reduced type 1 (Th1) cytokine responses in CC chemokine receptor 2 knockout mice. J Clin Invest 1997;100(10):2552–61.
[49] Bain CC, et al. Constant replenishment from circulating monocytes maintains the macrophage pool in the intestine of adult mice. Nat Immunol 2014;15(10):929–37.
[50] Scott CL, Henri S, Guilliams M. Mononuclear phagocytes of the intestine, the skin, and the lung. Immunol Rev 2014;262(1):9–24.
[51] Hoeffel G, Ginhoux F. Ontogeny of tissue-resident macrophages. Front Immunol 2015;6:486.
[52] Parakkal PF. Morphogenesis of the hair follicle during catagen. Z Zellforsch Mikrosk Anat 1970;107(2):174–86.
[53] Müller-Röver S, et al. Intercellular adhesion molecule-1 and hair follicle regression. J Histochem Cytochem 2000;48(4):557–68.

[54] Castellana D, Paus R, Perez-Moreno M. Macrophages contribute to the cyclic activation of adult hair follicle stem cells. PLoS Biol 2014;12(12), e1002002.
[55] Ghosh K, Capell BC. The senescence-associated secretory phenotype: critical effector in skin cancer and aging. J Investig Dermatol 2016;136(11):2133–9.
[56] Yunna C, et al. Macrophage M1/M2 polarization. Eur J Pharmacol 2020;877, 173090.
[57] Clebak KT, Malone MA. Skin infections. Prim Care 2018;45(3):433–54.
[58] Broz P. Recognition of intracellular bacteria by inflammasomes. In: Bacteria and intracellularity. Wiley; 2019. p. 287–97.
[59] David MZ, Daum RS. Treatment of *Staphylococcus aureus* infections. In: Staphylococcus aureus. Springer; 2017. p. 325–83.
[60] Chiller K, Selkin BA, Murakawa GJ. Skin microflora and bacterial infections of the skin. J Investig Dermatol Symp Proc 2001;6(3):170–4.
[61] Abtin A, et al. Perivascular macrophages mediate neutrophil recruitment during bacterial skin infection. Nat Immunol 2014;15(1):45–53.
[62] Feuerstein R, et al. MyD88 in macrophages is critical for abscess resolution in staphylococcal skin infection. J Immunol 2015;194(6):2735–45.
[63] Brandt SL, et al. Macrophage-derived LTB4 promotes abscess formation and clearance of *Staphylococcus aureus* skin infection in mice. PLoS Pathog 2018;14(8), e1007244.
[64] Cho JS, et al. Neutrophil-derived IL-1β is sufficient for abscess formation in immunity against *Staphylococcus aureus* in mice. PLoS Pathog 2012;8(11), e1003047.
[65] Italiani P, Della Camera G, Boraschi D. Induction of innate immune memory by engineered nanoparticles in monocytes/macrophages: from hypothesis to reality. Front Immunol 2020;11:2324.
[66] Chan LC, et al. Protective immunity in recurrent *Staphylococcus aureus* infection reflects localized immune signatures and macrophage-conferred memory. Proc Natl Acad Sci 2018;115(47):E11111–9.
[67] Feuerstein R, et al. Resident macrophages acquire innate immune memory in staphylococcal skin infection. elife 2020;9, e55602.
[68] Nurden AT, et al. Platelets and wound healing. Front Biosci 2008;13:3532.
[69] Eming SA, Martin P, Tomic-Canic M. Wound repair and regeneration: mechanisms, signaling, and translation. Sci Transl Med 2014;6(265):265sr6.
[70] Goren I, et al. A transgenic mouse model of inducible macrophage depletion: effects of diphtheria toxin-driven lysozyme M-specific cell lineage ablation on wound inflammatory, angiogenic, and contractive processes. Am J Pathol 2009;175(1):132–47.
[71] Mirza R, DiPietro LA, Koh TJ. Selective and specific macrophage ablation is detrimental to wound healing in mice. Am J Pathol 2009;175(6):2454–62.
[72] Ridiandries A, Tan J, Bursill CA. The role of chemokines in wound healing. Int J Mol Sci 2018;19(10):3217.
[73] Yasukawa K, Okuno T, Yokomizo T. Eicosanoids in skin wound healing. Int J Mol Sci 2020;21(22):8435.
[74] Aoki M, et al. Sphingosine-1-phosphate facilitates skin wound healing by increasing angiogenesis and inflammatory cell recruitment with less scar formation. Int J Mol Sci 2019;20(14):3381.
[75] Davies LC, et al. Tissue-resident macrophages. Nat Immunol 2013;14(10):986.
[76] Malissen B, Tamoutounour S, Henri S. The origins and functions of dendritic cells and macrophages in the skin. Nat Rev Immunol 2014;14(6):417–28.
[77] Minutti CM, et al. Tissue-specific contribution of macrophages to wound healing. Semin Cell Dev Biol 2017;61:3–11.
[78] Martin P, Leibovich SJ. Inflammatory cells during wound repair: the good, the bad and the ugly. Trends Cell Biol 2005;15(11):599–607.
[79] Stockmann C, et al. A wound size-dependent effect of myeloid cell-derived vascular endothelial growth factor on wound healing. J Investig Dermatol 2011;131(3):797–801.

[80] Willenborg S, et al. CCR2 recruits an inflammatory macrophage subpopulation critical for angiogenesis in tissue repair. Blood 2012;120(3):613–25.
[81] Ploeger DT, et al. Cell plasticity in wound healing: paracrine factors of M1/M2 polarized macrophages influence the phenotypical state of dermal fibroblasts. Cell Commun Signal 2013;11(1):29.
[82] Hou J, et al. M2 macrophages promote myofibroblast differentiation of LR-MSCs and are associated with pulmonary fibrogenesis. Cell Commun Signal 2018;16(1):1–14.
[83] Hinz B. The role of myofibroblasts in wound healing. Curr Res Transl Med 2016;64(4):171–7.
[84] Olczyk P, Mencner Ł, Komosinska-Vassev K. The role of the extracellular matrix components in cutaneous wound healing. Biomed Res Int 2014;2014, 747584.
[85] Xue M, Jackson CJ. Extracellular matrix reorganization during wound healing and its impact on abnormal scarring. Adv Wound Care 2015;4(3):119–36.
[86] Kono H, Onda A, Yanagida T. Molecular determinants of sterile inflammation. Curr Opin Immunol 2014;26:147–56.
[87] Muller L, et al. Increased risk of common infections in patients with type 1 and type 2 diabetes mellitus. Clin Infect Dis 2005;41(3):281–8.
[88] Esposito S, Ascione T, Pagliano P. Management of bacterial skin and skin structure infections with polymicrobial etiology. Expert Rev Anti-Infect Ther 2019;17(1):17–25.
[89] Brandt SL, et al. Excessive localized leukotriene B4 levels dictate poor skin host defense in diabetic mice. JCI Insight 2018;3(17), e120220.
[90] Rich J, Lee JC. The pathogenesis of *Staphylococcus aureus* infection in the diabetic NOD mouse. Diabetes 2005;54(10):2904–10.
[91] Dejani NN, et al. Topical prostaglandin E analog restores defective dendritic cell-mediated Th17 host defense against methicillin-resistant *Staphylococcus aureus* in the skin of diabetic mice. Diabetes 2016;65(12):3718–29.
[92] Baltzis D, Eleftheriadou I, Veves A. Pathogenesis and treatment of impaired wound healing in diabetes mellitus: new insights. Adv Ther 2014;31(8):817–36.
[93] Falanga V. Wound healing and its impairment in the diabetic foot. Lancet 2005;366(9498):1736–43.
[94] Ramalho T, et al. Impaired wound healing in type 1 diabetes is dependent on 5-lipoxygenase products. Sci Rep 2018;8(1):1–13.
[95] Yaseen H, Khamaisi M. Skin well-being in diabetes: role of macrophages. Cell Immunol 2020;356, 104154.
[96] Bannon P, et al. Diabetes induces stable intrinsic changes to myeloid cells that contribute to chronic inflammation during wound healing in mice. Dis Model Mech 2013;6(6):1434–47.
[97] Mirza RE, et al. Blocking interleukin-1β induces a healing-associated wound macrophage phenotype and improves healing in type 2 diabetes. Diabetes 2013;62(7):2579–87.
[98] Gallagher KA, et al. Epigenetic changes in bone marrow progenitor cells influence the inflammatory phenotype and alter wound healing in type 2 diabetes. Diabetes 2015;64(4):1420–30.
[99] Mirza R, Koh TJ. Dysregulation of monocyte/macrophage phenotype in wounds of diabetic mice. Cytokine 2011;56(2):256–64.
[100] Nassiri S, et al. Relative expression of proinflammatory and antiinflammatory genes reveals differences between healing and nonhealing human chronic diabetic foot ulcers. J Invest Dermatol 2015;135(6):1700.
[101] Okizaki S-I, et al. Vascular endothelial growth factor receptor type 1 signaling prevents delayed wound healing in diabetes by attenuating the production of IL-1β by recruited macrophages. Am J Pathol 2016;186(6):1481–98.
[102] Okonkwo UA, et al. Compromised angiogenesis and vascular integrity in impaired diabetic wound healing. PLoS One 2020;15(4), e0231962.

[103] Morey M, et al. Hyperglycemia acts in synergy with hypoxia to maintain the pro-inflammatory phenotype of macrophages. PLoS One 2019;14(8), e0220577.
[104] Khanna S, et al. Macrophage dysfunction impairs resolution of inflammation in the wounds of diabetic mice. PLoS One 2010;5(3), e9539.
[105] Kimball A, et al. Ly6CHi blood monocyte/macrophage drive chronic inflammation and impair wound healing in diabetes mellitus. Arterioscler Thromb Vasc Biol 2018;38(5):1102–14.
[106] Barman PK, Urao N, Koh TJ. Diabetes induces myeloid bias in bone marrow progenitors associated with enhanced wound macrophage accumulation and impaired healing. J Pathol 2019;249(4):435–46.
[107] Kruidenier L, et al. A selective jumonji H3K27 demethylase inhibitor modulates the proinflammatory macrophage response. Nature 2012;488(7411):404–8.

CHAPTER 7

Bone—From macrophage to osteoclast and osteolytic diseases

Erivan S. Ramos-Junior[a], Thaise M. Taira[b], and Sandra Y. Fukada[b]
[a]Department of Oral Biology & Diagnostic Science, The Dental College of Georgia, Augusta University, Augusta, GA, United States
[b]School of Pharmaceutical Sciences of RibeirãoPreto, Department of BioMolecular Sciences, University of São Paulo, Ribeirão Preto, SP, Brazil

Introduction and historical background

Bones are dynamic organs that undergo continuous remodeling throughout life involving the removal of mineralized bone followed by the formation of a bone matrix. Apart from the calcified matrix, four major cell types populate bone tissue: osteoprogenitor cells (stem cells), osteoblasts, osteocytes, and osteoclasts. All of them, except for the osteoclasts, are connective tissue cells (mesenchymal cells). Osteoclasts are professional bone-resorbing cells derived from precursors belonging to the monocyte-macrophage lineage [1]. The story behind the finding that osteoclasts could resorb bone started more than 200 years ago. The identification of giant multinucleated cells able to resorb bone was made in 1849 by C.H. Robin; however, the name "osteoclast" was created by Killiker only in 1873. All these historical facts about the beginning of the osteoclast discovery were marvelously reviewed and summarized by N.M. Hancox, in 1949, compiled over a century of publications written in several languages [2].

From 1949 until the 1970s, the osteoclast origin was still obscure, and the scientists postulated that osteoclasts and osteoblasts were from the same lineage. In the mid-1970s, Dr. Donald Walker shed some light on osteoclasts' origin after a series of experiments on mouse models of osteopetrosis. By joining the circulation of osteopetrotic mice to wild-type (parabiosis), osteopetrosis histologic signs were wiped out. Later, he showed that the osteopetrotic parameters were reduced by irradiation and by wild-type bone marrow or spleen cell injection [1,3,4].

In the following years, the myeloid origin of osteoclasts was confirmed by several scientists. Zambonin et al. [5] observed that osteoclasts arise from

Macrophages in the Human Body
https://doi.org/10.1016/B978-0-12-821385-8.00011-1

monocytes when using time-lapse microscopy; they showed, by direct visualization, that the fusion of several monocytes originated the osteoclasts. Moreover, significant advances in mouse models showed that osteopetrotic (op/op) mice induced by a mutation in the gene encoding the macrophage colony factor-1 (M-CSF1 or CSF1) exhibited very few macrophages, monocytes, and osteoclasts [6,7]. Similarly, mice deficient in PU.1, a crucial transcription factor for initial monocyte/macrophage differentiation, also presented the absence of osteoclasts showing that the genes required for myeloid development were essential for osteoclast formation. These studies revealed the beginning of the cytokines requirements for osteoclastogenesis in vitro.

The identity of the soluble factor responsible for osteoclast formation remained obscure for a long time. In late 1990, four independent groups identified the cytokine RANKL (receptor activator of nuclear factor-κappa B ligand) that is now widely recognized as the main element for osteoclast differentiation [8–11]. The observation that animals deficient in RANKL presented osteopetrotic, toothless, and deficiency in osteoclasts confirmed the fundamental role of RANKL for osteoclastogenesis [12]. Subsequently, Tsuda et al. identified one protein produced by stromal cells that inhibited osteoclast formation in vitro, which they named osteoclast inhibitor factor (OCIF) [13]. Simultaneously, Simonet et al. found that the sequence of OCIF was identical to a member of the TNF receptor family called osteoprotegerin (OPG) [14]. The role of OPG inhibiting osteoclast formation was identified in OPG-deficient mice that exhibited severe osteoporosis and fractures due to an increased number of osteoclasts [15].

Finally, in 2000 a committee from the American Society for Bone and Mineral Research (ASBMR) recommended a standard nomenclature. The committee recommended the receptor activator of NF-kappaB (RANK) for the membrane receptor, RANK ligand (RANKL) for the ligand, and osteoprotegerin (OPG) for the decoy receptor [16].

The identification of RANK/RANKL/OPG unveiled the critical molecular mechanism for osteoclastogenesis and allowed researchers to generate osteoclasts in vitro from myeloid precursors. Moreover, the results showing that RANKL can be produced by stromal and immune cells were essential to link bone biology and immunology, which has been explored as the osteoimmunology field [17]. In this chapter, we compile and discuss the crucial molecular signature for osteoclasts formation. We also review the evidence showing the critical role of osteoclasts in bone loss observed in some diseases.

Molecular signature for osteoclast differentiation

Osteoclastogenesis is a complex multi-step process where myeloid cells/macrophages are first committed as osteoclasts precursors near the bone surface. Osteoclasts differentiation requires two essential cytokines M-CSF and RANKL, which lead to the fusion of the precursor's cells. The initial phase is characterized by an increased expression of nuclear factor of activated T cell cytoplasmic 1 (NFATc1), the main transcription factor for osteoclast differentiation. Next, the pre-osteoclasts' fusion step is marked by the expression of dendritic cell-specific transmembrane protein (DC-STAMP), an indispensable element for the formation of large, multinucleated osteoclasts. Finally, the mature osteoclasts adhere to the bone surface by integrins and digest the underlying bone tissue by releasing hydrochloric acid and digestive enzymes such as Cathepsin K and Matrix metalloproteinase 9 (MMP-9) onto the bone (actin ring area ruffle border). Since discovering the main factors for osteoclast differentiation (M-CSF, OPG, RANK, and RANKL), the signaling pathway's elucidation and its role in osteoclast biology paved new avenues in the bone biology field. Also, it contributed to the progression of innovative therapeutic strategies for bone-related disorders. Here we will go deeper into the signaling molecules responsible for the commitment, fusion, and maturation of osteoclasts.

M-CSF/CSF1R (c-Fms)

Myeloid cells capable of differentiating into osteoclasts express the M-CSF receptor (M-CSF1R). Since the 1990s, it is known the role of M-CSF/M-CSF1R signaling on the survival, proliferation, and differentiation of osteoclasts (reviewed in Ref. [18]). The M-CSF receptor's activation phosphorylates six tyrosine residues in the cytoplasmic domain (Y559, Y697, Y706, Y721, Y807, and Y974). Still, only the tyrosine residues Y559 and Y807 are essential for osteoclast proliferation and differentiation, whereas the others play a minor role [19]. The phosphorylated Y559 interacts with c-Src, a non-receptor tyrosine kinase, recruits phosphatidylinositol 3-kinase (PI3K) complex, and activates the Akt pathway [20,21]. The PI3K/AKT complex plays an essential role in critical biological functions, including cellular proliferation, survival, and cytoskeletal reorganization in preosteoclasts [22]. On the other hand, phospho-Y697/Y974 recruits Grb2, a critical adaptor protein for Ras/Raf/MEK/ERK cascades. Moreover, numerous studies have shown that ERK and PI3K/AKT cascades are also involved in osteoclasts precursors survival (reviewed in Ref. [18]).

Additionally, Arai et al. demonstrated that M-CSF induces RANK mRNA expression in bone marrow cells enabling their subsequent response to the critical osteoclastogenic cytokine RANKL [23].

RANK/RANKL/OPG

The discovery of the molecular triad RANK/RANKL/OPG profoundly impacted the bone biology field in the 1990s. Chronologically, OPG was the first to be discovered, and Willian J. Boyle's team found that OPG was part of the tumor necrosis factor receptor (TNFR) superfamily [14]. Curiously, OPG was described as a secreted soluble molecule, in contrast to most members of the TNFR superfamily. The transgenic mice overexpressing OPG exhibited an increased bone density and reduced number of osteoclasts [14]. Whereas OPG-deficient mice developed severe osteoporosis [15]; however, the putative ligand for OPG was still obscure.

Subsequent in vitro experiments revealed that the ligand for OPG was a TNF-related cytokine. This cytokine could replace stromal cells, vitamin D3, and glucocorticoids in the coculture model for osteoclastogenesis. The molecule named OPG ligand (OPGL) directly activates osteoclasts in vitro and in vivo [9]. At almost the same time, another independent research group using an extensive cDNA library from dendritic cells and murine thymocytes cloned the receptor RANK and its ligand RANKL. The study also demonstrated that RANKL boosts the ability of dendritic cells to stimulate naive T cells [12].

Interestingly, it was shown that the OPGL sequence was identical to that of RANKL. Thus, RANK was considered to be the cell surface receptor on osteoclast precursors, and RANKL was its natural ligand. The studies mentioned above revealed that RANK, RANKL, and OPG belong to the TNF and its receptor superfamilies and settled a revolution in the bone biology field.

RANK downstream signaling pathway

Molecular analysis showed that RANK contains four cysteine-rich repeat domains at the amino-terminus and three TNFR-associated factors (TRAF) domains at the carboxy-terminus. The TRAF family comprises of seven members (TRAF1–7) that mainly mediate signals produced by TNF family cytokines and pathogen-associated molecular patterns (PAMPs). TRAF6 is an essential component of osteoclastogenesis. Mice deficient in TRAF6 are osteopetrotic with massive bone remodeling defects due to impaired osteoclast function and lymph-node organogenesis [24,25]. TRAF6 activation leads to two main signaling pathways, namely (i) the MAPK and (ii) NF-κB signaling pathways.

The MAPK signaling

M-CSF and RANKL trigger MAPKs (e.g., extracellular signal-regulated kinase (ERK), c-Jun N-terminal kinase (JNK), and p38) signaling during osteoclast differentiation. MAPK signaling activated by M-CSF is mainly involved in regulating osteoclast precursor proliferation, whereas RANKL-induced MAPK activation is implicated in osteoclast commitment. Here we will focus on the three most-studied MAPK pathways implicated in osteoclast differentiation, i.e., (i) p38, (ii) JNK, and (iii) ERK 1/2.

To date, four splice variants of the p38 family have been identified: p38α, p38β, p38γ (ERK6, SAPK3), and p38δ (SAPK4) (reviewed in Ref. [26]). After RANK activation, TRAF6 recruits and activates TAK1 that activates MKK6 and MKK7 to induce p38 and JNK MAPK pathways. TAK1-deficient monocytes failed to activate p38 and did not form multinucleated osteoclasts [27]. Furthermore, overexpression of a dominant-negative form of p38α and treatment with a specific p38 inhibitor in osteoclast precursors lead to a complete blockage of RANKL-induced osteoclastogenesis in vitro [28]. Also, the ablation of p38α in monocytic osteoclast precursors indirectly inhibited osteoblast proliferation and differentiation via decreased expression of coupling factors such as BMP-2 and platelet-derived growth factor, which are expressed via the p38 signaling in osteoclasts. These results demonstrated that p38α plays an essential role in linking osteoclastogenesis and osteoblastogenesis [29]. The functions of the other p38 isoforms (p38β, p38γ, and p38δ) in osteoclasts remain to be elucidated.

JNK signaling

Three genes encode the JNK proteins, MAPK8 (which encodes JNK1), MAPK9 (which encodes JNK2), and MAPK10 (which encodes JNK3), which are alternatively spliced, giving rise to at least 10 isoforms [30]. Experiments revealed that only JNK1, but not JNK2, is activated by RANKL [31]. The impairment of JNK signaling by overexpression of dominant-negative JNK1 or the JNK-specific inhibitor SP600125 abolishes the anti-apoptotic effect of RANKL/RANK/TRAF6 signaling in osteoclasts [32], resulting in low osteoclast differentiation.

ERK signaling

RANKL also activates ERK1/2 kinases. ERK signaling has been implicated in the survival, proliferation, apoptosis, formation, polarity, podosome disassembly, and differentiation of osteoclasts. ERK1 knockout mice present an increased bone density compared with WT mice. Furthermore, ERK-1 deletion reduced osteoclast progenitor cell numbers, compromised pit formation, and diminished M-CSF-mediated adhesion and migration [33]. This result suggests that the ERK pathway positively regulates osteoclastogenesis.

Canonical and non-canonical NF-κB signaling pathways

Many of the biological effects exerted by the RANK-RANKL axis are mediated through the transcription factor NF-κB. The NF-κB family includes five members: RelA (also known as p65), p50, p52, RelB, and c-Rel. Both p50 and p52 are cleaved forms, derived from precursor proteins p105 and p100, respectively. All family members share a Rel homology domain (RHD) in their N-terminus responsible for homo- and heterodimerization. RelB preferentially associates with p52, whereas RelA and c-Rel preferentially associate with p50 [34]. Studies had shown that NF-κB1/2 double knockout (dKO) mice had a failure of tooth eruption and osteopetrosis [35,36]. Curiously, dKO mice have increased numbers of Cd11b-positive and RANK-positive osteoclasts precursors, but the cells failed in differentiating into mature osteoclasts. The defect in osteoclast differentiation in NF-κB dKO osteoclast precursors is rescued by either Fos or NFATc1 retroviral constructs, indicating that they act downstream NF-κB [37]. Collectively, these findings suggest that NF-κB1/2 signaling is not only required for the differentiation of myeloid cells into osteoclast progenitors but is necessary for terminal differentiation [37,38].

During osteoclast differentiation, NF-κB activation depends on two distinct pathways, the canonical and the non-canonical NF-κB signaling pathway. Both pathways contain several activating steps that permit ubiquitination and proteasomal degradation of proteins that work as NF-κB signaling inhibitors. These inhibitors proteins are called IκB kinase (IKK complex) and include IKKα, IKKβ, and the NF-κB essential modulator (NEMO, or IKKγ). Throughout the activation of NF-κB canonical signaling in RANKL-stimulated cells, the IKK complex phosphorylates IKKα leading to its ubiquitination and proteasomal degradation. The heterodimer RelA/p50 translocates to the nucleus and induces Fos and NFATc1 expression, which has been called the master regulator of osteoclastogenesis. The central role of NFATc1 at this early stage is to overcome the expression of transcription factors responsible for limiting osteoclast differentiation and function such as Bcl6. For example, RANKL stimulation was shown to suppress the expression of Bcl6, which directly binds to the promoters of NFATc1, to suppress its autoamplification [39]. In later stages, Fos expression induced by NF-κB boosts NFATc1 to induce osteoclast-specific genes such as the Tartrate-resistant acid phosphatase (TRAP), Calcitonin receptor, Cathepsin K, β3 Integrin β3, ATPase H+ Transporting V0 Subunit D2 (Atp6v0d2), DC-STAMP, and OC-STAMP [37].

The non-canonical NF-κB pathway is initiated hours after the canonical signaling activation. In this pathway, NF-κB-induced kinase (NIK) activates IKKα leading to a degradation of p100 to form the processed product p52. Interestingly, this activation occurs in response to RANKL but not TNF, leading to RelB/p52 heterodimers' translocation to the nucleus [40]. Evidence suggests that osteoclast differentiation of $NIK^{-/-}$ osteoclast precursor cells are completely blocked in vitro. Besides, the trabecular bone volume is slightly but significantly increased in $NIK^{-/-}$ mice [40]. Interestingly, although NIK is thought to be essential for activation of RelB/p52, NIK's absence dampens the nuclear translocation of not only RelB but also RelA, indicating that NIK also regulates both canonical and non-canonical pathways in osteoclast differentiation [40].

Ca^{2+} and costimulatory signaling in osteoclasts

It is well accepted that Ca^{2+} serves as a ubiquitous second messenger that specifically mediates and regulates various downstream signaling pathways, including enzyme activation, cell survival, and differentiation. The two most common mechanisms of Ca^{2+} mobilization by extracellular stimuli during osteoclast differentiation involve the RANKL-induced and the immunoglobulin-like (Ig) receptors-induced phosphorylation of phospholipase C (PLC)-γ2. PLCγ2 activation leads to the production of inositol-1,4,5-trisphosphate (IP3) that binds to and activates the IP3 receptor (IP3R), resulting in Ca^{2+} release from the endoplasmic reticulum (ER). This continuous Ca^{2+} flux is crucial for the long-lasting activation of NFATc1 (reviewed in Ref. [41]). In osteoclast precursors cells, Ca^{2+} oscillation induced by RANKL requires the Grb2-associated binder 2 (Gab2). The finding showing that $Gab2^{-/-}$ mice are osteopetrotic due to poor osteoclast differentiation elucidates the crucial role of Gab2 in osteoclastogenesis [42]. Mechanistically Gab2 works as an adaptor protein for PLCγ2 (Fig. 1), and the increase of intracellular calcium levels stimulates NFATc1 and activates the osteoclastogenic machinery [42]. PLCγ2 also binds to Syk tyrosine kinase that is coupled with ITAM adaptors of Ig-like receptors. DAP12 and FcRγ are both transmembrane adapter proteins with ITAM adaptors that couple to downstream pathways through the Syk tyrosine kinase. A study demonstrated that Syk-deficient osteoclasts failed to organize their cytoskeleton and degrade mineralized matrix in vitro [43].

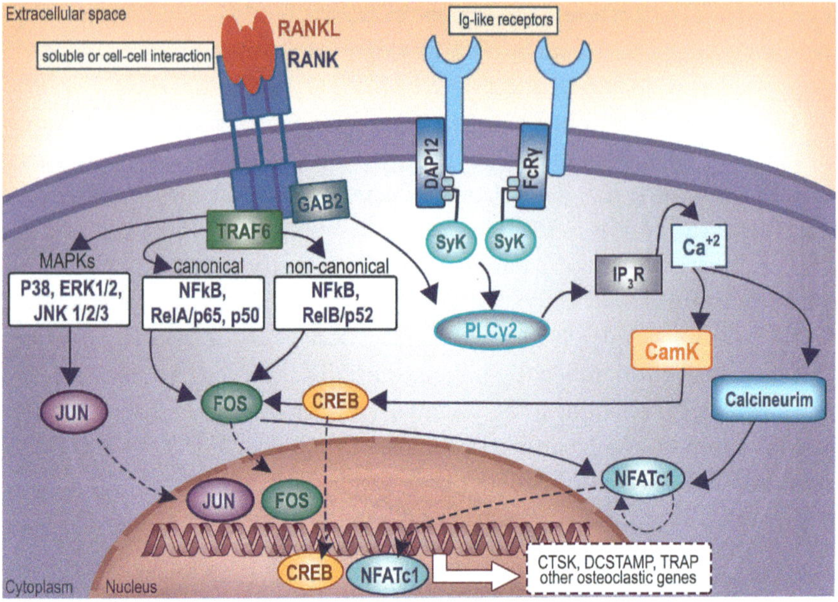

Fig. 1 RANK/RANKL signaling transduction in osteoclast differentiation. RANKL recruits TRAF6 to activate several signaling transduction pathways leading to MAPK, NFκB (canonical/non-canonical) activation. NFκB induces FOS to promote early NFATc1 expression. RANKL also recruits GAB2 in conjunction with Ig-like receptors to activate the Ca^{2+} pathway. NFATc1 expression is activated and sustained by the Ca^{2+} signal downstream of PLCγ2. Finally, NFATc1 translocates to the nucleus working together with other transcription factors, such as AP-1 (JUN/FOS) and CREB, to induce several osteoclast-specific genes.

These data suggest that RANK and Ig-like receptors form an osteoclastogenic signaling complex where PLCγ2 appears to be a key regulator, exerting its effects by modulating intracellular Ca^{2+}. After being released from the endoplasmic reticulum, calcium binds to calmodulin and activates the calmodulin-dependent phosphatase calcineurin. Calcineurin directly dephosphorylates serine residues in NFATc1, allowing its rapid translocation into the nucleus and subsequent activation. Once in the nucleus, NFATc1 binds directly to the fusion-mediating molecules' promoter regions, Atp6v0d2 (an osteoclast V-ATPase subunit) and DC-STAMP, and induces expression of these genes, thereby regulating osteoclast multinucleation [44]. NFATc1 also forms complexes with other transcription factors such as AP-1, PU.1, and MITF to regulate several osteoclastic genes. The NFATc1/AP-1 complex is essential for the induction of the TRAP and calcitonin receptor genes and the robust autoamplification of NFATc1

[45]. Fos, a member of the AP-1 transcription factor, is induced at an early stage during osteoclast differentiation and Fos-knockout mice develop osteopetrosis [46]. It has been shown that NFATc1 induction by RANKL is completely abolished in Fos-deficient cells [45]. NFATc1 also cooperates with PU.1 to activate the proteolytic enzyme cathepsin K at the terminal stage of differentiation [47]. It has been suggested that the transcription factor MITF also cooperates with NFATc1 to express OSCAR, an Ig-like receptor responsible for Ca^{2+} signaling activation [48]. Osteoclast differentiation can be divided into several stages in the context of NFATc1 activation (Fig. 1). (i) RANK/RANKL engagement results in MAPK signaling and TRAF6 recruitment, leading to NF-κB activation and Fos expression (as part of an AP-1 transcription factor). NFATc1 expression is initiated at a very early phase. (ii) Continuous calcium flux induced by RANKL and Ig-like receptors induces NFATc1 translocation to the nucleus and recruitment to its promoter (autoamplification). (iii) Several osteoclastic genes such as v-ATPase, DC-STAMP, TRAP, Calcitonin receptor, Cathepsin K, and Osteoclast-associated immunoglobulin-like receptor (OSCAR) are activated by NFATc1 in collaboration with other transcriptions factors such as AP-1, PU.1, and MITF.

Osteolytic diseases

The uncoupling of the bone remodeling process caused by hormonal dysregulation, genetic predisposition, or inflammatory conditions leads to several osteolytic diseases such as osteoporosis, rheumatoid arthritis (RA), and periodontitis. Therefore, the intricate relationship between bone remodeling and these factors, mainly the immune system, has been increasingly explored to unravel key molecular protagonists' role in osteoclast biology behind pathological bone loss. Here, we will describe how immunoinflammatory cells and cytokines trigger uncontrolled bone loss in some osteolytic diseases. A better knowledge of this cross-talk will clarify the unresolved challenge and help to develop novel therapeutic approaches.

Osteoporosis

The National Institutes of Health Consensus Development Panel on Osteoporosis defined, "Osteoporosis is a skeletal disorder characterized by compromised bone strength predisposing a person to an increased risk of fracture." This bone disorder affects both genders and is caused by an age-related decline of bone mass. However, this disease's spotlight has been on postmenopausal women due to estrogen levels' declination. Estrogen is considered a bone-protective hormone and is largely used to prevent the early stages of osteoporosis and reduce fracture risk in postmenopausal women [49].

The demonstration that osteoblasts, osteocytes, osteoclasts, and cells responsible for the bone remodeling, and their progenitors express receptors for estrogen pointed them as estrogen targets [50–52]. The activation of ERα, one estrogen receptor subtype, suppresses the RANKL and regulates OPG expression in bone lining cells [53]. The downregulation of RANKL/OPG expression by estrogen results in impaired osteoclasts differentiation. Chiefly, estrogen inhibits bone degradation by direct mechanisms that involve the restraining of key lysosomal enzymes secretion by osteoclasts [54] and the control of mature osteoclasts' life span by inducing apoptosis [55,56]. Interestingly, estrogen-induced osteoclasts death is mediated via FAS ligand (FASL)/FAS signaling on osteoblasts [57]. The bone protection effect of estrogen is not only controlling osteoclast survival and function; it also involves a positive outcome on osteoblasts' maturation, autophagy, and apoptosis [58,59]. Thus, the estrogen deprivation, observed during menopause, impacts osteoclasts' lifespan extension and osteoblasts apoptosis that combined, guide the effect toward marked bone resorption and osteoporosis.

In addition to playing a direct effect on bone cell regulation, estrogen also meddles in the immune system during osteoporosis. The lack of estrogen stimulates several inflammatory cytokines such as interleukin (IL)-1, tumor necrosis factor (TNF)-α, IL-6, IL-17, M-CSF, RANKL, and prostaglandins, all mediators known as osteoclastogenic inductors [50,60–62]. Indeed, different epidemiologic studies have reported the relationship between inflammatory conditions and susceptibility to osteoporosis [63–65]. The clinical and molecular studies observed several estrogen-regulated cytokines associated with accelerated bone loss in post-menopause [66]. Estrogen deficiency seen in post-menopause conditions induces IL-1 and IL-6 production from peripheral blood monocytes and bone marrow cells from humans [67,68]. IL-1 and IL-6 are potent stimulators of bone loss since it stimulates osteoclasts' activity and inhibits osteoblasts. Estrogen downregulates IL-1 receptor expression in human osteoclasts so that it minimizes the proinflammatory and pro-resorptive actions of IL-1 [69].

Besides being the precursor's cells for osteoclasts differentiation, macrophages release several cytokines that orchestrate osteoclastogenesis-associated bone loss. IL-23 is one of these players, and it directly mediates bone loss by inducing osteoclastogenesis and RANKL expression in T cells.

The blockage of IL-23 prevented trabecular bone loss in ovariectomized mice [70]. Moreover, other cells, including T and B cells, are also involved in osteoporosis's pathogenesis. Proliferation and expansion of activated T cells due to estrogen deficiency leads to high production of proinflammatory cytokines such as TNF-α and IL-17 and further stimulates M-CSF and RANKL production by bone marrow stem cells [62]. Early postmenopausal women treated with etanercept, a specific TNF-α blocker, reduced the rise of osteoclasts markers in the serum and urine evaluated 3 weeks after its administration [71]. These data show an essential role of TNF-α mediating bone loss during osteoporosis-induced estrogen deficiency in postmenopausal women; however, there are several other elements involved in the progression of this pathology as exposed above, and all of them should be considered.

Rheumatoid arthritis/osteoarthritis

Rheumatoid arthritis (RA) is a systemic autoimmune disorder characterized by joint inflammation and the destruction of cartilage and bone. Although it is not clear what triggers this disease, there are many risk factors related to RA development, such as smoking, obesity, periodontal disease, hormones, genetic predisposition, infections, and gut microbiota. RA is one among many osteolytic disorders that involve an immunoinflammatory response; however, the spotlight is on bone erosion mediated by osteoclasts activity, which causes severe deformations. Osteoclasts reside in the interface between inflamed synovial tissue and periarticular bone surface [72]. Although it is still unclear how the monocytes lineage or osteoclast precursors cells arrive at the joint, osteoclasts are differentiated and activated locally in the synovial tissue subsequently of M-CSF and RANKL expression. Inflammatory mediators such as TNF-α, IL-6, IL-1, and IL-17 are elevated in synovial joints and the serum of patients with RA [73–75]. The activation of these cells in the inflamed synovial tissue is associated with bone loss progression and the joint's severity. In chronic joint inflammation, T helper (Th) cells produce IL-17 and subsequently induce RANKL expression by osteoblasts [76]. Moreover, IL-17 upregulates IL-6 production by fibroblasts exacerbating the synovium inflammation.

Inflammatory cells can express RANKL in their membrane-associated form or induce the upregulation through NF-kB activation on synovial and T cells [77,78]. Poubelle and colleagues [79] observed that healthy peripheral blood neutrophils express RANKL. Simultaneously, synovial fluid neutrophils from RA patients express not only RANKL but also RANK and OPG, showing the dual involvement of neutrophils as immune and osteoclasts inductor cells in the RA inflammatory site. This triad is the main pathway for osteoclasts differentiation and, subsequently, bone resorption; however, TNF has also been targeted to control the inflammation, pain, and bone erosion in patients with RA [80]. In RA, TNF induces monocytes and synovial-fibroblast activation and survival, downregulates the collagen synthesis, to raise pain-receptor sensitization, pathognomonic signals of RA. Besides this, TNF, IL-1, and IL-6 released by synovial fibroblasts accelerate osteoclasts' catabolic activity by producing matrix enzymes, as MMPs, cathepsins, and aggrecanases [81]. This enzymatic milieu accelerates the joints' cartilage and bone degradation process, worsening the disease's prognosis. Monotherapy with anti-TNF showed to be effective in RA experimental models and, more importantly, in patients [82,83]; however, some patients presented only partial improvement or treatment failure. Therefore, current studies have investigated new therapeutic targets or associations of therapies.

Periodontitis

Periodontitis is a chronic inflammatory lesion that affects the tissues of protection and support of the teeth, which are the gingiva, periodontal ligament, cementum, and alveolar bone. Primarily, the dental biofilm represented by several periodontal pathogens around the teeth causes gingival inflammation called gingivitis. This scenario is still reversible with improved oral hygiene. However, if this gingival inflammation persists and is exacerbated, the tissues that support the teeth, mainly the alveolar bone, are reabsorbed, resulting in tooth loss [84]. Nowadays, it is very well established that dental biofilm is not the only factor that triggers periodontal disease. The immunopathogenesis of this disease involves several variances in the dental biofilm (bacteria), genetic predispositions [85], epigenetic factors [86], and host immune response [87]. These factors determine the inflammatory state, as well as the severity of the periodontal tissue damage.

During the periodontitis development, innate immune cells as neutrophils, macrophages are recruited to the local infection. Their surface receptors recognize specific molecules patterns of bacteria, allowing phagocytosis and host immune response activation. Besides, neutrophils express RANKL [79], the main ligand for osteoclast differentiation and bone resorption. In contrast to neutrophils, macrophages act indirectly on bone resorption, producing TNF in response to infection [88]. In an inflammatory context, TNF-α induces mononuclear osteoclasts precursors to differentiate by stimulating the bone marrow macrophages to produce RANKL [89]. Both cytokines play a critical role in driving inflammatory bone destruction during periodontitis [90].

In an infected milieu, the pathogens induce immune cells to produce and release effector molecules such as granular proteins, chemokines, and cytokines [91]. A significant amount of cytokines like TNF, interferon (IFN)-γ, and transforming growth factor-β (TGF-β) as well as matrix metalloproteinases (MMPs) and reactive oxygen species (ROS) are found in inflamed periodontal tissue, as well in the gingival crevicular fluid from patients with periodontitis [92,93]. These mediators are essential drivers for modulating leukocytes, osteoblasts, and osteoclasts activity, displaying an intimate relationship between osteoclastogenesis and bone resorption [94]. The link between immune cells and bone resorption involves MMPs, collagenolytic enzymes activated by IL-1, IL-6, IL-8, and TNF-α [95]. In addition to cytokines, bacterial plaque may induce the neutrophils to release collagenases such as MMPs during phagocytosis or cell lysis [96]. Among MMPs, MMP-2 and MMP-9 are strongly expressed in the inflamed pocket epithelium and can rapidly degrade the collagen in the bone matrix. Mäkelä and colleagues [97] found a significant increase of both MMP-9 and MMP-2 levels evaluated in oral fluids from periodontitis patients when compared to healthy patients. Notably, these collagenases are responsible for the collagen and bone degradation of periodontal tissues seen in periodontal disease.

Fracture healing

The bone healing process implies an intense bone remodeling to repair bone damage mainly caused by a fracture. Immediately after injury, a quick inflammatory response is activated, which will conduct the initial healing process. This inflammatory phase comprises removing necrotic tissues, hematoma formation, recruitment of mesenchymal stem cells to differentiate into chondrocytes and osteoblasts, and concomitant vascular ingrowth and neoangiogenesis [98,99]. This cascade aiming the tissue repair with cartilage and bone formation is guided by macrophages and some immune cells, including the release of cytokines such as IL-6, IL-1, and TNF-α [100]. Although TNF-α is an osteoclastogenic inductor, in this primary scenario, it plays an essential role in the recruitment of mesenchymal stem cells and on the hypertrophic chondrocytes apoptosis during endochondral bone formation [100]. Simultaneously in the fracture site, the platelets release growth factors as TGF-β, platelet-derived growth factor (PDGF), and osteoprogenitor cells express the bone morphogenic protein (BMP-2).

These factors initiate cartilaginous callus formation [98]. Besides, infiltrating macrophages and other inflammatory cells secrete fibroblast growth factor (FGF) and additional PDGF and TGF to form the hematoma [101]. These factors allow immediate fracture stabilization to further endochondral and intramembranous bone ossification.

After this inflammatory stage, new cartilage tissue is formed. With this, angiopoietins and vascular endothelial growth factors (VEGFs) stimulate vascular ingrowth in the periosteum to supply the new bone, an essential element for the fracture repair process progress. Following the healing process, the cartilage extracellular matrix and the first mineralized tissue start to be reabsorbed by osteoclasts to give a place for secondary bone formation. During the intense bone remodeling phase, the undifferentiated mesenchymal stem cells and osteoprogenitor cells are stimulated to differentiate in osteoblasts by BMPs, TGF-β2, and TGF-β3 [100,102]. RANKL and M-CSF are upregulated immediately after fracture injury, and their level is highly kept during the entire healing process [103]. Together with several cytokines as IL-1, IL-6, IL-11, IL-12, TNF-α, and IFN-γ, these factors modulate the renewing and resorptive actions given by the osteoblasts and osteoclasts undergoing the replacement of cartilage tissue to the woven bone [104]. Despite bone repair being concluded in some weeks, molecular and cellular signaling in the fracture site remains active for many years until recovery.

References

[1] Walker DG. Control of bone resorption by hematopoietic tissue. The induction and reversal of congenital osteopetrosis in mice through use of bone marrow and splenic transplants. J Exp Med 1975;142(3):651–63.

[2] Hancox NM. The osteoclast. Biol Rev Camb Philos Soc 1949;24(4):448–71.

[3] Walker DG. Congenital osteopetrosis in mice cured by parabiotic union with normal siblings. Endocrinology 1972;91(4):916–20.

[4] Walker DG. Spleen cells transmit osteopetrosis in mice. Science 1975;190(4216):785–7.

[5] Zambonin Zallone A, Teti A, Primavera MV. Monocytes from circulating blood fuse in vitro with purified osteoclasts in primary culture. J Cell Sci 1984;66:335–42.

[6] Wiktor-Jedrzejczak W, Bartocci A, Ferrante Jr AW, Ahmed-Ansari A, Sell KW, Pollard JW, et al. Total absence of colony-stimulating factor 1 in the macrophage-deficient osteopetrotic (op/op) mouse. Proc Natl Acad Sci USA 1990;87(12):4828–32.

[7] Yoshida H, Hayashi S, Kunisada T, Ogawa M, Nishikawa S, Okamura H, et al. The murine mutation osteopetrosis is in the coding region of the macrophage colony stimulating factor gene. Nature 1990;345(6274):442–4.

[8] Anderson DM, Maraskovsky E, Billingsley WL, Dougall WC, Tometsko ME, Roux ER, et al. A homologue of the TNF receptor and its ligand enhance T-cell growth and dendritic-cell function. Nature 1997;390(6656):175–9.

[9] Lacey DL, Timms E, Tan HL, Kelley MJ, Dunstan CR, Burgess T, et al. Osteoprotegerin ligand is a cytokine that regulates osteoclast differentiation and activation. Cell 1998;93(2):165–76.

[10] Wong BR, Rho J, Arron J, Robinson E, Orlinick J, Chao M, et al. TRANCE is a novel ligand of the tumor necrosis factor receptor family that activates c-Jun N-terminal kinase in T cells. J Biol Chem 1997;272(40):25190–4.
[11] Yasuda H, Shima N, Nakagawa N, Yamaguchi K, Kinosaki M, Mochizuki S, et al. Osteoclast differentiation factor is a ligand for osteoprotegerin/osteoclastogenesis-inhibitory factor and is identical to TRANCE/RANKL. Proc Natl Acad Sci USA 1998;95(7):3597–602.
[12] Kong YY, Yoshida H, Sarosi I, Tan HL, Timms E, Capparelli C, et al. OPGL is a key regulator of osteoclastogenesis, lymphocyte development and lymph-node organogenesis. Nature 1999;397(6717):315–23.
[13] Tsuda E, Goto M, Mochizuki S, Yano K, Kobayashi F, Morinaga T, et al. Isolation of a novel cytokine from human fibroblasts that specifically inhibits osteoclastogenesis. Biochem Biophys Res Commun 1997;234(1):137–42.
[14] Simonet WS, Lacey DL, Dunstan CR, Kelley M, Chang MS, Luthy R, et al. Osteoprotegerin: a novel secreted protein involved in the regulation of bone density. Cell 1997;89(2):309–19.
[15] Bucay N, Sarosi I, Dunstan CR, Morony S, Tarpley J, Capparelli C, et al. Osteoprotegerin-deficient mice develop early onset osteoporosis and arterial calcification. Genes Dev 1998;12(9):1260–8.
[16] American Society for B, Mineral Research President's Committee on N. Proposed standard nomenclature for new tumor necrosis factor family members involved in the regulation of bone resorption. The American Society for Bone and Mineral Research President's Committee on Nomenclature. J Bone Miner Res 2000;15(12):2293–6.
[17] Rho J, Takami M, Choi Y. Osteoimmunology: interactions of the immune and skeletal systems. Mol Cell 2004;17(1):1–9.
[18] Mun SH, Park PSU, Park-Min KH. The M-CSF receptor in osteoclasts and beyond. Exp Mol Med 2020;52(8):1239–54.
[19] Feng X, Takeshita S, Namba N, Wei S, Teitelbaum SL, Ross FP. Tyrosines 559 and 807 in the cytoplasmic tail of the macrophage colony-stimulating factor receptor play distinct roles in osteoclast differentiation and function. Endocrinology 2002;143(12):4868–74.
[20] Lee AW, States DJ. Both src-dependent and -independent mechanisms mediate phosphatidylinositol 3-kinase regulation of colony-stimulating factor 1-activated mitogen-activated protein kinases in myeloid progenitors. Mol Cell Biol 2000;20(18):6779–98.
[21] Xiong Y, Song D, Cai Y, Yu W, Yeung YG, Stanley ER. A CSF-1 receptor phosphotyrosine 559 signaling pathway regulates receptor ubiquitination and tyrosine phosphorylation. J Biol Chem 2011;286(2):952–60.
[22] Gyori DS, Mocsai A. Osteoclast signal transduction during bone metastasis formation. Front Cell Dev Biol 2020;8:507.
[23] Arai F, Miyamoto T, Ohneda O, Inada T, Sudo T, Brasel K, et al. Commitment and differentiation of osteoclast precursor cells by the sequential expression of c-Fms and receptor activator of nuclear factor kappaB (RANK) receptors. J Exp Med 1999;190(12):1741–54.
[24] Lomaga MA, Yeh WC, Sarosi I, Duncan GS, Furlonger C, Ho A, et al. TRAF6 deficiency results in osteopetrosis and defective interleukin-1, CD40, and LPS signaling. Genes Dev 1999;13(8):1015–24.
[25] Naito A, Azuma S, Tanaka S, Miyazaki T, Takaki S, Takatsu K, et al. Severe osteopetrosis, defective interleukin-1 signalling and lymph node organogenesis in TRAF6-deficient mice. Genes Cells 1999;4(6):353–62.
[26] Zarubin T, Han J. Activation and signaling of the p38 MAP kinase pathway. Cell Res 2005;15(1):11–8.

[27] Lamothe B, Lai Y, Xie M, Schneider MD, Darnay BG. TAK1 is essential for osteoclast differentiation and is an important modulator of cell death by apoptosis and necroptosis. Mol Cell Biol 2013;33(3):582–95.
[28] Matsumoto M, Sudo T, Saito T, Osada H, Tsujimoto M. Involvement of p38 mitogen-activated protein kinase signaling pathway in osteoclastogenesis mediated by receptor activator of NF-kappa B ligand (RANKL). J Biol Chem 2000;275(40):31155–61.
[29] Cong Q, Jia H, Li P, Qiu S, Yeh J, Wang Y, et al. p38alpha MAPK regulates proliferation and differentiation of osteoclast progenitors and bone remodeling in an aging-dependent manner. Sci Rep 2017;7:45964.
[30] Wagner EF, Nebreda AR. Signal integration by JNK and p38 MAPK pathways in cancer development. Nat Rev Cancer 2009;9(8):537–49.
[31] David JP, Sabapathy K, Hoffmann O, Idarraga MH, Wagner EF. JNK1 modulates osteoclastogenesis through both c-Jun phosphorylation-dependent and -independent mechanisms. J Cell Sci 2002;115(Pt 22):4317–25.
[32] Ikeda F, Matsubara T, Tsurukai T, Hata K, Nishimura R, Yoneda T. JNK/c-Jun signaling mediates an anti-apoptotic effect of RANKL in osteoclasts. J Bone Miner Res 2008;23(6):907–14.
[33] He Y, Staser K, Rhodes SD, Liu Y, Wu X, Park SJ, et al. Erk1 positively regulates osteoclast differentiation and bone resorptive activity. PLoS One 2011;6(9):e24780.
[34] Boyce BF, Xiu Y, Li J, Xing L, Yao Z. NF-kappaB-mediated regulation of osteoclastogenesis. Endocrinol Metab 2015;30(1):35–44.
[35] Franzoso G, Carlson L, Xing L, Poljak L, Shores EW, Brown KD, et al. Requirement for NF-kappaB in osteoclast and B-cell development. Genes Dev 1997;11(24):3482–96.
[36] Iotsova V, Caamano J, Loy J, Yang Y, Lewin A, Bravo R. Osteopetrosis in mice lacking NF-kappaB1 and NF-kappaB2. Nat Med 1997;3(11):1285–9.
[37] Yamashita T, Yao Z, Li F, Zhang Q, Badell IR, Schwarz EM, et al. NF-kappaB p50 and p52 regulate receptor activator of NF-kappaB ligand (RANKL) and tumor necrosis factor-induced osteoclast precursor differentiation by activating c-Fos and NFATc1. J Biol Chem 2007;282(25):18245–53.
[38] Xing L, Bushnell TP, Carlson L, Tai Z, Tondravi M, Siebenlist U, et al. NF-kappaB p50 and p52 expression is not required for RANK-expressing osteoclast progenitor formation but is essential for RANK- and cytokine-mediated osteoclastogenesis. J Bone Miner Res 2002;17(7):1200–10.
[39] Miyauchi Y, Ninomiya K, Miyamoto H, Sakamoto A, Iwasaki R, Hoshi H, et al. The Blimp1-Bcl6 axis is critical to regulate osteoclast differentiation and bone homeostasis. J Exp Med 2010;207(4):751–62.
[40] Novack DV, Yin L, Hagen-Stapleton A, Schreiber RD, Goeddel DV, Ross FP, et al. The IkappaB function of NF-kappaB2 p100 controls stimulated osteoclastogenesis. J Exp Med 2003;198(5):771–81.
[41] Hwang SY, Putney Jr JW. Calcium signaling in osteoclasts. Biochim Biophys Acta 2011;1813(5):979–83.
[42] Wada T, Nakashima T, Oliveira-dos-Santos AJ, Gasser J, Hara H, Schett G, et al. The molecular scaffold Gab2 is a crucial component of RANK signaling and osteoclastogenesis. Nat Med 2005;11(4):394–9.
[43] Faccio R, Teitelbaum SL, Fujikawa K, Chappel J, Zallone A, Tybulewicz VL, et al. Vav3 regulates osteoclast function and bone mass. Nat Med 2005;11(3):284–90.
[44] Kim K, Lee SH, Ha Kim J, Choi Y, Kim N. NFATc1 induces osteoclast fusion via up-regulation of Atp6v0d2 and the dendritic cell-specific transmembrane protein (DC-STAMP). Mol Endocrinol 2008;22(1):176–85.
[45] Takayanagi H, Kim S, Koga T, Nishina H, Isshiki M, Yoshida H, et al. Induction and activation of the transcription factor NFATc1 (NFAT2) integrate RANKL signaling in terminal differentiation of osteoclasts. Dev Cell 2002;3(6):889–901.

[46] Wang ZQ, Ovitt C, Grigoriadis AE, Mohle-Steinlein U, Ruther U, Wagner EF. Bone and haematopoietic defects in mice lacking c-fos. Nature 1992;360(6406):741–5.
[47] Matsumoto M, Kogawa M, Wada S, Takayanagi H, Tsujimoto M, Katayama S, et al. Essential role of p38 mitogen-activated protein kinase in cathepsin K gene expression during osteoclastogenesis through association of NFATc1 and PU.1. J Biol Chem 2004;279(44):45969–79.
[48] Kim Y, Sato K, Asagiri M, Morita I, Soma K, Takayanagi H. Contribution of nuclear factor of activated T cells c1 to the transcriptional control of immunoreceptor osteoclast-associated receptor but not triggering receptor expressed by myeloid cells-2 during osteoclastogenesis. J Biol Chem 2005;280(38):32905–13.
[49] Dobbs MB, Buckwalter J, Saltzman C. Osteoporosis: the increasing role of the orthopaedist. Iowa Orthop J 1999;19:43–52.
[50] Manolagas SC, Jilka RL. Bone marrow, cytokines, and bone remodeling. Emerging insights into the pathophysiology of osteoporosis. N Engl J Med 1995;332(5):305–11.
[51] Girasole G, Jilka RL, Passeri G, Boswell S, Boder G, Williams DC, et al. 17 beta-estradiol inhibits interleukin-6 production by bone marrow-derived stromal cells and osteoblasts in vitro: a potential mechanism for the antiosteoporotic effect of estrogens. J Clin Invest 1992;89(3):883–91.
[52] Oursler MJ, Osdoby P, Pyfferoen J, Riggs BL, Spelsberg TC. Avian osteoclasts as estrogen target cells. Proc Natl Acad Sci USA 1991;88(15):6613–7.
[53] Streicher C, Heyny A, Andrukhova O, Haigl B, Slavic S, Schuler C, et al. Estrogen regulates bone turnover by targeting RANKL expression in bone lining cells. Sci Rep 2017;7(1):6460.
[54] Kremer M, Judd J, Rifkin B, Auszmann J, Oursler MJ. Estrogen modulation of osteoclast lysosomal enzyme secretion. J Cell Biochem 1995;57(2):271–9.
[55] Kameda T, Mano H, Yuasa T, Mori Y, Miyazawa K, Shiokawa M, et al. Estrogen inhibits bone resorption by directly inducing apoptosis of the bone-resorbing osteoclasts. J Exp Med 1997;186(4):489–95.
[56] Nakamura T, Imai Y, Matsumoto T, Sato S, Takeuchi K, Igarashi K, et al. Estrogen prevents bone loss via estrogen receptor alpha and induction of Fas ligand in osteoclasts. Cell 2007;130(5):811–23.
[57] Krum SA, Miranda-Carboni GA, Hauschka PV, Carroll JS, Lane TF, Freedman LP, et al. Estrogen protects bone by inducing Fas ligand in osteoblasts to regulate osteoclast survival. EMBO J 2008;27(3):535–45.
[58] Allison H, McNamara LM. Inhibition of osteoclastogenesis by mechanically stimulated osteoblasts is attenuated during estrogen deficiency. Am J Physiol Cell Physiol 2019;317(5):C969–82.
[59] Yang YH, Chen K, Li B, Chen JW, Zheng XF, Wang YR, et al. Estradiol inhibits osteoblast apoptosis via promotion of autophagy through the ER-ERK-mTOR pathway. Apoptosis 2013;18(11):1363–75.
[60] Ammann P, Rizzoli R, Bonjour JP, Bourrin S, Meyer JM, Vassalli P, et al. Transgenic mice expressing soluble tumor necrosis factor-receptor are protected against bone loss caused by estrogen deficiency. J Clin Invest 1997;99(7):1699–703.
[61] Kimble RB, Vannice JL, Bloedow DC, Thompson RC, Hopfer W, Kung VT, et al. Interleukin-1 receptor antagonist decreases bone loss and bone resorption in ovariectomized rats. J Clin Invest 1994;93(5):1959–67.
[62] Pacifici R. Role of T cells in ovariectomy induced bone loss—revisited. J Bone Miner Res 2012;27(2):231–9.
[63] Haugeberg G, Lodder MC, Lems WF, Uhlig T, Ørstavik RE, Dijkmans BA, et al. Hand cortical bone mass and its associations with radiographic joint damage and fractures in 50-70 year old female patients with rheumatoid arthritis: cross sectional Oslo-Truro-Amsterdam (OSTRA) collaborative study. Ann Rheum Dis 2004;63(10):1331–4.

[64] Mikuls TR, Saag KG, Curtis J, Bridges Jr SL, Alarcon GS, Westfall AO, et al. Prevalence of osteoporosis and osteopenia among African Americans with early rheumatoid arthritis: the impact of ethnic-specific normative data. J Natl Med Assoc 2005;97(8):1155–60.
[65] Mitra D, Elvins DM, Speden DJ, Collins AJ. The prevalence of vertebral fractures in mild ankylosing spondylitis and their relationship to bone mineral density. Rheumatology (Oxford) 2000;39(1):85–9.
[66] Brincat SD, Borg M, Camilleri G, Calleja-Agius J. The role of cytokines in postmenopausal osteoporosis. Minerva Ginecol 2014;66(4):391–407.
[67] Bismar H, Diel I, Ziegler R, Pfeilschifter J. Increased cytokine secretion by human bone marrow cells after menopause or discontinuation of estrogen replacement. J Clin Endocrinol Metab 1995;80(11):3351–5.
[68] Sato F, Ouchi Y, Masuyama A, Nakamura T, Hosoi T, Okamoto Y, et al. Effects of estrogen replacement on insulin-like growth factor I concentrations in serum and bone tissue and on interleukin 1 secretion from spleen macrophages in oophorectomized rats. Calcif Tissue Int 1993;53(2):111–6.
[69] Sunyer T, Lewis J, Collin-Osdoby P, Osdoby P. Estrogen's bone-protective effects may involve differential IL-1 receptor regulation in human osteoclast-like cells. J Clin Invest 1999;103(10):1409–18.
[70] Shukla P, Mansoori MN, Singh D. Efficacy of anti-IL-23 monotherapy versus combination therapy with anti-IL-17 in estrogen deficiency induced bone loss conditions. Bone 2018;110:84–95.
[71] Charatcharoenwitthaya N, Khosla S, Atkinson EJ, McCready LK, Riggs BL. Effect of blockade of TNF-alpha and interleukin-1 action on bone resorption in early postmenopausal women. J Bone Miner Res 2007;22(5):724–9.
[72] Schett G, Gravallese E. Bone erosion in rheumatoid arthritis: mechanisms, diagnosis and treatment. Nat Rev Rheumatol 2012;8(11):656–64.
[73] Eklund KK, Leirisalo-Repo M, Ranta P, Mäki T, Kautiainen H, Hannonen P, et al. Serum IL-1beta levels are associated with the presence of erosions in recent onset rheumatoid arthritis. Clin Exp Rheumatol 2007;25(5):684–9.
[74] Marti L, Golmia R, Golmia AP, Paes AT, Guilhen DD, Moreira-Filho CA, et al. Alterations in cytokine profile and dendritic cells subsets in peripheral blood of rheumatoid arthritis patients before and after biologic therapy. Ann NY Acad Sci 2009;1173:334–42.
[75] Tchetverikov I, Ronday HK, Van El B, Kiers GH, Verzijl N, TeKoppele JM, et al. MMP profile in paired serum and synovial fluid samples of patients with rheumatoid arthritis. Ann Rheum Dis 2004;63(7):881–3.
[76] Kuwabara T, Ishikawa F, Kondo M, Kakiuchi T. The role of IL-17 and related cytokines in inflammatory autoimmune diseases. Mediat Inflamm 2017;2017:3908061.
[77] Horwood NJ, Elliott J, Martin TJ, Gillespie MT. Osteotropic agents regulate the expression of osteoclast differentiation factor and osteoprotegerin in osteoblastic stromal cells. Endocrinology 1998;139(11):4743–6.
[78] Tanaka Y, Ohira T. Mechanisms and therapeutic targets for bone damage in rheumatoid arthritis, in particular the RANK-RANKL system. Curr Opin Pharmacol 2018;40:110–9.
[79] Poubelle PE, Chakravarti A, Fernandes MJ, Doiron K, Marceau AA. Differential expression of RANK, RANK-L, and osteoprotegerin by synovial fluid neutrophils from patients with rheumatoid arthritis and by healthy human blood neutrophils. Arthritis Res Ther 2007;9(2):R25.
[80] Williams RO, Feldmann M, Maini RN. Anti-tumor necrosis factor ameliorates joint disease in murine collagen-induced arthritis. Proc Natl Acad Sci USA 1992;89(20):9784–8.

[81] McInnes IB, Schett G. Cytokines in the pathogenesis of rheumatoid arthritis. Nat Rev Immunol 2007;7(6):429–42.
[82] Asquith DL, Miller AM, McInnes IB, Liew FY. Animal models of rheumatoid arthritis. Eur J Immunol 2009;39(8):2040–4.
[83] Elliott MJ, Maini RN, Feldmann M, Long-Fox A, Charles P, Katsikis P, et al. Treatment of rheumatoid arthritis with chimeric monoclonal antibodies to tumor necrosis factor alpha. Arthritis Rheum 1993;36(12):1681–90.
[84] Lamster IB, Pagan M. Periodontal disease and the metabolic syndrome. Int Dent J 2017;67(2):67–77.
[85] Borrell LN, Papapanou PN. Analytical epidemiology of periodontitis. J Clin Periodontol 2005;32(Suppl. 6):132–58.
[86] Larsson L, Castilho RM, Giannobile WV. Epigenetics and its role in periodontal diseases: a state-of-the-art review. J Periodontol 2015;86(4):556–68.
[87] Kinane DF, Stathopoulou PG, Papapanou PN. Periodontal diseases. Nat Rev Dis Primers 2017;3:17038.
[88] Papadopoulos G, Weinberg EO, Massari P, Gibson 3rd FC, Wetzler LM, Morgan EF, et al. Macrophage-specific TLR2 signaling mediates pathogen-induced TNF-dependent inflammatory oral bone loss. J Immunol 2013;190(3):1148–57.
[89] Lam J, Takeshita S, Barker JE, Kanagawa O, Ross FP, Teitelbaum SL. TNF-alpha induces osteoclastogenesis by direct stimulation of macrophages exposed to permissive levels of RANK ligand. J Clin Invest 2000;106(12):1481–8.
[90] Kitaura H, Sands MS, Aya K, Zhou P, Hirayama T, Uthgenannt B, et al. Marrow stromal cells and osteoclast precursors differentially contribute to TNF-alpha-induced osteoclastogenesis in vivo. J Immunol 2004;173(8):4838–46.
[91] Nauseef WM. How human neutrophils kill and degrade microbes: an integrated view. Immunol Rev 2007;219:88–102.
[92] Berthelot JM, Le Goff B. Rheumatoid arthritis and periodontal disease. Joint Bone Spine 2010;77(6):537–41.
[93] Liao F, Li Z, Wang Y, Shi B, Gong Z, Cheng X. Porphyromonas gingivalis may play an important role in the pathogenesis of periodontitis-associated rheumatoid arthritis. Med Hypotheses 2009;72(6):732–5.
[94] Birkedal-Hansen H. Role of cytokines and inflammatory mediators in tissue destruction. J Periodontal Res 1993;28(6 Pt 2):500–10.
[95] Uitto VJ, Overall CM, McCulloch C. Proteolytic host cell enzymes in gingival crevice fluid. Periodontol 2000;2003(31):77–104.
[96] Mancini S, Romanelli R, Laschinger CA, Overall CM, Sodek J, McCulloch CA. Assessment of a novel screening test for neutrophil collagenase activity in the diagnosis of periodontal diseases. J Periodontol 1999;70(11):1292–302.
[97] Mäkelä M, Salo T, Uitto VJ, Larjava H. Matrix metalloproteinases (MMP-2 and MMP-9) of the oral cavity: cellular origin and relationship to periodontal status. J Dent Res 1994;73(8):1397–406.
[98] Fazzalari NL. Bone fracture and bone fracture repair. Osteoporos Int 2011;22(6):2003–6.
[99] Ozaki A, Tsunoda M, Kinoshita S, Saura R. Role of fracture hematoma and periosteum during fracture healing in rats: interaction of fracture hematoma and the periosteum in the initial step of the healing process. J Orthop Sci 2000;5(1):64–70.
[100] Gerstenfeld LC, Cho TJ, Kon T, Aizawa T, Tsay A, Fitch J, et al. Impaired fracture healing in the absence of TNF-alpha signaling: the role of TNF-alpha in endochondral cartilage resorption. J Bone Miner Res 2003;18(9):1584–92.
[101] Barnes GL, Kostenuik PJ, Gerstenfeld LC, Einhorn TA. Growth factor regulation of fracture repair. J Bone Miner Res 1999;14(11):1805–15.

[102] Cho TJ, Gerstenfeld LC, Einhorn TA. Differential temporal expression of members of the transforming growth factor beta superfamily during murine fracture healing. J Bone Miner Res 2002;17(3):513–20.
[103] Kon T, Cho TJ, Aizawa T, Yamazaki M, Nooh N, Graves D, et al. Expression of osteoprotegerin, receptor activator of NF-kappaB ligand (osteoprotegerin ligand) and related proinflammatory cytokines during fracture healing. J Bone Miner Res 2001;16(6):1004–14.
[104] Mountziaris PM, Mikos AG. Modulation of the inflammatory response for enhanced bone tissue regeneration. Tissue Eng B Rev 2008;14(2):179–86.

CHAPTER 8

Microglia and border-associated macrophages in the central nervous system

N.G. Zanluqui[a,b,c], C.M. Polonio[a,c], M.G. de Oliveira[c], L.G. Oliveira[a,c], L.C. Faria[c], and J.P.S. Peron[a,b,c]

[a]Laboratory of Neuroimmunology of Arboviruses, Scientific Platform Pasteur-USP (SPPUSP), São Paulo, SP, Brazil
[b]Immunopathology and Allergy Post Graduate Program, School of Medicine, University of São Paulo, São Paulo, SP, Brazil
[c]Neuroimmune Interactions Laboratory, Department of Immunology, Institute of Biomedical Sciences, University of São Paulo, São Paulo, SP, Brazil

CNS macrophages heterogeneity and beyond

Microglia cells in the parenchyma and pvMΦ are long-living cells with a low rate of renewal and insignificant replacement by circulating monocytes [1–3]. Studies using ^{14}C [4] analysis evidenced a very heterogeneous microglial population in human CNS according to survival (around two decades) and renewal (28% annually) rates [1]. Microglia cells originate during embryonic development from erythromyeloid progenitors (EMP) located in yolk-sac [5], as observed for most of the resident macrophages in the body [6], without any replenish by circulating monocytes under physiological conditions [7–10]. Interestingly, intracerebral injected hematopoietic stem cell-derived macrophages acquire an abnormal morphology in comparison with resident microglia and a microglia-like gene signature is elicited only during neurodegeneration [11].

The colonization of brain parenchyma occurs in different stages of embryonic development. In mice, the first wave of CNS colonization by microglia happens at day 9.5 post-conception (dpc), giving rise to the first microglia population [12]. The progenitors seed in different CNS regions around dpc 12.5 when CNS structures start to emerge [6]. In contrast, mMΦ and cpMΦ are short-lived and it is still unclear whether they are originated from yolk-sac and then replenish by bone-marrow-derived cells [6,13].

Macrophages in the Human Body
https://doi.org/10.1016/B978-0-12-821385-8.00013-5

Different cellular populations of microglia emerge during the development and are distinct in each brain area. The features of microglia populations in specific brain regions during health and disease have been well-described. Microglia diversity was considered by density, morphology, self-renewal, molecular signatures, metabolism, and ontogenesis [14].

CD11b and CD11c integrins have important roles in regulating the migration and localization of microglia in the CNS. The first one is constitutively active, while the second is highly expressed in activated microglia [15]. Studies concerning microglial density agree that the highest cellular abundancy is within the forebrain in comparison to other regions, as the hippocampus and midbrain, with the lowest numbers in the cerebellum [16–18]. Different concentrations within the same area were also described [14]. In the cerebellum the most populated region is the nucleus, with no differences among medial, interposed, and lateral areas, contrasting with the cortex, which has its own density variations with more cells in the granular than in the molecular layer [16,19]. Similarly, the dentated gyrus and CA1 regions within the hippocampus are more populated than CA3 area [20]. However, it is not well-elucidated whether these cellular concentrations are due to contrasting cellular migration, turnover, or cell death.

The regional variety of microglial phenotypes was mostly evidenced by the expression of immunoregulatory genes and metabolic processes [21]. Concerning this, cerebellar and hippocampal microglia have greater energetic metabolism and immune-alert state than microglia in the striatum and cortex. These changes were evidenced by transcriptome analysis from different brain regions in which most upregulated pathways are related to electron-chain transport and generation of metabolic precursors in addition to a striking expression of immune effector response. Interestingly, sometimes the same area in the brain may also present a diversity of microglia phenotypes. However, independent of the brain region, the overall profile of CNS microglia differs from tissue-resident macrophages outside the brain [21].

There are four types of microglial morphology: (1) amoeboid, i.e., globular without ramifications; (2) primed, with enlarged processes, increased proliferation and decreased secondary branching; (3) ramified, with multiple thin and radial processes; and (4) reactive, thickened processes and decreased ramifications [22]. Microglia shape is affected brain areas and if the cells are near to blood vessels, neuronal cell bodies, or axons and dendrites [23–25]. Additionally, the composition of white and gray matter, where microglia is located, also affects the concentration and morphology. In the cerebellum, microglial cells are less branched with augmented cytoplasm, contrasting with cells in the hippocampus and frontal cortex. Curiously, regions that lack complete BBB, as subventricular zone (SVZ) and other regions around the ventricle, exhibit rounded cells with fewer ramifications [23,26].

In addition to parenchymal cells varieties, there is also a great heterogeneity among BAMs. BAMs are considered an isolated population of brain phagocytes regarding their ontogenesis and location [27]. During the development, two distinct populations of CNS macrophages emerge, which notably diverge in their phenotypes [28]. For instance, microglia cells differentiation and accumulation are dependent on TGF-β receptor whereas BAMs are established independently [28].

BAMs display different morphology and are localized in totally distinct anatomical areas. This is the reason why they were further characterized as border-associated, as they sit on interfaces of blood- or ventricle-brain barriers. Mostly studied are mMΦ, pvMΦ and cpMΦ. Aside from anatomical location, BAMs have both discrepancies and similarities to microglial cells, for instance, sharing the expression of CD11b and Iba-1 [6]. cpMΦ have a radial morphology, whereas mMΦ and pvMΦ are mostly stretched and less branched [29,30]. Similar to parenchymal microglia, pvMΦ are sessile and can extend their processes to distant perivascular areas [30,31], as well as mMΦ can elongate their processes along their environment to scan the region [32,33]. Little is known concerning cpMΦ's motility in vivo [34].

Macrophage's phenotype in the brain

Early microglia precursors in yolk-sac are $CD45^+$ and $c\text{-}Kit^+$ but they completely abrogate c-Kit expression after migration to the brain and start to acquire immature macrophages markers, as CD11b, CD115, F4/80, CX3CR1, and CSF-1R [9,12,35]. Additionally, immature microglia depend on PU.1 and IRF-8 transcriptional factors for full maturation [36]. As they mature, microglial cells greatly resemble macrophages, sharing markers as CD11b, F4/80, and Fc receptors. However, they diverge from macrophages on lower levels of CD45 and the expression of CSF1R, CD200, SIRPA, CX3CR1, IBA1. Moreover, a more specific transcriptional signature has been proposed, with *P2ry12*, *Tmem119*, *Siglec-h*, *Gpr34*, *Socs3*, *Hexb*, *Olfml3*, and *Fcrls* [37–44].

In fact, recent studies using single-cell technologies as CyTOF and RNA sequencing have evidenced many different microglial signatures in mouse and human brains. In vivo comparisons of mice and human microglia, heterogeneity was carried out by combined single-cell RNA sequencing with immunofluorescence. During mice embryo development, six clusters of mature microglia were found in all brain regions, with different expressions of the genes *Tmem119*, *Selplg*, *Slc2a5*, *Malat1* (a long non-coding RNA), *Ctsb*, *Ctsd*, *Lamp1*, and *Apoe*. Biopsies from healthy human brain evidenced four clusters of cortical microglia with similarities to murine postnatal microglial profile, although with a distinct expression of *Cst3*, *Pr2y13* and *Ccl4* [45].

Single-cell RNA sequencing comparing male and female mice microglia at embryonic day 14.5 and postnatal days 4/5, 30, 100, and 540 clustered 9 different microglia signatures. The greatest microglial diversity was found at the earliest stages (E14.5 and P5), whereas considerably less diversity was observed in juveniles (P30) and adults (P100). Gene expression analysis showed that canonical microglial genes (*Fcrls*, *P2ry12*, *Cx3cr1*, *Trem2*, and *C1qa*) were highly expressed by most of the analyzed cells, whereas only *C1qa*, *Fcrls*, and *Trem2* were uniformly expressed in all clusters. Only a small cluster (around 0.5% of microglia) in P4/5 was enriched in females, with high levels of *CD74*, *Ccl24*, and *Arg1* genes. In summary, sex has little or no impact on microglial diversity whereas aging has a slight impact on microglial phenotype with an increase in inflammatory and interferon-responsive microglia [46].

A study using CyTOF analysis of adult mice microglia, confirmed the heterogeneity of microglial cells by the $CD45^{lo}CD11b^{lo}F4/80^{+}CD64^{+}MerTK^{+}CX3CR1^{hi}Siglec\text{-}H^{+}CD88^{lo}$ phenotype in steady state [4]. During neuroinflammation, however, some of these markers are decreased, whereas others increase, generating a phenotype referred to as disease-associated microglia (DAM) [47]. Another population of cells with distinct macrophage-associated markers MerTK, CD64, F4/80, and CD16/32 expression were also described, with a $CD45^{lo/+}CD11b^{lo}F4/80^{hi}CD64^{hi}MerTK^{+}CX3CR1^{+}CD88^{hi}$, evidencing the complexity and the dynamic of CNS phagocytes' phenotype. However, these cells lacked typical microglial Siglec-H marker and *Sall1* transcriptional factor but instead expressed CD206, CD38, and Lyve1.

BAMs are the second most abundant immune cells in the mice CNS after microglia [4]. Worthy to remember that, whereas microglia reside in the brain parenchyma, BAM locate within the borders of the CNS, as the meninges and choroid plexus. Four BAM subsets were described based on differential expression of CD38, MHC-II, and CCR2: subset 1 ($CD38^{+}MHC\text{-}II^{-}CCR2^{-}$), subset 2 ($CD38^{+}MHC\text{-}II^{+}CCR2^{-}$), subset 3 ($CD38^{-}MHC\text{-}II^{+}CCR2^{-}$), and subset 4 ($CD38^{-}MHC\text{-}II^{+}CCR2^{+}$). Subset 1 is located in all brain-border regions, subset 2 is preferentially enriched in the pia mater and perivascular spaces, and less in the dura mater, whereas subsets 3 and 4 were enriched in the dura mater and only subset 4 was described in the choroid plexus [4].

Other markers can be used to distinguish BAMs subtypes and to identify their location. CD206, a receptor of non-parenchymal macrophages in the CNS, and CD163, a scavenger receptor cysteine-rich superfamily group B, are present in pvMΦ between the laminin-positive endothelial (CD31) and glial basement membranes [48,49]. ERTR7, a fibroblast-like cell marker can be used to localize subdural mMΦ lining the meninges and meningeal vasculature [50]. Finally, CCR2 is a marker for cpMΦ found exclusively in the stroma and epithelial layer of the choroid plexus. It is consistent with the high turnover rate and replacement by $CCR2^+$ bone marrow-derived cells within the choroid plexus.

Altogether, BAMs are a heterogeneous resident macrophage population that can be divided into subsets based on their phenotype and transcriptional signature. Also, they are specifically enriched in different CNS border compartments, shaped by cues from their specific local microenvironments that may reflect in different functional roles during health or disease.

General functions of microglia and BAMs

Due to their aforementioned complexity, it is a challenge to understand how microglia and BAMs behave under different contexts, as neurodevelopment, aging, and brain diseases, either autoimmune or infectious. Clearly, microglia and BAMs greatly vary the expression of transcription factors during these conditions [51]. Moreover, their location next to endothelial vessels, meningeal spaces, and choroid plexus, greatly difficult access, and the isolation of purified cellular populations [52].

During both resting and activation states, microglia secrete trophic factors, such as brain-derived neurotrophic factor (BDNF) [53], nerve growth factor (NGF) [54], fibroblast growth factor (FGF), hepatocyte growth factor (HGF), platelet-derived growth factor (PDGF) and epidermal growth factor (EGF) [55] that contribute for appropriate CNS function [56]. Likewise, microglia also express neurotransmitters receptors, such as glutamate ionotropic receptors (AMPA and NMDA) and glutamate metabotropic receptors (mGlutR2) that work in a two-way street in neuroimmunology modulation [15].

Microglial cells are also responsible for tissue repair and antigen sampling by prolonging cellular branches and interacting with neuronal synapses [57], identified by electron microscopy [58] using primary cerebral rat cultures and time-lapse video microscopy [59]. Microglia are highly mobile and pinocytic [60], making a brief (approximately 5 min) and recurrent (about once per hour) contact with neuronal synapses [58]. More specifically, microglia is in connection with neuronal cell body, dendritic spines [61], and also making microglia-neuronal junctions, as shown with transfected HEK-cells. Interestingly, the sites of interaction between microglia and neuronal somatic membrane are abundant of Kv2.2 clusters, involved in cell-to-cell communication and monitoring neuronal activity [62].

Microglia is largely responsible for immunosurveillance in the CNS and rapidly responds against infections, dead cells, and protein aggregates [63]. They express many innate immunity receptors, such as Toll-like receptor (TLR), NOD-like receptors (NLRs), and scavenger receptors (CD36, SR1, and MARCO) that may recognize either pathogen-associated molecular pattern (PAMPS) [64] as well as Danger-associated molecular patterns (DAMPs). Although with a more reparative and modulatory function at a steady state, after stimulation, microglia may produce diverse pro-inflammatory molecules, such as cytokines, chemokines, nitric oxide (NO), and reactive oxygen species (ROS) [65].

NOD-like receptors are important players in microglial function for the induction of IL-1β release. The NLRP3 inflammasome (NOD-, LRR- and pyrin domain-containing 3), promotes activation of caspase-1 and consequently cleavage of pro-IL-1β/IL-18 in the activated form of IL-1β and IL-18 [66]. IL-1β can interact with the hypothalamus-pituitary-adrenal axis, an important neuroendocrine system, acting in the induction of "sickness behaviors," responsible for the constant sensation of loss appetite, lethargy, depressed mood, sleepiness, and increase in body temperature [67,68]. IL-1β is also related to the increase of MHC expression in microglia and recruitment of leucocytes, such as monocytes, neutrophils, dendritic cells, and $CD4^+$ and $CD8^+$ T-cells, that is mediated by IL-1R1 receptor [69]. Additionally, blood-brain barrier leakage is due to IL-1β modulating the expression of ICAM-1 (intercellular adhesion molecule 1) in endothelial cells and reducing tight junctions leading to increase permeability [70].

The ionotropic receptors P2X4 and P2X7, and the metabotropic receptors P2Y6, P2Y12, P2Y13 are expressed in microglia and can be activated by their ligand, ATP [71]. When activated, this signaling alters membrane polarization and increases intracellular Ca^{2+} levels and inducing inflammasome assembling, increasing cytokine production, metabolic changes, and increased phagocytosis by these cells [72]. ATP is important to mediate communication between microglial processes and neuronal cell bodies and are linked to mitochondrial metabolic activity, being important to sustaining the cells' contact, and mediate a neuroprotective response [62], but also may be involved in neurodegenerative diseases, mediated by P2X7R and increase of IL-1β cytokine [73]. Interestingly, microglia express a bioenergetic profile related to both glycolytic and oxidative metabolism, which results in the production of ATP molecules. These quiescent microglia mainly use oxidative phosphorylation to produce ATP that can be changed when microglia turns activated, when they start to use aerobic glycolysis as the main source for ATP production [74], evidenced a fine tunned immune metabolic autocrine loop.

Other classes of receptors include LDL receptors (LDLR, ApoER2, and VLDL) and tyrosine kinase receptors (Tyro3, Axl, and Mertk), whose ligands are the annexin V binding proteins growth arrest-specific 6 (GAS6) and protein-S [75]. Fc receptors, C-type lectin receptors (CLEC7A), and complement receptors interact with their respective ligands to promote phagocytosis, endocytosis, and cytokine release in the brain, regulating homeostasis and pathogen response [76]. Markers as colony-stimulating factor 1 receptor (CSF1R), a tyrosine kinase receptor, and the chemokine receptor CX3CR1, are restricted to microglia and can be used as a tool for studying microglial biology [77,78]. The binding of CX3CR1 with its ligand, neuron-derived fractalkine (CX3CL1), greatly accounts for maintaining microglial resting state [79]. Moreover, after systemic inflammation by LPS injection, activated microglia increase FcγR mRNA levels with the increase of IgG in the cerebral vasculature. Also, it is demonstrated that microglia express either IFNRI or IL-17r, demonstrating a clear connection with adaptive immunity through either B and T lymphocytes to mediate neuroinflammation [80].

During brain development from embryo to adulthood, microglia also regulates neural precursors cells (NPCs) synapses formation [81]. During adulthood, microglia promotes phagocytosis of precursors cells, mostly in the subgranular and SVZ, whereas the absence of microglia increases the number of apoptotic NPCs, resulting in cognitive impairment [82]. During synaptic pruning, synaptic elements are phagocytosed by resident microglial cells in the thalamus, cerebral cortex, and hippocampus. To alert microglia that these cells need to be phagocytosed, "find-me" or "eat-me" signals are released, to complement protein C3, eliciting microglial migration and activation through CR3 [83,84]. Lastly, microglia is also important for neurite formation, myelination, and synchronized synaptic activity [85]. This evidences that microglia controls neuronal death and survival, clearance, and thus shaping the synapses and refining neural networks of the developing brain [86].

As for macrophages, microglia used to be simply defined by two distinct phenotypes: the classic "M1 profile" (pro-inflammatory) and the alternative "M2 profile" (anti-inflammatory). However, although this didactic segregation has been useful, due to the huge heterogeneity of microglia shown through single-cell RNA-seq analysis, these nomenclatures have been deeply questioned and suggested to be abandoned [87].

Not only microglial cells are capable of communicating with CNS resident cells, but also they may send and receive stimuli from all over the body. The BBB and autonomic nervous system also tightly control the passage of molecules and ions, such as cytokines and other molecules from the periphery to the brain and vice versa [88]. For example, microbiota secrets different mediators in the gut, that will act on CNS-infiltrating cells, establishing this communication, therefore promoting the gut-brain-axis communication [86,89]. RNA-seq analyses identified microglial impairment in germ-free mice (GF; i.e., lacking all commensal and pathogenic microbes) compared to specific pathogen-free animals. Indeed, GF mice mRNA profile indicates downregulation of activation markers (Mapk8, Fcgr2β, IL-1α, Ly86, Cd86, and Hif1α), accompanied by reduction in type I IFN-related genes (jak3 and stat1) suggesting a dysregulated microglia in the absence of microbiota [90] and evidencing the importance of environmental cues in shaping microglial biology.

Moreover, behavioral and environmental stimuli are important protagonists in orchestrating microglial functions. In fact, stress and neurological disorders in children can disrupt the normal development of microglia and contribute to pro-inflammatory responses throughout life [91]. Children neglected by the family have abnormal hippocampal development, synaptic maturation, and pruning, axonal growth, neurogenesis, and myelination. These modifications correlate to microglia alteration of morphology, density, and inflammatory gene expression [92]. Early live stress in Wistar rats increase glial cells numbers and the expression of IL-6 mRNA [93], IL-1β, and TNF-α in the hippocampus and amygdala [94]. This was consistent with an epigenetic reprogramming of microglia through DNA methylation [95].

Microglial cells are also in constant interaction with other glial cells, such as astrocytes. For instance, microglia secret TNF-α that leads to astrocytes activation [96]. Corroborating this, microglial depletion leads to a decrease in pro-inflammatory cytokine production by astrocytes. In addition, IL-1β-producing microglia stimulates the expression of tissue inhibitors of matrix metalloproteinases 1 (TIMP-1), resulting in inhibition of proteolysis and tissue damage induced by activated microglia [15,97].

Concerning BAMs and their location, it is expected that their functions also present some peculiarities. pvMΦ perform phagocytoses, lymphatic clearance [98], antigen presentation [99], as well as BBB maintenance [100]. In the context of BBB integrity, macrophages act in conjunction with pericytes in the early brain to establish vascular anastomosis and blood flow and share similar markers with macrophages in subventricular vascular plexus (SVP) [101]. Antigen presentation is related to CNS lymphatic pathways, responsible for draining CSF, directly communicating BAMs, and patrolling or meningeal resident T cells [102,103]. Thereby, MHC-II$^+$ macrophages in dura mater have been identified as functional APCs [103].

Dural macrophages may be infected by viruses, increasing the levels of IFN-β and TNF-α in the meninges, and thus upregulating genes associated with immune functions, such as antiviral activity (Interferon Stimulated Genes), phagocytosis, endocytosis, and antigen-presentation [104].

Moreover, the single-cell atlas of mouse brain macrophages identified that mostly genes of BAMs are related to cholesterol and lipid metabolism, immune response, and blood vessel development [105]. These macrophages located in the apical surface of the choroid plexus epithelium present transcriptional signatures of lipid metabolism and phagocytosis (Cst7, Apoe, Clec7a, and Lpl), that when compared to microglia, shows a profile very similar to disease-associated microglia (DAM).

Concluding, microglia, and macrophages in the CNS have many important biological functions, that may correlate either with homeostasis, as neurodevelopment and tissue repair, as well in the control of invading pathogens. Thus, it is expected that these cells converge with many different molecular mechanisms, as contact-dependent phagocytosis, or the secretion and capture of soluble mediators, as neurotransmitters, cytokines, and chemokines. The intrinsic contact with resident cells in CNS, such as neurons and astrocytes, allows microglia to regulate the biology of these cell populations, greatly impacting their function. Lastly, these cells under different circumstances, may either promote or dampen disease progression, depending on the context and modifications present in the environment.

Macrophages in CNS disorders

As discussed, tissue-resident parenchymal and non-parenchymal macrophages help during many brain functions, either during physiological processes or disease. Further on, we will briefly discuss the role of macrophage subsets during aging and common CNS-related disorders.

Aging

Aging of the CNS mostly results from loss of both gray and white matter with consequent cognitive decline. Morphological evidence of structural deterioration of microglia and macrophages in the brain indicates that these cells are also submitted to senescence [106,107]. Aged microglia have been characterized with a dystrophic morphology, including deramification, shortening, gnarling, spheroid shaping, and cytoplasmic fragmentation with an increase in pro-inflammatory cytokine signature, ROS production, and reduced migration and phagocytosis [108,109].

Aging brain experiences significant metabolic changes, with glucose and lipids metabolism are impaired. Conversely, reduced glucose utilization in the temporal and parietal lobes correlates to cognitive impairment [110,111]. Additionally, the neuronal glucose transporter GLUT3 is vulnerable to oxidative stress and inflammation [112]. Neuronal glucose metabolism is compromised during aging, the metabolism of ketone bodies β-hydroxybutyrate and acetoacetate is maintained, which may be detrimental [113]. Interestingly, in aged microglia and macrophages, prostaglandin E2 and EP2 receptor signaling promote the sequestration of glucose into glycogen, reducing glucose flux and mitochondrial respiration resulting in a cognitive deficit. The blockage or inhibition of EP2 signaling restores cellular bioenergetics as well as hippocampal synaptic plasticity, spatial memory, and cognition [114].

RNA sequencing revealed that aged microglia display decreased expression of genes associated with endogenous ligand recognition and up-regulated genes associated with microbe recognition and host defence. To counterbalance these insults, aging was associated with an overall increase in the expression of microglial genes involved in neuroprotection [115]. Markers usually expressed by activated microglia are elevated in aged microglia, including MHC-II, CD11b, CD14, CD68, and TLRs [116]. CX3CR1, an important marker in suppressing excessive microglia activation, and CD47, with function in Alzheimer's disease, are down-regulated in aged microglia [117,118].

Fe^{2+} serves as a catalytic cofactor in a variety of biochemical reactions, especially in oxidative stress. This may become a threat when promoting exacerbated production of hydroxyl radicals, which induce intracellular pro-oxidant conditions. Conversely, iron deposits in the brain increase with age mainly in the basal ganglia of the putamen, globus pallidus, and substantia nigra [119–121]. Aged dystrophic microglia were found to store iron through elevated levels of ferritin, and this storage and metabolism have contributed to microglia degeneration through increased oxidative stress [122].

Hormones can influence microglial function by showing antioxidant properties. Hormone decline during aging or menopause is associated with mitochondrial dysfunction, neuroinflammation, synaptic and cognitive impairment, favoring the development of age-related disorders [123]. On the other hand, hormones may have dimorphic effects on regulating neuroimmune functions. Whereas oestradiol treatment in male microglia results in an anti-inflammatory effect, inhibiting phagocytosis and decreasing ROS production, iNOS expression, and PGE2 release, female microglia display increased IL-1β expression [124].

Microglia senescence greatly reflects in changes of NF-κB signaling pathway activation. It is evidenced in apro-inflammatory skewed polarization, with increased production of complement molecules, ROS production, oxidative stress, and lipid peroxidation. Consequently, it may cause mitochondrial DNA damage as well as telomere shortening and decrease in telomerase activity [125–128]. Additionally, aged microglial cells have impairment in the ability to clear misfolded proteins and reduced expression of genes related to the actin cytoskeleton and cell surface sensor receptors [129,130]. Aged and chronic inflammation of microglia and macrophages are triggers to neurodegenerative diseases development.

Alzheimer's disease

Alzheimer's disease (AD) is the most common cause of dementia and the sixth cause of death in the United States and the fifth among Americans older than 65 years [131]. Amyloid-β (Aβ) accumulation and neurofibrillary pathology formed by the aggregation of tau protein are central events in the pathogenesis of AD. The cleavage of amyloid precursor protein generates Aβ peptides, which form soluble Aβ oligomers, polymers, and insoluble amyloid plaques. In this context, microglia activation has a pivotal function on AD development and many mechanisms have been described [132,133].

Physiologically, tau protein maintains DNA structure and stabilizes microtubules to regulate axonal transport. Under pathological conditions, modification of tau protein leads to its detachment from microtubules, resulting in synaptic loss, neuronal dysfunction, and tau aggregation [134,135]. Compared with Aβ deposition, tau accumulation in the cortex occurs later and correlates with cognitive decline in patients with AD [136]. The colocalization of reactive microglia with Aβ plaques and then the emergence of neurofibrillary tangles raised the amyloid-cascade inflammation hypothesis, suggesting that sequential events occur, in which microglial activation acts as a bridge between Aβ plaques formation and tau aggregation [137–139]. This was supported by the chronic activation of microglia and secretion of cytokines and reactive molecules that exacerbated plaque pathology as well as enhanced the hyperphosphorylation of tau protein and subsequently the development of neurofibrillary tangle [139].

Plenty of receptors expressed in microglia surface cooperate in the recognition and clearance of Aβ, leading to intracellular activation, cytokine production, and phagocytic alterations. Among them, TLR-2 interaction with Aβ triggers microglia activation with induction of, IL-1β, IL-8, and TNF-α. The inhibition of TLR-2 shifts M1- to M2-polarization, which can be beneficial in AD progression [140]. The interaction with scavenger receptors, CD36, α6β1 integrin, and CD47 is required for induction of Aβ phagocytosis [141]. The complement system and Fc receptors also regulate neuroinflammation by complement activation, opsonization, and recognition by CR1 and, subsequently, Aβ phagocytosis. Conversely, polymorphisms in the *CR1* locus are considered a risk factor for the development of AD and correlate with increased levels of Aβ in the CSF [142].

Triggering receptor expressed on myeloid cells 2 (TREM2) is a broad sensor for lipids and lipoproteins (e.g., APOE) that triggers intracellular protein tyrosine phosphorylation [143]. TREM2 senses damage-associated lipids and fibrillar Aβ complex, facilitating phagocytosis and limiting diffusion of Aβ plaques throughout the brain [143–145]. Besides that, TREM2 signaling promotes microglia survival, proliferation, and the maintenance of phagocytosis [143]. Conversely, the human polymorphism *R47H* of TREM2correlates with increased susceptibility to develop AD by impairing lipid detection and uptake [143]. TREM2 activation also induces ROS-dependent calcium influx for NLRP3 assembly and consequently activation of caspase-1 and maturation of IL-1β, resulting in enhanced Aβ clearance [146]. TREM2 seems to have both beneficial and detrimental functions on microglia during AD.

Although the beneficial effect of NLRP3, its deficiency skewed microglial cells fromM1- to M2-phenotype resulting in decreased deposition of Aβ in AD [147,148]. Moreover, pro-inflammatory cytokines produced by microglia upregulate, via NF-κB, the production of β-secretase (BACE) and consequently increase in Aβcleavage [149,150]. Microglia also releases C-terminal caspase recruitment domain (ASC) specks, an adaptor protein of NLRP3, which binds to Aβ and induces the increase of Aβ oligomers and aggregates, acting as an inflammation-driven seed for Aβ [151]. Additionally, microglia actively spreads tau through phagocytosis and exosome secretion, providing a propagation and trans-synaptic transmission of tau [152].

Recent studies using single-cell RNA sequencing analysis characterized immune cells at different stages of AD in an experimental model. Microglial early-response showed enrichment of cell cycle genes involved in DNA replication and repair indicating that microglia expansion occurs at the early stages of AD [153]. Interestingly, two distinct DAMs were found in AD and were absent in normal brains. These DAMs showed reduced expression of P2ry12/P2ry13, *Cx3cr1*, and *Tmem119*, and enrichment of genes known as common risk factors for AD, including *Apoe*, *Ctsd*, *Lpl*, *Tyrobp*, and *Trem2* [154].

Progression of AD causes inflammation and recruitment of monocytes to the CNS. Although microglia can phagocyte and degrade Aβ aggregates, macrophages degrade fibrillary Aβ deposition more efficiently [155]. Patrolling monocytes have been found crawling on Aβ-positive veins to internalize it, and then circulate back to the bloodstream [156]. Also, pvMΦ are cells enriched with scavenger receptors and CCR2 and have been pivotal in the clearance of these Aβ plaques, as the depletion of this population drastically impaired Aβ clearance and amplified Aβ vascular deposition [157]. Interestingly, the recognition of Aβ by pvMΦ can be highly pathogenic due to neurovascular damage through the activation of CD36 and NADPH oxidase leading to oxidative stress [158].

Multiple sclerosis

Multiple Sclerosis (MS) is a chronic autoimmune neuroinflammatory and neurodegenerative disease with great emotional and functional impact in people's lives [159]. MS is characterized by episodes of demyelination caused by self-reactive T responses against myelin-derived protein antigens, mostly played by cytotoxic CD8 lymphocytes [160]. After clonal expansion and differentiation, lymphocytes infiltrate the CNS and secrete many mediators that initiate the local inflammatory process by activating glial cells and, subsequently, recruiting infiltrating cells from the periphery [161–163].

In the steady state, microglial cells and CNS-associated macrophages are in a quiescent state and present M2-phenotype, associated with tissue repair and immune surveillance [31]. As microglial cells actively monitor the brain parenchyma and spinal cord, they present low expression of CD80, CD86, CD40, CD45, and MHC-II molecules, and release several anti-inflammatory cytokines, such as IL-4, IL-10, IL-13, IL-33, and TGF-β [164,165]. Also, its activation is related to structural modifications, once it was revealed that ramified macrophages-like turn to amoeboid morphology [166].

On the other hand, CNS resident cells are also widely involved in the pathogenesis of MS and EAE, the murine model for MS. In fact, analysis of active lesions from MS patients showed that 45% of macrophages-like were microglia with a significant reduction of P2RY12 expression, a homeostatic microglial marker [40]. In the early stages, microglial cells assume a pro-inflammatory phenotype, whereas in later stages the phenotype can be skewed to anti-inflammatory. In inactive lesions from relapsing-remitting patients, the density of microglia is reduced, with increased P2RY12 expression. Curiously, during EAE other homeostatic markers, such as CXCR1, MertK and Siglec-H were reduced, whereas CD44, CD86, PDL1, CD11c, MHC-II, and Sca-1 were increased. This was associated with a decreased number of BAMs and increased infiltrating monocytes-derived cells (MdCs), that expressed activation markers, like CD44, MHC-II, Sca-1, PDL1, CD117, and CD11c [4].

To better understand the role of microglial cells during EAE, it was created *CD11b-HVTK* transgenic mice, which express herpes simplex thymidine kinase in microglia, to further deplete these cells after ganciclovir treatment [167]. It was shown that microglia paralysis leads to an ameliorating EAE, associated with a decreased TNF-α, NO, and MIP-1 β production, leading to reduced infiltration of $CD4^+$, $CD8^+$, and $CD11b^+$ cells. Interestingly, it has been shown that microglial cells may be activated even before the onset of EAE symptoms [168]. In the acute phase, microglia proliferate, increase the expression of CD80, CD86, and MHC-II, and migrate to the injury site, favoring the antigen presentation and consequently activating encephalitogenic T lymphocytes [169,170]. As microglia assume an M1-phenotype, characterized by high ROS and inflammatory cytokines production, it is clear that they assist in the oxidative damage of CNS cells, the opening of BBB, and leukocyte extravasation, as well as Th1 polarization [165,171].

Cytokines as GM-CSF and IL-17 A are extremely important for MS and EAE pathogenesis [172–175]. CNS resident cells, such as astrocytes and microglia, express IL-17R constitutively [176]. It was shown that these cytokines lead to overactivation of astrocytes with increased metalloproteinase-9 (MMP-9), IL-6, CXCL2, and CXCL9 secretion, associated with augmented ICAM-1 and VCAM [177]. Moreover, GM-CSF induces IL-23 production by macrophages, which increases Th17 cells polarization and pathogenicity [178] and consequently more GM-CSF production [179]. Thus, there is a recruitment of different immune cells, such as infiltrating macrophages into the CNS that sustain positive feedback of inflammatory process, promoting gradual myelin sheath degradation and axonal degeneration.

Microglial cells may also have an ambiguous role during astrocyte activation. Microglia produces TGF-α that limits pathogenic astrocytes via ErbB1, whereas VEGF-β-producing microglia triggers FLT-1 in astrocytes worsening EAE [180]. The expression of TGF-α and VEGF-β in microglial cells correlates with multiple sclerosis progression. Surprisingly, both are controlled by dietary tryptophan metabolites produced by the commensal flora.

Curiously, myelin internalization induces an M1-phenotype in microglia and macrophages, characterized by increased TNF-α and nitric oxide (NO) production [181]. At the same time, these cells scavenge myelin debris and assume immunosuppressive phenotype increasing ARG1, CD206, and insulin-like growth factor-1 (IGF-1), which promotes oligodendrocytes differentiation and CNS repair [182–184]. Also, during severe EAE, the activation of microglia is inversely proportional to miR-124 expression, a brain-specific miRNA. Interestingly, the exogenous administration of miR-124 ameliorated EAE promoting microglial quiescence via C/EBP-α-PU.1 pathway and inducing an M2-phenotype [185,186]. Therefore, M1/M2 balance has a huge relevance during MS and EAE progression.

Lastly, the role of pvMΦ during EAE has already been addressed. pvMΦ express high levels of lactate dehydrogenase A, an enzyme that converts pyruvate to lactate. Also, these macrophages express monocarboxylate transporter-4 (MCT-4) and CD147. Together, these molecules are responsible to export lactate from glycolytic cells. The inhibition of MCT-4 and CD147 interaction by α-cyano-4-hydroxy-cinnamic acid (CHCA) showed to decrease lactate production by macrophages, impairing their migration to CNS, evidencing an important role for glycolytic metabolism of infiltrating macrophages during severe EAE [187].

Brain cancer

Brain tumors are a benign or malignant mass of abnormal cells that can originate from brain tissue, as meningiomas and gliomas, specifically glioblastoma [188], or disseminate into the brain, mostly from systemic metastatic cancers. Tumors are composed of a plethora of cells, including non-tumoral cells, as resident and bone marrow-derived macrophages and also lymphocytes. Macrophages represent the most abundant stromal cell in many cancers and comprise up to 50% of the cellular content of the tumor [189,190].

Tumor-associated macrophages (TAM) are mostly monocyte-derived (around 85%), whereas just 15% are microglial cells that predominate in the peritumoral areas [191]. Some factors are reported to be crucial for the development, seeding, and characteristics of TAM. Tumor cells can produce stromal-derived factor-1 (SDF-1) that acts on CXCR4 and is involved in the accumulation of bone-marrow-derived TAM within hypoxic areas of the tumor to promote angiogenesis [192]. CCL2 can also be released by glioma cells, acting on CCR2-expressing microglia, triggering IL-6 production and promoting invasiveness [193]. The interaction of glioma with TLR-2 in TAM, induces membrane-type 1-matrix metalloproteinase expression, promoting growth and degradation of extracellular matrix for invasiveness [194].

TAM within glioma normally expresses M2 markers, as CD163 and CD204. Their abundance positively correlates with the pathological grade, and M-CSF seems to have an important role in the development of glioma-associated M2 TAM [195]. Blood monocytes may support immunosuppression and promote the growth of malignant cells by releasing high amounts of EGF [196]. Release of stress-inducible protein 1 (STI1) and TGF-β by microglia increases the proliferation and migration of glioblastoma cells, whereas blockage of TGF-β signaling impairs glioma growth [197,198].

Using RNA sequencing, it was possible to evidence the heterogeneity of brain tumor cells. Analysis of glioma demonstrated a higher representation of macrophage programs (CD163, IFITM2, TGFB1) over microglia signatures (CX3CR1, P2RY12) [199]. Analysis of murine brain malignancy revealed distinct clusters of TAM populations, in which macrophages showed distinct transcriptional profiles and activation states from microglia. While TAM macrophages were enriched for chemokines genes involved in wound healing, including *Ccl22*, *Ccl17*, *Cxcl2*, *Cxcl3*, and *Cxcl16*, TAM microglia were enriched for expression of *Ccl4* and *Tnf* genes, associated with pro-inflammatory response [200].

Additionally, differences in TAM gene expression between tumor center and surrounding cells showed more pro-inflammatory markers expressed in the tumor periphery accompanied by the expression of microglial signature genes. At the same time, more anti-inflammatory and pro-angiogenic factors were expressed in the tumor core associated with macrophages gene signature [201]. Although TAM characteristics are variable and sometimes related to poor prognostic, some researchers demonstrated that TAM also has tumor-killing activities in some models [202,203]. The design of immunotherapies targeting macrophages in cancers is a potentially novel avenue in the treatment of incurable malignancies in the brain.

Stroke

Stroke is an acquired brain injury caused by a thrombotic or embolic event, cerebral hypoperfusion, venous thrombosis, and intracerebral hemorrhage due to trauma, inflammation, or intracerebral vascular malformations. There are two different types: ischemic, the most common type that is caused by lack of blood flow, and hemorrhagic, which leads to bleeding [204]. Over the years, it has been shown that neuroinflammation helps in the pathogenesis and progression of ischemic cerebral injury, and in this context, microglia and infiltrating macrophages have a pivotal role.

Following ischemic stroke, tissue damage is caused by cell death by necrosis, apoptosis, and autophagy, mainly in the brain tissue core. In the surrounding tissue, known as the penumbra, cell death is more controlled and less evident [205]. Consequently, DAMPS are released, which are recognized by microglia through PRRs. For example, the activation of TLR-2/4 culminates in NF-κB nuclear translocation, leading to pro-inflammatory cytokines secretion [206]. In addition, NLRP3 and NLRC4 inflammasome activation lead to IL-1β and IL-18 production, and further apoptosis and pyroptosis, greatly contributing to tissue damage [207,208].

Microglial cells produce mediators that also cause the positive feedback of the inflammatory process, such as TAK1 (transforming growth factor beta-activated kinase 1), NADPH oxidase, and iNOS (induced nitric oxide synthase). TAK1 inhibition in microglia decreased infarct volume and neuronal damage, at the same time that ameliorated neurological deficits and decreased IL-1β and TNF-α gene expression [209–211]. NADPH oxidase is mainly expressed by microglia in the brain and is the major source of ROS, leading to microglial activation and proliferation, and contributing to neuronal apoptosis [212,213]. During ischemic stroke, it has been observed iNOS upregulation and further TNF-α production by microglia. NADPH and iNOS seem to have synergistic effects, as both can cause neuronal damage during ischemic stroke [214,215].

Microglial cells may also have a neuroprotective effect during ischemic stroke. Purinergic receptors are triggered by purine and pyrimidine nucleotides, being ATP the most common ligand [216]. They have a neuroprotective role in microglia during ischemic stroke injury, once P2X4R mediates calcium entry leading to membrane potential changes and inducing anti-inflammatory phenotype [217]. Also, it is known that macrophages express P2X7R and its absence increases brain edema [218]. Moreover, muscarinic acetylcholine receptors (MR3) also play a neuroprotective role. The lack of MR3 on microglia increased neurological damage after ischemic stroke [219]. Using an oxygen-glucose-deprivation (OGD) ischemic cell model, it has been demonstrated that vinpocetine treatment, a phosphodiesterase 1b (PDE1B) inhibitor that hydrolysis cAMP to AMP, promotes microglial polarization to M2-phenotype, and the release of exosomes reducing neuronal damage. Also, vinpocetine-treated microglial exosomes regulate cortical neuronal survival under cerebral ischemic conditions in vivo [220]. Altogether, these evidence point to an ambiguous role for microglial cells during ischemic stroke.

Other brain's disease

Over the years, several psychiatric disorders are being associated with inflammation, including depression [221], bipolar disorder [222], and schizophrenia [223]. In this context, macrophages and microglia are very important cells [224]. Interestingly, an increase of microglial activation and macrophage recruitment was observed in the dorsal anterior cingulate matter in depressive patients [225] and in ventral prefrontal white matter in suicidal individuals [226]. Also, IL-1β, IL-6, and TNF-α production by microglia inhibited neurogenesis in depressive young rats [227]. These pro-inflammatory cytokines decrease glutamate transporter expression inducing glutamate release that contributes to neuronal damage during depression [228]. Thus, ketamine, an

N-methyl-D-aspartate (NMDA) antagonist with antidepressant effects, also showed anti-inflammatory capacity by decreasing TNF-α and IL-6 gene expression in activated macrophages [229].

Curiously, stressed Sprague-Dawley rats presented microglial activation and hyper-ramification in the prefrontal cortex [230], whereas minocycline antibiotic-treated rats decreased its activation [231]. Also, acute stress promotes morphological microglial activation in different parts of the brain, including the thalamus, hypothalamus, hippocampus, substantia nigra, and central gray [232]. Another mechanism demonstrated was that IL-8 production by microglia in the brain attracts neutrophils, increasing the pro-inflammatory immune response, which reduces the bioavailability of neurotransmitters [233] and affects neurogenesis in different neurological diseases, such as depression and schizophrenia [234,235].

In addition, chronic cerebral hypoperfusion caused by low blood pressure, elevated inflammatory response, and oxidative stress, leads to brain's white matter damage and consequently vascular impaired cognitive function, once activated microglia produces TNF-α, IL-1β, IL-6, MCP-1 and complement components [236–239]. Recent studies showed the ability of nonsteroidal anti-inflammatory drugs (NSAID) as a good candidate for age-associated mild cognitive impairment and dementia, once reduces glycolysis and polarizes macrophages to anti-inflammatory phenotype restoring memory capability [240].

Concluding remarks

Microglia and BAMs are critical components in the establishment and maintenance of the CNS. They not only help on damaged or unnecessary neurons and synapses during neurodevelopment but are also the primary active immune defense against injury and infection of the brain. Here, we summarized the origin, phenotypes, and functions of these phagocytic cells in health and disease, pointing to some of their functions as protective or detrimental for brain disease (Fig. 1). Microglia and BAMs functions may converge in a functional unit, acquiring their respective signatures and boosting functions in different phases of the disease. However, in some CNS disorders, an important question is raised whether macrophages dysfunction is the primary-driven or a consequence. A better molecular understanding that regulates the homeostasis and function of these macrophages would be targeted for effective therapeutic strategies.

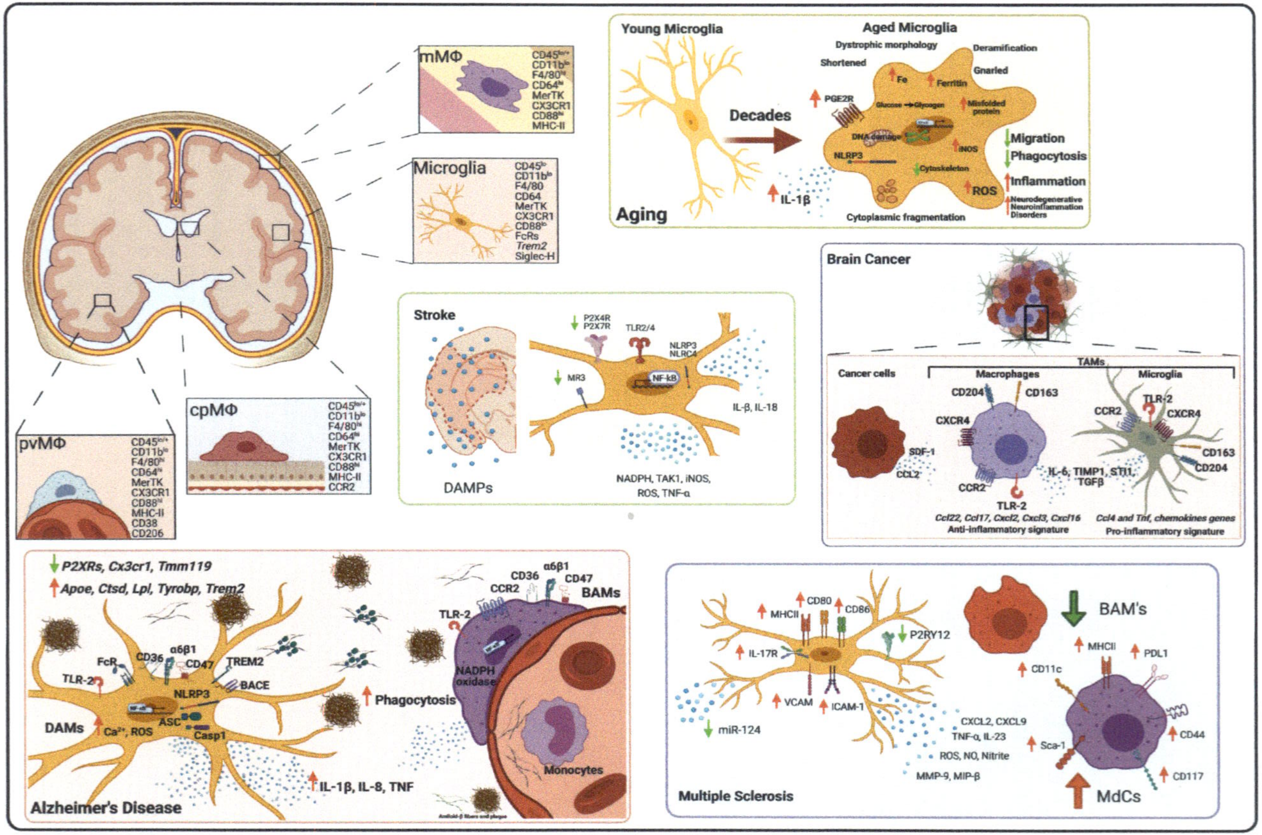

Fig. 1 See the legend in facing page

Fig. 1 Microglia and border-associated macrophages in homeostasis, aging and CNS-related disorders. (A) The brain is populated with several different macrophage subsets. Microglial cells are localized in the brain parenchyma and express peculiar markers that distinguish them from macrophages, whereas BAMs are located in blood-brain or CSF-brain interfaces. Meningeal macrophages (mMΦ) are situated in the dura mater, whereas perivascular (pvMΦ) and choroid plexus (cpMΦ) macrophages are adjacent to the vasculature and in the choroid plexus, respectively. During homeostasis, these cells express characteristic markers that significantly change during tissue repair or brain disease. (B) Aging has direct effects on microglial phenotype, shape, and function. For example, secreting pro-inflammatory cytokines and chemokines, changing metabolic status, and accumulating high concentrations of iron and ferritin intracellularly, compromising mitochondrial DNA. Additionally, there is an increase of inflammasome assembling with IL-1β release, a decrease of migration, and phagocytosis promoting the development of neurodegenerative and/or neuroinflammatory disease. (C) Stroke is an acquired brain injury caused by ischemic or hemorrhagic bleeding in the brain, which leads to the release of DAMPs that are recognized by microglial cells for further pro-inflammatory cytokines production. (D) Multiple sclerosis, is an autoimmune and neurodegenerative disorder caused by self-reactive T cell responses against myelin-derived protein antigens. Microglia and MdCs have increased expression of activation markers and production of pro-inflammatory mediators. (E) Brain tumors are composed of many different cells, including TAMs. High numbers of macrophages with anti-inflammatory signatures can be found in the tumor core, whereas microglia with pro-inflammatory signatures are found in the periphery. (F) Alzheimer's disease is a neurodegenerative disease caused by amyloid-β and tau protein accumulation in the brain. Aβ can be recognized by a range of receptors expressed by microglia, BAMs, and monocytes, which phagocyte Aβ proteins induce inflammasome activation and release of inflammatory cytokine and ASC specks.

References

[1] Réu P, et al. The lifespan and turnover of microglia in the human brain. Cell Rep 2017;20:779–84.
[2] Füger P, et al. Microglia turnover with aging and in an Alzheimer's model via long-term in vivo single-cell imaging. Nat Neurosci 2017;20:1371–6.
[3] Bechmann I, et al. Turnover of rat brain perivascular cells. Exp Neurol 2001;168:242–9.
[4] Mrdjen D, et al. High-dimensional single-cell mapping of central nervous system immune cells reveals distinct myeloid subsets in health, aging, and disease. Immunity 2018;48:380–395.e6.
[5] Stratoulias V, Venero JL, Tremblay M-È, Joseph B. Microglial subtypes: diversity within the microglial community. EMBO J 2019;38(17):1–18. https://doi.org/10.15252/embj.2019101997.
[6] Goldmann T, et al. Origin, fate and dynamics of macrophages at central nervous system interfaces. Nat Immunol 2016;17:797–805.
[7] Ginhoux F, Jung S. Monocytes and macrophages: developmental pathways and tissue homeostasis. Nat Rev Immunol 2014;14:392–404.
[8] Xu J, et al. Temporal-spatial resolution fate mapping reveals distinct origins for embryonic and adult microglia in zebrafish. Dev Cell 2015;34:632–41.
[9] Ginhoux F, et al. Fate mapping analysis reveals that adult microglia derive from primitive macrophages. Science 2010. https://doi.org/10.1126/science.1194637.
[10] Bennett ML, Bennett FC. The influence of environment and origin on brain resident macrophages and implications for therapy. Nat Neurosci 2020;23:157–66.
[11] Bennett FC, et al. A combination of ontogeny and CNS environment establishes microglial identity. Neuron 2018;98:1170–1183.e8.
[12] Kierdorf K, et al. Microglia emerge from erythromyeloid precursors via Pu. 1-and Irf8-dependent pathways. Nat Neurosci 2013;16:273–80.
[13] Liddelow SA. Development of the choroid plexus and blood-CSF barrier. Front Neurosci 2015;9:1–13.
[14] Yi YT, Li Y. Microglial regional heterogeneity and its role in the brain. Mol Psychiatry 2020;25:351–67. https://doi.org/10.1038/s41380-019-0609-8.
[15] Colonna M, Butovsky O. Microglia function in the central nervous system during health and neurodegeneration. Annu Rev Immunol 2017;35:441–68.
[16] Lawson LJ, Perry VH, Dri P, Gordon S. Heterogeneity in the distribution and morphology of microglia in the normal adult mouse brain. Neuroscience 1990;39:151–70.
[17] Savchenko VL, Nikonenko IR, Skibo GG, McKanna JA. Distribution of microglia and astrocytes in different regions of the normal adult rat brain. Neurophysiology 1997;29:343–51.
[18] Stowell RD, et al. Cerebellar microglia are dynamically unique and survey Purkinje neurons in vivo. Dev Neurobiol 2018;78:627–44.
[19] Vela JM, Dalmau I, González B, Castellano B. Morphology and distribution of microglial cells in the young and adult mouse cerebellum. J Comp Neurol 1995;361:602–16.
[20] Jinno S, Fleischer F, Eckel S, Schmidt V, Kosaka T. Spatial arrangement of microglia in the mouse hippocampus: a stereological study in comparison with astrocytes. Glia 2007;55:1334–47.
[21] Grabert K, et al. Microglial brain regionâ'dependent diversity and selective regional sensitivities to aging. Nat Neurosci 2016;19:504–16.
[22] del Fernández-Arjona MM, Grondona JM, Granados-Durán P, Fernández-Llebrez P, López-Ávalos MD. Microglia morphological categorization in a rat model of neuroinflammation by hierarchical cluster and principal components analysis. Front Cell Neurosci 2017;11:1–22.
[23] Abbott NJ, Rönnbäck L, Hansson E. Astrocyte–endothelial interactions at the blood–brain barrier. Nat Rev Neurosci 2006;7:41–53.

[24] Baalman K, et al. Axon initial segment–associated microglia. J Neurosci 2015;35:2283–92.
[25] Ajami B, Bennett JL, Krieger C, McNagny KM, Rossi FMV. Infiltrating monocytes trigger EAE progression, but do not contribute to the resident microglia pool. Nat Neurosci 2011;14:1142–50.
[26] Ribeiro Xavier AL, Kress BT, Goldman SA, De Lacerda Menezes JR, Nedergaard M. A distinct population of microglia supports adult neurogenesis in the subventricular zone. J Neurosci 2015;35:11848–61.
[27] Hammond TR, et al. Single-cell RNA sequencing of microglia throughout the mouse lifespan and in the injured brain reveals complex cell-state changes. Immunity 2019;50:253–271.e6.
[28] Utz SG, et al. Early fate defines microglia and non-parenchymal brain macrophage development. Cell 2020;181:557–573.e18.
[29] Zeisel A, et al. Cell types in the mouse cortex and hippocampus revealed by single-cell RNA-seq. Science 2015;347:1138–42.
[30] Nayak D, Zinselmeyer BH, Corps KN, McGavern DB. In vivo dynamics of innate immune sentinels in the CNS. IntraVital 2012;1:95–106.
[31] Nimmerjahn A, Kirchhoff F, Helmchen F. Neuroscience: resting microglial cells are highly dynamic surveillants of brain parenchyma in vivo. Science 2005;308:1314–8.
[32] Moeller HV, Peay KG, Fukami T. Activation of pial and dural macrophages and dendritic cells by CSD (67 chrs). FEMS Microbiol Ecol 2014;87:797–806.
[33] Russo MV, Latour LL, McGavern DB. Distinct myeloid cell subsets promote meningeal remodeling and vascular repair after mild traumatic brain injury article. Nat Immunol 2018;19:442–52.
[34] Joana M, Jordão C, Prinz M. Macrophages at CNS interfaces: ontogeny and function in health and disease. Nat Rev Neurosci 2019;20.
[35] Harrison JK, et al. Role for neuronally derived fractalkine in mediating interactions between neurons and CX3CR1-expressing microglia. Proc Natl Acad Sci USA 1998;95:10896–901.
[36] Neumann H, Wekerle H. Brain microglia: watchdogs with pedigree. Nat Neurosci 2013;16:253–5.
[37] Butovsky O, et al. Identification of a unique TGF-β-dependent molecular and functional signature in microglia. Nat Neurosci 2014;17:131–43.
[38] Satoh J-I, et al. TMEM119 marks a subset of microglia in the human brain. Neuropathology 2016;36:39–49.
[39] Zhu C, et al. Expression site of P2RY12 in residential microglial cells in astrocytomas correlates with M1 and M2 marker expression and tumor grade. Acta Neuropathol Commun 2017;5:1–12.
[40] Zrzavy T, et al. Loss of 'homeostatic' microglia and patterns of their activation in active multiple sclerosis. Brain 2017;140:1900–13.
[41] Butovsky O, Weiner HL. Microglial signatures and their role in health and disease. Nat Rev Neurosci 2018;19:622–35.
[42] Bennett ML, et al. New tools for studying microglia in the mouse and human CNS. Proc Natl Acad Sci USA 2016;113:E1738–46.
[43] Ginhoux F, Prinz M. Origin of microglia: current concepts and past controversies. Cold Spring Harb Perspect Biol 2015;7:1–15.
[44] Sousa C, Biber K, Michelucci A. Cellular and molecular characterization of microglia: a unique immune cell population. Front Immunol 2017;8.
[45] Masuda T, et al. Spatial and temporal heterogeneity of mouse and human microglia at single-cell resolution. Nature 2019. https://doi.org/10.1038/s41586-019-0924-x.
[46] Hammond TR, Marsh SE, Stevens B, et al. Immune signaling in neurodegeneration. Immunity 2019;50(4):955–74.

[47] Jordão MJC, et al. Neuroimmunology: single-cell profiling identifies myeloid cell subsets with distinct fates during neuroinflammation. Science 2019. https://doi.org/10.1126/science.aat7554.
[48] Galea I, et al. Mannose receptor expression specifically reveals perivascular macrophages in normal, injured, and diseased mouse brain. Glia 2005;49:375–84.
[49] Kim WK, et al. CD163 identifies perivascular macrophages in normal and viral encephalitic brains and potential precursors to perivascular macrophages in blood. Am J Pathol 2006;168:822–34.
[50] Kim JV, Kang SS, Dustin ML, McGavern DB. Myelomonocytic cell recruitment causes fatal CNS vascular injury during acute viral meningitis. Nature 2009;457:191–5.
[51] Norris GT, Kipnis J. Immune cells and CNS physiology: microglia and beyond. J Exp Med 2019;216:60–70.
[52] Lapenna A, De Palma M, Lewis CE. Perivascular macrophages in health and disease. Nat Rev Immunol 2018;18:689–702.
[53] Batchelor PE, et al. Activated macrophages and microglia induce dopaminergic sprouting in the injured striatum and express brain-derived neurotrophic factor and glial cell line-derived neurotrophic factor. J Neurosci 1999;19:1708–16.
[54] Barde Y. Microglia-derived nerve growth factor causes cell death in the developing retina. Neuron 1998;20:35–41.
[55] Yin J, Valin KL, Dixon ML, Leavenworth JW. The role of microglia and macrophages in CNS homeostasis, autoimmunity, and cancer. J Immunol Res 2017;2017.
[56] Patterson SL, Life B, Building S. Immune dysregulation and cognitive vulnerability in the aging brain: interactions of microglia, IL-1β, BDNF and synaptic plasticity. Neuropharmacology 2016;96:11–8.
[57] Parkhurst CN, et al. Microglia promote learning-dependent synapse formation through BDNF. Cell 2014;155:1596–609.
[58] Wake H, Moorhouse AJ, Jinno S, Kohsaka S, Nabekura J. Resting microglia directly monitor the functional state of synapses in vivo and determine the fate of ischemic terminals. J Neurosci 2009;29:3974–80.
[59] Thomas WE. Characterization of the dynamic nature of microglial cells. Brain Res Bull 1990;25:351–4.
[60] Booth PL, Thomas WE. Evidence for motility and pinocytosis in ramified microglia in tissue culture. Brain Res 1991;548:163–71.
[61] Lowery RL, Majewska AK. Microglial interactions with synapses are modulated by visual experience. PLoS One 2010;8:e1000527.
[62] Zedadra O, et al. Microglia monitor and protect neuronal function via specialized somatic purinergic junctions. Sustainability 2019;11:1–14.
[63] Hu Y, et al. mTOR-mediated metabolic reprogramming shapes distinct microglia functions in response to lipopolysaccharide and ATP. Glia 2020;68:1031–45.
[64] Mariani MM, Kielian T. Microglia in infectious diseases of the central nervous system. J NeuroImmune Pharmacol 2009;4:448–61.
[65] Janova H, et al. CD14 is a key organizer of microglial responses to CNS infection and injury. Glia 2016;64:635–49.
[66] Heneka MT, McManus RM, Latz E. Inflammasome signalling in brain function and neurodegenerative disease. Nat Rev Neurosci 2018;19:610–21.
[67] Dinarello CA. Interleukin-l. Rev Infect Dis 1984;6:51–95.
[68] Shaftel SS, Griffin WST, Kerry KM. The role of interleukin-1 in neuroinflammation and Alzheimer disease: an evolving perspective. J Neuroinflammation 2008;5:1–12.
[69] Shaftel SS, et al. Chronic interleukin-1β expression in mouse brain leads to leukocyte infiltration and neutrophil-independent blood-brain barrier permeability without overt neurodegeneration. J Neurosci 2007;27:9301–9.

[70] Anthony DC, Bolton SJ, Fearn S, Perry VH. Age-related effects of interleukin-1β on polymorphonuclear neutrophil-dependent increases in blood-brain barrier permeability in rats. Brain 1997;120:435–44.
[71] Calovi S, Mut-Arbona P, Sperlágh B. Microglia and the purinergic signaling system. Neuroscience 2019;405:137–47.
[72] Illes P, Norenberg W, Gebicke-Haerter PJ. Molecular mechanisms of microglial activation. B. Voltage- and purinoceptor-operated channels in microglia. Neurochem Int 1996;29:13–24.
[73] Facci L, Barbierato M, Zusso M, Skaper SD, Giusti P. Serum amyloid A primes microglia for ATP-dependent interleukin-1β release. J Neuroinflammation 2018;15:1–11.
[74] Yang G, Sau C, Lai W, Cichon J, Li W. Bioenergetic regulation of microglia. Glia 2015;344:1173–8.
[75] Kettenmann H, Hanisch U, Noda M, Verkhratsky A. Physiology of microglia. Physiol Rev 2021;461–553. https://doi.org/10.1152/physrev.00011.2010.
[76] Li Q, Barres BA. Microglia and macrophages in brain homeostasis and disease. Nat Rev Immunol 2018;18:225–42.
[77] Dai X, et al. Targeted disruption of the mouse colony-stimulating factor 1 receptor gene results in osteopetrosis, mononuclear phagocyte deficiency, increased primitive progenitor cell frequencies, and reproductive defects. Blood 2002;99:111–20.
[78] Gyoneva S, et al. Cx3cr1-deficient microglia exhibit a premature aging transcriptome. Life Sci Alliance 2019;2:1–16.
[79] Limatola C, Ransohoff RM, Harrison JK. Modulating neurotoxicity through CX-3CL1/CX3CR1 signaling. Front Cell Neurosci 2014;8:1–8.
[80] Lunnon K, et al. Systemic inflammation modulates fc receptor expression on microglia during chronic neurodegeneration. J Immunol 2011;186:7215–24.
[81] Galloway DA, Phillips AEM, Owen DRJ, Moore CS. Phagocytosis in the brain: homeostasis and disease. Front Immunol 2019;10:1–15.
[82] Wolf SA, Boddeke HWGM, Kettenmann H. Microglia in physiology and disease. Annu Rev Physiol 2017;79:619–43.
[83] Mosser CA, Baptista S, Arnoux I, Audinat E. Microglia in CNS development: shaping the brain for the future. Prog Neurobiol 2017;149–150:1–20.
[84] Allison RB, Stevens B. TGF-β signaling regulates neuronal C1q expression and developmental synaptic refinement. Nat Neurosci 2014;16:1773–82.
[85] Christopher AMLS. Microglia as dynamic cellular mediators of brain function. Physiol Behav 2016;176:100–6.
[86] Tay TL, Savage JC, Hui CW, Bisht K, Tremblay MÈ. Microglia across the lifespan: from origin to function in brain development, plasticity and cognition. J Physiol 2017;595:1929–45.
[87] Ransohoff RM. A polarizing question: do M1 and M2 microglia exist. Nat Neurosci 2016;19:987–91.
[88] Obermeier B, Daneman R, Ransohoff RM. Development, maintenance and disruption of the blood-brain barrier. Nat Med 2013;19:1584–96.
[89] Haq RA, Schlachetzki JCM, Glass CK, Mazmanian SK. Microbiome—microglia connections via the gut – brain axis. J Exp Med 2018;216:41–59.
[90] Erny D, et al. Host microbiota constantly control maturation and function of microglia in the CNS. Nat Neurosci 2015;18:965–77.
[91] Lenz KM, Nelson LH. Microglia and beyond: innate immune cells as regulators of brain development and behavioral function. Front Immunol 2018;9.
[92] Delpech JC, et al. Early life stress perturbs the maturation of microglia in the developing hippocampus. Brain Behav Immun 2016;57:79–93.
[93] Banqueri M, Méndez M, Gómez-Lázaro E, Arias JL. Early life stress by repeated maternal separation induces long-term neuroinflammatory response in glial cells of male rats. Stress 2019;22:563–70.

[94] Ferle V, et al. Synergistic effects of early life mild adversity and chronic social defeat on rat brain microglia and cytokines. Physiol Behav 2020;215:112791.
[95] Catale C, et al. Exposure to different early-life stress experiences results in differentially altered DNA methylation in the brain and immune system. Neurobiol Stress 2020;13:100249.
[96] Liddelow SA, et al. Neurotoxic reactive astrocytes are induced by activated microglia. Nature 2017;541:481–7.
[97] Jha MK, Jo M, Kim JH, Suk K. Microglia-astrocyte crosstalk: an intimate molecular conversation. Neuroscientist 2019;25:227–40.
[98] Carare RO, et al. Solutes, but not cells, drain from the brain parenchyma along basement membranes of capillaries and arteries: significance for cerebral amyloid angiopathy and neuroimmunology. Neuropathol Appl Neurobiol 2008;34:131–44.
[99] Hickey WF, Kimura H. Perivascular microglial cells of the CNS are bone marrow-derived and present antigen in vivo. Science 1988;239:290–2.
[100] Willis CL, Garwood CJ, Ray DE. A size selective vascular barrier in the rat area postrema formed by perivascular macrophages and the extracellular matrix. Neuroscience 2007;150:498–509.
[101] Yamamoto S, et al. A subset of cerebrovascular pericytes originates from mature macrophages in the very early phase of vascular development in CNS. Sci Rep 2017;7:1–16.
[102] Sun BL, et al. Lymphatic drainage system of the brain: a novel target for intervention of neurological diseases. Prog Neurobiol 2018;163–164:118–43.
[103] Rustenhoven J, et al. Functional characterization of the dural sinuses as a neuroimmune interface. Cell 2020;184:1–17.
[104] Rua R, et al. Infection drives meningeal engraftment by inflammatory monocytes that impairs CNS immunity. Nat Immunol 2019;20:407–19.
[105] Van Hove H, et al. A single-cell atlas of mouse brain macrophages reveals unique transcriptional identities shaped by ontogeny and tissue environment. Nat Neurosci 2019;22:1021–35.
[106] Rawji KS, et al. Immunosenescence of microglia and macrophages: impact on the ageing central nervous system. Brain 2016;139:653–61.
[107] Streit WJ. Microglial senescence: does the brain's immune system have an expiration date? Trends Neurosci 2006;29:506–10.
[108] Streit WJ, Xue QS, Tischer J, Bechmann I. Microglial pathology. Acta Neuropathol Commun 2014;2:1–17.
[109] Damani MR, et al. Age-related alterations in the dynamic behavior of microglia. Aging Cell 2011;10:263–76.
[110] Goyal MS, et al. Loss of brain aerobic glycolysis in normal human aging. Cell Metab 2017;26:353–360.e3.
[111] Kato T, Inui Y, Nakamura A, Ito K. Brain fluorodeoxyglucose (FDG) PET in dementia. Ageing Res Rev 2016;30:73–84.
[112] Mattson MP. Roles of the lipid peroxidation product 4-hydroxynonenal in obesity, the metabolic syndrome, and associated vascular and neurodegenerative disorders. Exp Gerontol 2009;44:625–33.
[113] Cunnane SC, et al. Can ketones help rescue brain fuel supply in later life? Implications for cognitive health during aging and the treatment of Alzheimer's disease. Front Mol Neurosci 2016;9:1–21.
[114] Minhas PS, et al. Restoring metabolism of myeloid cells reverses cognitive decline in ageing. Nature 2021;590:122–8.
[115] Hickman SE, et al. The microglial sensome revealed by direct RNA sequencing. Nat Neurosci 2013;16:1896–905.
[116] Letiembre M, et al. Innate immune receptor expression in normal brain aging. Neuroscience 2007;146:248–54.

[117] Bachstetter AD, et al. Fractalkine and CX 3CR1 regulate hippocampal neurogenesis in adult and aged rats. Neurobiol Aging 2011;32:2030–44.
[118] Floden AM, Combs CK. Microglia demonstrate age-dependent interaction with amyloid-β fibrils. J Alzheimers Dis 2011;25:279–93.
[119] Hohenberger J, Ray K, Meyer K. The biology and chemistry of high-valent iron-oxo and iron-nitrido complexes. Nat Commun 2012;3:1–13.
[120] Bresgen N, Eckl PM. Oxidative stress and the homeodynamics of iron metabolism. Biomolecules 2015;5:808–47.
[121] Connor JR, Snyder BS, Arosio P, Loeffler DA, LeWitt P. A quantitative analysis of isoferritins in select regions of aged, parkinsonian, and Alzheimer's diseased brains. J Neurochem 1995;65:717–24.
[122] Lopes KO, Sparks DL, Streit WJ. Microglial dystrophy in the aged and Alzheimer's disease brain is associated with ferritin immunoreactivity. Glia 2008;56:1048–60.
[123] Zárate S, Stevnsner T, Gredilla R. Role of estrogen and other sex hormones in brain aging. Neuroprotection and DNA repair. Front Aging Neurosci 2017;9:1–22.
[124] Loram LC, et al. Sex and estradiol influence glial pro-inflammatory responses to lipopolysaccharide in rats. Psychoneuroendocrinology 2012;37:1688–99.
[125] Hayashi Y, et al. Reverse of age-dependent memory impairment and mitochondrial DNA damage in microglia by an overexpression of human mitochondrial transcription factor A in mice. J Neurosci 2008;28:8624–34.
[126] Flanary BE, Streit WJ. Progressive telomere shortening occurs in cultured rat microglia, but not astrocytes. Glia 2004;45:75–88.
[127] Ma W, et al. Gene expression changes in aging retinal microglia: relationship to microglial support functions and regulation of activation. Neurobiol Aging 2013;34:2310–21.
[128] Koellhoffer EC, McCullough LD, Ritzel RM. Old maids: aging and its impact on microglia function. Int J Mol Sci 2017;18:1–25.
[129] Sarlus H, Heneka MT. Microglia in Alzheimer's disease. J Clin Investig 2017;127:3240–9.
[130] Galatro TF, et al. Transcriptomic analysis of purified human cortical microglia reveals age-associated changes. Nat Neurosci 2017;20:1162–71.
[131] Anonymous. 2020 Alzheimer's disease facts and figures. Alzheimers Dement 2020;16:391–460.
[132] Yang T, Li S, Xu H, Walsh DM, Selkoe DJ. Large soluble oligomers of amyloid β-protein from Alzheimer brain are far less neuroactive than the smaller oligomers to which they dissociate. J Neurosci 2017;37:152–63.
[133] Hansen DV, Hanson JE, Sheng M. Microglia in Alzheimer's disease. J Cell Biol 2018;217:459–72.
[134] Wang Y, Mandelkow E. Tau in physiology and pathology. Nat Rev Neurosci 2016;17:5–21.
[135] Braak H, Braak E. Neuropathological stageing of Alzheimer-related changes. Acta Neuropathol 1991;82:239–59.
[136] Jack CR, et al. Hypothetical model of dynamic biomarkers of the Alzheimer's pathological cascade. Lancet Neurol 2010;9:119–28.
[137] McGeer PL, Itagaki S, Tago H, McGeer EG. Reactive microglia in patients with senile dementia of the Alzheimer type are positive for the histocompatibility glycoprotein HLA-DR. Neurosci Lett 1987;79:195–200.
[138] Dani M, et al. Microglial activation correlates in vivo with both tau and amyloid in Alzheimer's disease. Brain 2018;141:2740–54.
[139] Kitazawa M, Yamasaki TR, LaFerla FM. Microglia as a potential bridge between the amyloid β-peptide and tau. In: Annals of the New York academy of sciences, vol. 1035. New York Academy of Sciences; 2004. p. 85–103.
[140] Liu S, et al. TLR2 is a primary receptor for Alzheimer's amyloid β peptide to trigger Neuroinflammatory activation. J Immunol 2012;188:1098–107.

[141] Koenigsknecht J, Landreth G. Microglial phagocytosis of fibrillar β-amyloid through a β1 integrin-dependent mechanism. J Neurosci 2004;24:9838–46.
[142] Brouwers N, et al. Alzheimer risk associated with a copy number variation in the complement receptor 1 increasing C3b/C4b binding sites. Mol Psychiatry 2012;17:223–33.
[143] Wang Y, et al. TREM2 lipid sensing sustains the microglial response in an Alzheimer's disease model. Cell 2015;160:1061–71.
[144] Yeh FL, Wang Y, Tom I, Gonzalez LC, Sheng M. TREM2 binds to Apolipoproteins, including APOE and CLU/APOJ, and thereby facilitates uptake of amyloid-beta by microglia. Neuron 2016;91:328–40.
[145] Krasemann S, et al. The TREM2-APOE pathway drives the transcriptional phenotype of dysfunctional microglia in neurodegenerative diseases. Immunity 2017;47:566–581.e9.
[146] Zhong Z, et al. TRPM2 links oxidative stress to NLRP3 inflammasome activation. Nat Commun 2013;4:1–11.
[147] Heneka MT, et al. NLRP3 is activated in Alzheimer's disease and contributes to pathology in APP/PS1 mice. Nature 2013;493:674–8.
[148] Alawieyah Syed Mortadza S, Sim JA, Neubrand VE, Jiang LH. A critical role of TRPM2 channel in Aβ42-induced microglial activation and generation of tumor necrosis factor-α. Glia 2018;66:562–75.
[149] Chen CH, et al. Increased NF-κB signalling up-regulates BACE1 expression and its therapeutic potential in Alzheimer's disease. Int J Neuropsychopharmacol 2012;15:77–90.
[150] Deng Y, et al. Amyloid-β protein (Aβ) Glu 11 is the major β-secretase site of β-site amyloid-β precursor protein-cleaving enzyme 1(BACE1), and shifting the cleavage site to Aβ Asp 1 contributes to Alzheimer pathogenesis. Eur J Neurosci 2013;37:1962–9.
[151] Venegas C, et al. Microglia-derived ASC specks crossseed amyloid-β in Alzheimer's disease. Nature 2017;552:355–61.
[152] Asai H, et al. Depletion of microglia and inhibition of exosome synthesis halt tau propagation. Nat Neurosci 2015;18:1584–93.
[153] Mathys H, et al. Temporal tracking of microglia activation in neurodegeneration at single-cell resolution. Cell Rep 2017;21:366–80.
[154] Keren-Shaul H, et al. A unique microglia type associated with restricting development of Alzheimer's disease. Cell 2017;169:1276–1290.e17.
[155] Majumdar A, et al. Degradation of fibrillar forms of Alzheimer's amyloid β-peptide by macrophages. Neurobiol Aging 2008;29:707–15.
[156] Michaud JP, Bellavance MA, Préfontaine P, Rivest S. Real-time in vivo imaging reveals the ability of monocytes to clear vascular amyloid beta. Cell Rep 2013;5:646–53.
[157] Mildner A, et al. Distinct and non-redundant roles of microglia and myeloid subsets in mouse models of Alzheimer's disease. J Neurosci 2011;31:11159–71.
[158] Park L, et al. Brain perivascular macrophages initiate the neurovascular dysfunction of Alzheimer Aβ peptides. Circ Res 2017;121:258–69.
[159] Rajaiah R, Moudgil KD. Animal models. Rheum Art Ther 2009;218–24. https://doi.org/10.1016/B978-032305475-1.50031-8.
[160] Course C. Clinical course and diagnosis; 2004.
[161] Miller DH, Leary SM. Primary-progressive multiple sclerosis. Lancet Neurol 2007;6:903–12.
[162] Rivers TM, Sprunt DH, Berry GP. Observations on attempts to produce acute disseminated encephalomyelitis in monkeys. J Exp Med 1933;58(1):39–53.
[163] Hemmer B, Archelos JJ, Hartung H-P. New concepts in the Immunopathogenesis of multiple sclerosis. Nat Rev Neurosci 2002;3:291–301.
[164] Jiang HR, et al. IL-33 attenuates EAE by suppressing IL-17 and IFN-γ production and inducing alternatively activated macrophages. Eur J Immunol 2012;42:1804–14.

[165] Chu F, et al. The roles of macrophages and microglia in multiple sclerosis and experimental autoimmune encephalomyelitis. J Neuroimmunol 2018;318:1–7.
[166] Stence N, Waite M, Dailey ME. Dynamics of microglial activation: a confocal time-lapse analysis in hippocampal slices. Glia 2001;33:256–66.
[167] Heppner FL, et al. Experimental autoimmune encephalomyelitis repressed by microglial paralysis. Nat Med 2005;11:146–52.
[168] Ponomarev ED, Shriver LP, Maresz K, Dittel BN. Microglial cell activation and proliferation precedes the onset of CNS autoimmunity. J Neurosci Res 2005;81:374–89.
[169] Wlodarczyk A, et al. Pathologic and protective roles for microglial subsets and bone marrow- and blood-derived myeloid cells in central nervous system inflammation. Front Immunol 2015;6:1–11.
[170] Juedes AE, Ruddle NH. Resident and infiltrating central nervous system APCs regulate the emergence and resolution of experimental autoimmune encephalomyelitis. J Immunol 2001;166:5168–75.
[171] Dudvarski Stankovic N, Teodorczyk M, Ploen R, Zipp F, Schmidt MHH. Microglia–blood vessel interactions: a double-edged sword in brain pathologies. Acta Neuropathol 2016;131:347–63.
[172] Codarri L, et al. RORγ3t drives production of the cytokine GM-CSF in helper T cells, which is essential for the effector phase of autoimmune neuroinflammation. Nat Immunol 2011;12:560–7.
[173] Ivanov II, et al. The orphan nuclear receptor ROR??T directs the differentiation program of Proinflammatory IL-17+ T helper cells. Cell 2006;126:1121–33.
[174] Korn T, et al. IL-21 initiates an alternative pathway to induce proinflammatory T H17 cells. Nature 2007;448:484–7.
[175] Mangan PR, et al. Transforming growth factor-β induces development of the T H17 lineage. Nature 2006;441:231–4.
[176] Das Sarma J, et al. Functional interleukin-17 receptor A is expressed in central nervous system glia and upregulated in experimental autoimmune encephalomyelitis. J Neuroinflammation 2009;6:1–12.
[177] Kang Z, et al. Astrocyte-restricted ablation of interleukin-17-induced Act 1-mediated signaling ameliorates autoimmune encephalomyelitis. Immunity 2010;32:414–25.
[178] Lee Y, et al. Induction and molecular signature of pathogenic T H 17 cells. Nat Immunol 2012;13:991–9.
[179] El-Behi M, et al. The encephalitogenicity of TH 17 cells is dependent on IL-1- and IL-23-induced production of the cytokine GM-CSF. Nat Immunol 2011;12:568–75.
[180] Rothhamer V, Tjon EC, Takenaka MC, Chao CC, Ardura-Fabregat A, de Lima KA, et al. Microglial control of astrocytes in response to microbial metabolites. Nature 2018;557(7707):724–8. https://doi.org/10.1038/s41586-018-0119-x.
[181] Bogie JFJ, Stinissen P, Hendriks JJA. Macrophage subsets and microglia in multiple sclerosis. Acta Neuropathol 2014;128:191–213.
[182] Boven LA, et al. Myelin-laden macrophages are anti-inflammatory, consistent with foam cells in multiple sclerosis. Brain 2006;129:517–26.
[183] Miron VE, et al. M2 microglia and macrophages drive oligodendrocyte differentiation during CNS remyelination. Nat Neurosci 2013;16:1211–8.
[184] Guerrero BL, Sicotte NL. Microglia in multiple sclerosis: friend or foe? Front Immunol 2020;11:1–8.
[185] Ponomarev ED, Veremeyko T, Barteneva N, Krichevsky AM, Weiner HL. MicroRNA-124 promotes microglia quiescence and suppresses EAE by deactivating macrophages via the C/EBP-α-PU.1 pathway. Nat Med 2011;17:64–70.
[186] Ponomarev ED, Veremeyko T, Weiner HL. MicroRNAs are universal regulators of differentiation, activation, and polarization of microglia and macrophages in normal and diseased CNS. Glia 2013;61:91–103.

[187] Kaushik DK, et al. Enhanced glycolytic metabolism supports transmigration of brain-infiltrating macrophages in multiple sclerosis. J Clin Invest 2019;129:3277–92.
[188] McFaline-Figueroa JR, Lee EQ. Brain Tumors. Am J Med 2018;131:874–82.
[189] Gutmann DH, Kettenmann H. Microglia/brain macrophages as central drivers of brain tumor pathobiology. Neuron 2019;104:442–9.
[190] Charles NA, Holland EC, Gilbertson R, Glass R, Kettenmann H. The brain tumor microenvironment. Glia 2012;60:502–14.
[191] Chen Z, et al. Cellular and molecular identity of tumor-associated macrophages in glioblastoma. Cancer Res 2017;77:2266–78.
[192] Wang SC, Hong JH, Hsueh C, Chiang CS. Tumor-secreted SDF-1 promotes glioma invasiveness and TAM tropism toward hypoxia in a murine astrocytoma model. Lab Investig 2012;92:151–62.
[193] Zhang J, et al. A dialog between glioma and microglia that promotes tumor invasiveness through the CCL2/CCR2/interleukin-6 axis. Carcinogenesis 2012;33:312–9.
[194] Hu F, et al. Glioma-derived versican promotes tumor expansion via glioma-associated microglial/macrophages toll-like receptor 2 signaling. Neuro-Oncology 2015;17:200–10.
[195] Komohara Y, Ohnishi K, Kuratsu J, Takeya M. Possible involvement of the M2 anti-inflammatory macrophage phenotype in growth of human gliomas. J Pathol 2008;216:15–24.
[196] Fries G, Perneczky A, Kempski O. Glioblastoma-associated circulating monocytes and the release of epidermal growth factor. J Neurosurg 1996;85:642–7.
[197] da Fonseca ACC, et al. Increased expression of stress inducible protein 1 in glioma-associated microglia/macrophages. J Neuroimmunol 2014;274:71–7.
[198] Wesolowska A, et al. Microglia-derived TGF-β as an important regulator of glioblastoma invasion—an inhibition of TGF-β-dependent effects by sh RNA against human TGF-β type II receptor. Oncogene 2008;27:918–30.
[199] Venteicher AS, et al. Decoupling genetics, lineages, and microenvironment in IDH-mutant gliomas by single-cell RNA-seq. Science 2017;355.
[200] Bowman RL, et al. Macrophage ontogeny underlies differences in tumor-specific education in brain malignancies. Cell Rep 2016;17:2445–59.
[201] Darmanis S, et al. Single-cell RNA-Seq analysis of infiltrating neoplastic cells at the migrating front of human glioblastoma. Cell Rep 2017;21:1399–410.
[202] Kirsch M, Fischer H, Schackert G. Activated monocytes kill malignant brain tumor cells in vitro. J Neuro-Oncol 1994;20:35–45.
[203] Wang C, et al. Monocyte-mediated chemotherapy drug delivery in glioblastoma. Nanomedicine 2018;13:157–78.
[204] Amarenco P, Bogousslavsky J, Caplan LR, Donnan GA, Hennerici MG. Classification of stroke subtypes. Cerebrovasc Dis 2009;27:493–501.
[205] Jung S, et al. Factors that determine penumbral tissue loss in acute ischaemic stroke. Brain 2013;136:3554–60.
[206] Anttila JE, Whitaker KW, Wires ES, Harvey BK, Airavaara M. Role of microglia in ischemic focal stroke and recovery: focus on toll-like receptors. Prog Neuro-Psychopharmacol Biol Psychiatry 2017;79:3–14.
[207] Poh L, et al. Evidence that NLRC4 inflammasome mediates apoptotic and pyroptotic microglial death following ischemic stroke. Brain Behav Immun 2019;75:34–47.
[208] Habib P, Harms J, Zendedel A, Beyer C, Slowik A. Gonadal hormones E2 and p mitigate cerebral ischemia-induced upregulation of the AIM2 and NLRC4 inflammasomes in rats. Int J Mol Sci 2020;21:1–24.
[209] Wang R, et al. Transforming growth factor beta-activated kinase 1-dependent microglial and macrophage responses aggravate long-term outcomes after ischemic stroke. Stroke 2020;975–85. https://doi.org/10.1161/STROKEAHA.119.028398.

[210] Zeyen T, et al. Microglial-specific depletion of TAK1 is neuroprotective in the acute phase after ischemic stroke. J Mol Med 2020;98:833–47.
[211] Neubert M, Ridder DA, Bargiotas P, Akira S, Schwaninger M. Acute inhibition of TAK1 protects against neuronal death in cerebral ischemia. Cell Death Differ 2011;18:1521–30.
[212] Mander P, Brown GC. Activation of microglial NADPH oxidase is synergistic with glial iNOS expression in inducing neuronal death: a dual-key mechanism of inflammatory neurodegeneration. J Neuroinflammation 2005;2:1–15.
[213] Tang XN, Cairns B, Kim JY, Yenari MA. NADPH oxidase in stroke and cerebrovascular disease. Neurol Res 2012;34:338–45.
[214] Zhou M, Wang CM, Yang WL, Wang P. Microglial CD14 activated by iNOS contributes to neuroinflammation in cerebral ischemia. Brain Res 2013;1506:105–14.
[215] Brown GC. Mechanisms of inflammatory neurodegeneration: INOS and NADPH oxidase. Biochem Soc Trans 2007;35:1119–21.
[216] Anwar S, Pons V, Rivest S. Microglia purinoceptor P2Y6: an emerging therapeutic target in CNS diseases. Cell 2020;9:1–13.
[217] Nguyen HM, et al. Biophysical basis for Kv1.3 regulation of membrane potential changes induced by P2X4-mediated calcium entry in microglia. Glia 2020;68:2377–94.
[218] Bergamin S, et al. Differential macrophage activation alters the expression profile of NTPDase and ecto-5 9-nucleotidase. PLoS One 2012;7.
[219] Costa A, et al. Deletion of muscarinic acetylcholine receptor 3 in microglia impacts brain ischemic injury. Brain Behav Immun 2021;91:89–104.
[220] Zang J, et al. Inhibition of PDE1-B by Vinpocetine regulates microglial exosomes and polarization through enhancing autophagic flux for neuroprotection against ischemic stroke. Front Cell Dev Biol 2021;8:1–19.
[221] Hoyo-Becerra C, Schlaak JF, Hermann DM. Insights from interferon-α-related depression for the pathogenesis of depression associated with inflammation. Brain Behav Immun 2014;42:222–31.
[222] Kim YK, Jung HG, Myint AM, Kim H, Park SH. Imbalance between pro-inflammatory and anti-inflammatory cytokines in bipolar disorder. J Affect Disord 2007;104:91–5.
[223] Girgis RR, Kumar SS, Brown AS. The cytokine model of schizophrenia: emerging therapeutic strategies. Biol Psychiatry 2014;75:292–9.
[224] Réus GZ, et al. The role of inflammation and microglial activation in the pathophysiology of psychiatric disorders. Neuroscience 2015;300:141–54.
[225] Torres-Platas SG, Cruceanu C, Chen GG, Turecki G, Mechawar N. Evidence for increased microglial priming and macrophage recruitment in the dorsal anterior cingulate white matter of depressed suicides. Brain Behav Immun 2014;42:50–9.
[226] Schnieder TP, et al. Microglia of prefrontal white matter in suicide. J Neuropathol Exp Neurol 2014;73:880–90.
[227] Zhao Q, et al. Maternal sleep deprivation inhibits hippocampal neurogenesis associated with inflammatory response in young offspring rats. Neurobiol Dis 2014;68:57–65.
[228] Barger SW, Goodwin ME, Porter MM, Beggs ML. Glutamate release from activated microglia requires the oxidative burst and lipid peroxidation. J Neurochem 2007;101:1205–13.
[229] Wu GJ, Chen TL, Ueng YF, Chen RM. Ketamine inhibits tumor necrosis factor-α and interleukin-6 gene expressions in lipopolysaccharide-stimulated macrophages through suppression of toll-like receptor 4-mediated c-Jun N-terminal kinase phosphorylation and activator protein-1 activation. Toxicol Appl Pharmacol 2008;228:105–13.
[230] Tynan RJ, et al. Chronic stress-induced disruption of the astrocyte network is driven by structural atrophy and not loss of astrocytes. Acta Neuropathol 2013;126:75–91.
[231] Soczynska JK, et al. Novel therapeutic targets in depression: minocycline as a candidate treatment. Behav Brain Res 2012;235:302–17.

[232] Sugama S, Takenouchi T, Fujita M, Kitani H, Hashimoto M. Cold stress induced morphological microglial activation and increased IL-1β expression in astroglial cells in rat brain. J Neuroimmunol 2011;233:29–36.
[233] Jeon SW, Kim YK. Neuroinflammation and cytokine abnormality in major depression: cause or consequence in that illness? World J Psychiatry 2016;6:283.
[234] Kim YK, Na KS, Myint AM, Leonard BE. The role of pro-inflammatory cytokines in neuroinflammation, neurogenesis and the neuroendocrine system in major depression. Prog Neuro-Psychopharmacol Biol Psychiatry 2016;64:277–84.
[235] Steiner J, et al. Immunological aspects in the neurobiology of suicide: elevated microglial density in schizophrenia and depression is associated with suicide. J Psychiatr Res 2008;42:151–7.
[236] Hou X, Liang X, Chen JF, Zheng J. Ecto-5′-nucleotidase (CD73) is involved in chronic cerebral hypoperfusion-induced white matter lesions and cognitive impairment by regulating glial cell activation and pro-inflammatory cytokines. Neuroscience 2015;297:118–26.
[237] Manso Y, et al. Minocycline reduces microgliosis and improves subcortical white matter function in a model of cerebral vascular disease. Glia 2018;66:34–46.
[238] Fowler JH, et al. Dimethyl fumarate improves white matter function following severe hypoperfusion: involvement of microglia/macrophages and inflammatory mediators. J Cereb Blood Flow Metab 2018;38:1354–70.
[239] Zhang LY, et al. Microglia exacerbate white matter injury via complement C3/C3aR pathway after hypoperfusion. Theranostics 2020;10:74–90.
[240] Mendelsohn AR, Larrick J. Metabolic reprogramming rejuvenates aged myeloid cells restoring cognition. Rejuvenation Res 2021;24:65–7.

CHAPTER 9

Intestines—Inflammatory and digestive system

Eloisa Martins da Silva[a], Renan Willian Alves[b], Lorena Doretto-Silva[b], and Vinicius Andrade-Oliveira[b]
[a]Paulista Medical School, Federal University of São Paulo, São Paulo, Brazil
[b]Federal University of ABC, Center for Human and Natural Sciences, São Paulo, Brazil

Introduction

The immune system, composed of several cell types distributed throughout the tissues, acts as a protective barrier of the body to avoid the entry of invasive microorganisms and it is influenced by some stress, such as infections, allergies, inflammation, genetic, and metabolic changes [1,2]. The immune response is classified into two large groups, the innate immune system, composed of macrophages, neutrophils, eosinophils, monocytes, NK cells, and the adaptive immune system, composed mainly of T and B lymphocytes [3]. These cell types are generated from precursors of the bone marrow, a primary lymphoid organ, and undergo maturation either in bone marrow or in the thymus (T cells) or spleen (B cells). After maturation, immune cells migrate to specific tissues through signaling molecules. Besides, some tissue-resident immune cells, such as macrophages, originate from precursors of the yolk sac or fetal liver [4].

Macrophages have the great ability to ingest and destroy microorganisms and therefore exert the initial barrier of the body [5]. In addition, the crosstalk with other cells is crucial to recruit other immune cells to sites of infection, and unlike other immune system cells, they also help in the clearance of apoptotic and cellular debris at the sites of inflammation being a dominant effector cell in the final stages of the innate immune response [6]. They have high plasticity and heterogeneity and respond to many environmental signals, these signals altering their phenotype. Macrophages have two major phenotypes of classification, named M1 and M2, associated with inflammatory and anti-inflammatory responses, respectively [7]. However, inside this classification (M1 and M2) exists a gradient of phenotypes, demonstrating the high heterogeneity of this population. Classic markers

Macrophages in the Human Body
https://doi.org/10.1016/B978-0-12-821385-8.00007-X

for M1 at homeostasis are INF-γ+ and STAT1, whereas M2 markers are IL-4+ and STAT6 [8].

Different tissues may have different immune cell composition, including tissue-resident immune cells, perhaps by the inherent function and specific demand of each tissue. The gastrointestinal (GI) system is particularly challenging due to the fact that it is in intimate contact with the external environment. In this sense, the GI tract has to deal with microorganisms, nutrients from diet among other factors while still exerting its function. It has been reported that several GI-related diseases are directly related to immune cell activity, such as inflammations, infections, and cancer [9,10]. Specifically, intestinal diseases such as Ulcerative Colitis (UC), Crohn's diseases (CD), and colorectal cancer (CRC) are experimental models used to investigate pathways and mechanisms of actions of cells and signaling molecules [11,12]. Macrophages, as well as other tissue-resident immune cells, are continuously working for the maintenance of homeostasis [11]. In the intestinal context, there is a huge range of macrophage subtypes, which are classified according to their metabolic profile, by the most active signaling pathways, cytokine production, and also their location in the intestine, may be present in the lamina propria or even in the intestinal muscle [13–15].

Although the classification of macrophages relies on the features aforementioned, The profile of macrophages residing in the GI tract is not yet fully evidenced regarding the metabolic, activation, and signaling pathways already described in other contexts. In this review, we grouped the main studies in which the crosstalk between intestinal microenvironment and GI diseases is evidenced and its influence in macrophages.

Macrophages: Ontogeny, subtypes, and classic markers

For intestinal homeostasis, macrophages have an important role related to immune regulation and maintenance of gastrointestinal function, in health and disease [16,17] in part by their functional plasticity and the ability to interpret signals from food, signals from the commensal bacterial, pathogen, and host-microbiota [16].

Macrophages in the GI tract are identified with protein markers associated with resting conditions, proinflammatory, and anti-inflammatory states, and there are some differences between the surface markers to identify mouse and human macrophages, for instance, F4/80 and inducible nitric oxide synthase (iNOS) markers are utilized to identify mouse but not human macrophages, whereas there are some shared markers such as major histocompatibility complex II (MHCII) and CD206 markers for both human and mouse [17]. A summary of the main markers employed to identify macrophages and their precursors in different organs and tissues are shown in Table 1.

Macrophages are also resident in the intestinal lamina propria [20] and it presents itself with two subtypes: a resident macrophage, the one that inhabits all the time in the gut, and a recruited macrophage which can be divided into M1 and M2 [21]. The resident macrophage can arise from the yolk sac and fetal liver precursors [22] and share some features from both M1 and M2, leaving aside this M1/M2 classification, making its main feature of resident macrophages the adaptation to the gut environment, playing

Table 1 Different markers during macrophage maturation and polarization.

Local	Cell type	Markers	
Bone marrow	Monocyte-macrophage DC progenitor	CD117/c-kit, CD115/M-CSF, CD135/Fit3, CD13, CX3CR1	[18]
	Common monocyte progenitor	CD117/c-kit, CD115/M-CSF, Ly6C,CD13 CX3CR1	[18]
	Proliferating monocytes	CD115/M-CSF, Ly6C, CD11b^{low}, CX3CR1	[18]
	Ly6C high	CD115/M-CSF, Ly6C, CD11b, CX3CR1	[18]
	Ly6C low	CD115/M-CSF, Ly6C, CD11b, CX3CR1	[18]
Peripheral blood	Monocyte	Ly6C^{hi}, CX3CR1^{int}, CD115+, CCR2+	[11]
	Patrolling monocyte	CD11b+, CX3CR1^{low}, LY6C^{low}, CCR2-, CD115+	[19]
Tissues	Resident macrophage	CX3CR1^{hi}, F4/80^{hi}, CD11b+, MHCII^{hi}, CD64+, TREM2, CD33	[11]
	Macrophage M1	IFN-γ, STAT1 CD64, CXCL9, CXCL10, NOS2	[7,8]
	Macrophage M2	IL-4, STAT6 CD23, CCL22, Mrc1, Arg1	[7,8]

M-CSF, macrophage colony-stimulating factor; *CX3CR1*, CX3C chemokine receptor 1; *Ly6C*, lymphocyte antigen 6C; *CCR2*, C-C chemokine receptor type 2; *MHCII*, major histocompatibility complex class 2; *TREM2*, triggering receptor expressed on myeloid cells 2; *IFN-γ*, interferon gamma; *STAT*, signal transducer and activator of transcription; *CXCL*, chemokine (C-X-C motif) ligand; *NOS2*, nitric oxide synthase 2; *CCL22*, C-C motif chemokine ligand 22; *Mrc1*, mannose receptor C-type 1; *Arg1*, arginase 1.

housekeeping functions, such as epithelial barrier integrity and tissue remodeling, high phagocytic capacity, capturing and destroying pathogens, apoptotic and senescent cells [11]. It was demonstrated in both mouse and human intestines that the colonic inflammation modifies the heterogeneous composition of the local macrophage pool. Thus, the same macrophage type can occupy a healthy and inflamed colon, presumably, by recruiting cells most similar to lymphocyte antigen 6C high ($Ly6C^{hi}$) monocytes [13]. The different signals in each condition, healthy and during inflammation, will dictate the type of macrophage the Ly6C cells become.

Like T helper for Th1 and Th2, macrophage also presents a similar dichotomy classification when we look at the cytokines involved: M1 suffers the influence of IFN-γ, like T helper 1 and M2 suffers the influence of IL-4, like T helper 2 [7]. As we have two types of macrophages, their polarization occurs differentially, by their stimulation and consequent populations. M1, activated by IFN-γ + LPS or TNF, presents the classical Th1 response, with anti-tumor and type 1 inflammation response. However, M2 can be activated and generates three different response: IL-4 and IL-13 induce M2a alternative response; immune complex + TLR or IL-1R ligands induce M2b type II; and IL-10 induces M2c deactivated which present, respectively, the Th2 responses, Th2 activation, and matrix deposition and tissue remodeling [23].

Interestingly, in germ-free (GF) mice, two macrophage populations, $CD11c+CD206^{int}CD121b+$ and $CD11c-CD206^{hi}CD169+$, appeared reduced in the colon lamina propria and microbiota recolonization re-emerged these populations. This shows that commensal microbiota is important to support the development of some colon macrophage populations [24]. It would be interesting to identify to which extent the absence of the microbiota impacts other macrophage subpopulations in the GI tract as well as if there are specific organisms or substances released by the microbiota, responsible for macrophage differentiation and survival.

Gastrointestinal inflammation and colorectal cancer

Inflammatory bowel disease (IBD), commonly divided into CD and UC, are disorders that affect millions worldwide impacting the patient's life [11]. These conditions are remarkably noticed by chronic inflammation that affects the GI tract and may result from genetic, environmental, immunological, and gut microbiota interactions. CD can affect any part of the GI tract, disrupting the epithelial intestinal wall in a discontinuous manner [25,26], whereas UC is located mostly in the colon and rectum, disturbing the mucosal layer in a continuous pattern [27,28].

Although the etiology of both IBD and UC remains unknown, several studies in the last years provided enough evidence to support that the pathogenesis of IBD and UC is due to a combination of genetic host conditions, intestinal microbiota, immune response, and environmental factors that lead to an exacerbated inflammatory response [29]. Regarding the immunological risks, the main factor observed in IBD and UC is the epithelial damage causing a disruption of the epithelial barrier. This epithelial rupture is the source of bacterial translocation, increasing antigen exposure that leads to persistent activation of immune cells, such as dendritic cells, neutrophils, macrophages, B and T cells, production of proinflammatory cytokines along with dysregulation of immune control of inflammatory response [30,31].

Intestinal-resident macrophages are the cells responsible for maintaining homeostasis in this organ, located in the mucosa throughout the GI tract residing mainly in the LP near the epithelial monolayer [23,32]. These cells are housekeeping components of the LP characterized by intensified bactericidal and phagocytic activity, clearance of apoptotic cells and tissue remodeling, and reduced pro-inflammatory cytokine secretion [30]. Also, it was described a small macrophage subpopulation located in the smooth muscle layers of the gut wall regulating intestinal motility [33]. This group, denominated *muscularis macrophages* (MM), a distinct macrophage population reside near the myenteric plexus and regulate enteric neuron activity and colon peristalsis by secretion of Bone morphogenetic protein 2 (BMP2), in experiments with mice treated with antibiotics, a decrease in this factor has been seen and also a decrease in the number of MMs in conjunction with GI dysmotility [15].

At steady-state in the mucosa, macrophages secrete cytokines that maintain tissue homeostasis and stimulate the epithelial progenitors' cell proliferation in intestinal crypts and by doing so regulate epithelial barrier integrity [32]. When these cells are selectively depleted by blockade of CSF1R, an important receptor involved in macrophages growth and differentiation, increases the susceptibility of DSS-induced colitis in a murine model, reducing intestinal stem cell viability [31,32]. Yet, the main feature of resident intestinal macrophages is their enhanced phagocytic activity, due to high levels of TREM2 phagocytic receptors, allowing them to capture and destroy any bacteria or material that breach the epithelial barrier [34]. Importantly, intestinal macrophages when exposed to intestinal bacteria or inflammatory stimuli do not produce pro-inflammatory responses, such as the production of TNFα, IL-1, IL-6, or other inflammatory cytokines and chemokines. At metabolic levels, bacterial exposure does not increase respiratory activity or nitric acid generation in intestinal macrophages [11]. In general, intestinal macrophages reveal an anti-inflammatory, pro-tolerogenic, and M2-like phenotype [35]. Despite that, intestinal macrophages are not in an anergic stage of cytokine production, some recent studies in mice revealed that they produce basal levels of IL-10 and TNF-α [36]. Interestingly, TNF-α regulates enterocyte growth and alters the permeability of the epithelial barrier [36]. Due to the expression of MHC II and the capacity to engulf bacteria and protein antigen, intestinal macrophages act as antigen-presenting cells modulating CD4+ T cell differentiation [37]. In addition, LPS-stimulated IL-10-production modulates epithelial cell growth and Treg proliferation [36,38].

Although intestine macrophages have an anti-inflammatory profile, paradoxically macrophages are the main cell type related to the pro-inflammatory profile in acute and chronic IBD, hence driving disease progression [39]. Under such conditions of homeostasis perturbance, the composition of intestinal macrophages alters. In the mucosa of patients with CD or UC pro-inflammatory (M1) macrophages become the predominant macrophage population [14]. Despite the challenge of characterizing intestinal macrophages, some studies used pan-myeloid cell markers, such as CD13, CD14, and CD33 [40,41]. Compared to macrophages found in the healthy intestine, pro-inflammatory macrophages produce a great amount of IL-1, IL-6, TNF-α, nitric oxide (NO), and reactive oxygen intermediates, indicating that these cells may be a therapeutic target to regulate inflammation in both diseases. Besides, pro-inflammatory macrophages present a high expression of CD14 when compared to resident macrophages in steady conditions, facilitating their identification [40].

Recent studies using CD patient samples revealed an increase in macrophages infiltration with pro-inflammatory phenotype, with enhanced CD14+ macrophages in inflamed samples when compared to normal tissue from control patients [40]. In addition, a similar study showed an increase in CD14+CD33+ proinflammatory colonic macrophages from CD patients, which are able to secrete more IL-23 and TNF-α, besides induce Th1 producer of IFN-γ response when challenged with commensal bacterial such as, *E. coli* or *Enterococcus faecalis* [42]. In the same fashion, macrophage infiltration is observed in samples from UC patients. Inflamed colon areas from UC revealed a similar profile to CD samples, with an increase in CD14+CD33+ proinflammatory macrophages [40]. Interestingly, in this case, CD14+CD33+ macrophages from colon samples produced higher levels of IL-6 [40].

Regarding CRC, macrophages play a crucial role in the tumor microenvironment. Tumor-associated macrophages (TAMs) are myeloid cells recruited to the site of the tumor and in most cases present a M2-like phenotype [43]. This anti-inflammatory feature enables TAMs to inhibit immune response against tumor cells allowing tumor growth [44]. However, some in vitro studies have demonstrated that TAMs can present an antitumor response in CRC [45]. In CRC tissue of patients, a decrease in markers was seen CD14+CD11c+CD163−, markers for TAMs M1, also, these patients had an increase in specific genera of bacteria *Ruminococcaceae*, *Faecalibacterium*, *Eubacterium* (phylum *Firmicutes*) and *Bacteroides* [46]. Furthermore, the intestinal microbiota stimulates the secretion of the enzyme cathepsin K (CTSK) promoting metastasis in colorectal cancer, IL-10 secretion, and polarization to M2 TAMs [47], suggesting the association of the microbiota with the polarization of TAMs in CRC [46,47]. New evidence indicates that the intense production of acid lactic through glycolysis by tumor cells is the main stimuli to modulate macrophages polarization toward M2 phenotype [48,49]. It is still controversial but the presence of TAMs can dictate a better or worse CRC prognosis. Edin et al. in an extensive study, analyzed 485 samples for CRC tumors and evaluated for the presence of M1 and M2 macrophages. They found that the M2 phenotype was strongly associated with invasive tumors while M1 macrophages confer a better prognosis [50]. In addition, McLean et al. observed that in colonic adenomatous polyps there is an increase in macrophages presence when compared to normal areas. Thus, this rise of macrophages numbers in polyps is related to the gravity of the injury and cancerous polyps revealed a higher quantity of M2 macrophages [51]. Different from BMDMs, the metabolism of resident macrophages of the intestine and the microbiota as a modulator of these cells need further studies for a complete characterization of their metabolic profile.

Overall macrophages are crucial cells, in the intestinal microenvironment, for the maintenance of homeostasis state. During the inflammation and cancer process, this state is dysregulated and can modulate the disease condition. The comprehension of the different phenotypes and functions of macrophages can deepen our understanding of their essential role in homeostasis, inflammation, and cancer. In light of that, macrophages represent an alternative and potential therapeutic target to treat patients with IBD, UC, or even to optimize CRC treatment (Fig. 1).

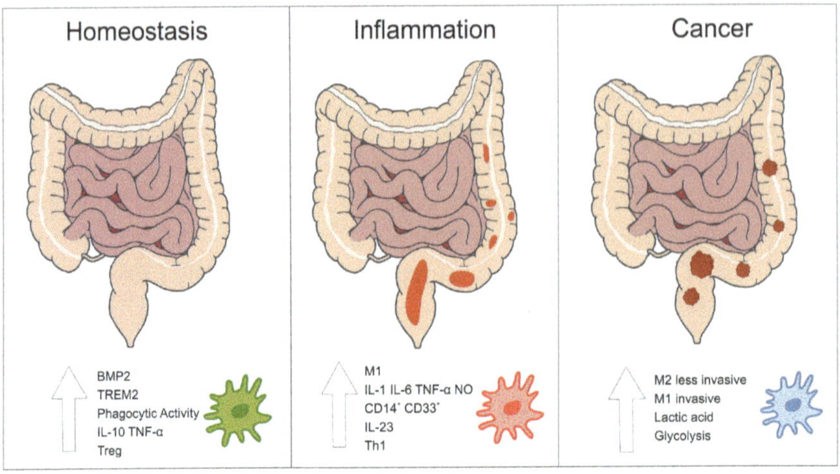

Fig. 1 Macrophages phenotypes in gastrointestinal inflammation and colorectal cancer. During homeostasis, lamina propria macrophages secret BMP2, which regulates intestinal motility. These cells express high levels of TREM2, which confers enhanced phagocytic activity. Besides intestinal macrophages produce basal levels of TNF-α and IL-10 modulating Treg proliferation. However, this process is dysregulated in patients with intestinal inflammation, leading to an accumulation of pro-inflammatory macrophages (M1) in the inflamed site. These CD14hi cells express CD33 and produce a large amount of IL-1, IL-6, TNF-α, and NO. In cancer, TAMs inhibit immune response resembling a M2-like phenotype. In that manner, the intense rate of glycolysis and acid lactic production modulates macrophages' polarization toward the M2 phenotype. *BMP2*, bone morphogenetic protein 2; *IL-1*, interleukin-1; *IL-6*, interleukin-6; *IL-10*, interleukin-10; *IL-23*, interleukin-23; *NO*, nitric oxide; *TNF-α*, tumor necrosis factor; *Treg*, regulatory T cells; *TREM2*, triggering receptor expressed on myeloid cells 2.

Metabolic pathways macrophages

Recent studies with immune cells show the important interconnection between the metabolic profile and cellular function, directly associating the type of metabolism demanded by the cell, with its effector or immunosuppressive function [52]. For example, macrophages have high plasticity and heterogeneity and respond to many environmental signals, and these signals alter your phenotype [53]. Moreover, metabolic enzymes can also influence immune cells by exerting non-metabolic function, mostly via epigenetics mechanisms [54]. As mentioned before, as much as there is the classic classification phenotype in M1/M2, it is known that there is a gradient of differences/polarization, that can be best classified in M1, M2a, M2b, M2c, and M2d (TAM). Looking for the metabolic demand of these macrophages, there are some differences about metabolic enzymes expression and activity, in which M1 increases iNOS (inducible Nitric Oxide Synthase), PFKFB3 (6-phosphofructo-2-kinase 3), PKM2 (pyruvate kinase M2), ACOD1 (aconitate decarboxylase 1), whereas M2a and M2b have increased ARG1 (Arginase 1) and CARKL (carbohydrate kinase-like), M2c ARG1 and GS (glutamine synthetase), and M2d AGR1 and IDO (indoleamine dioxygenase) [55]. Thus, this suggests that different macrophages and subtypes may have different metabolic demand to exert their function. Studies have focused on understanding how the metabolic pathways are triggered in different macrophage subsets since there are differences in metabolic pathways in M1 compared to M2 macrophages [52]. In vitro stimulation of bone marrow-derived macrophages (BMM) by LPS + IFNy, which differentiates these cells in M1 phenotype, increases the pentose phosphate pathway (PPP) and glycolysis, while the oxidative phosphorylation (OXPHOS) is attenuated, which culminates with increased reactive oxygen species (ROS) production [56–58]. The ROS is important for host defense, in homeostasis is required for cellular growth and proliferation, and a key player of adaptive immunity [56]. In contrast, stimulation of BMM by IL-4, which differentiates in M2 phenotype, increases fatty acid oxidation (FAO) and drives the produced pyruvates to generate acetyl-CoA, which are directed for Krebs cycle, increasing OXPHOS [57]. Knowing about all the plasticity of macrophages, it is clear that it may not be possible to extend the findings observed in vitro polarization studies to in vivo, since specific tissue possesses different microenvironments that might impact the macrophage metabolism specifically (Fig. 2).

Microbial metabolites in intestinal macrophages

It is extensively known the importance of the gut microbiota presence and function to maintain intestinal homeostasis. In a situation of disruption of gut microbiota composition, also known as intestinal dysbiosis, some factors produced and influenced by the microbiota are commonly altered, such as the profile of infiltrating immune cells in the lamina propria. These microorganisms present in the intestine contribute to the development and regulation of the immune system [59,60]. GF animal studies have shown that these animals have a small amount of resident immune cell populations, including macrophages [61]. Furthermore, the phenotype of innate and adaptive immune cells can be modulated by the microbiota [59].

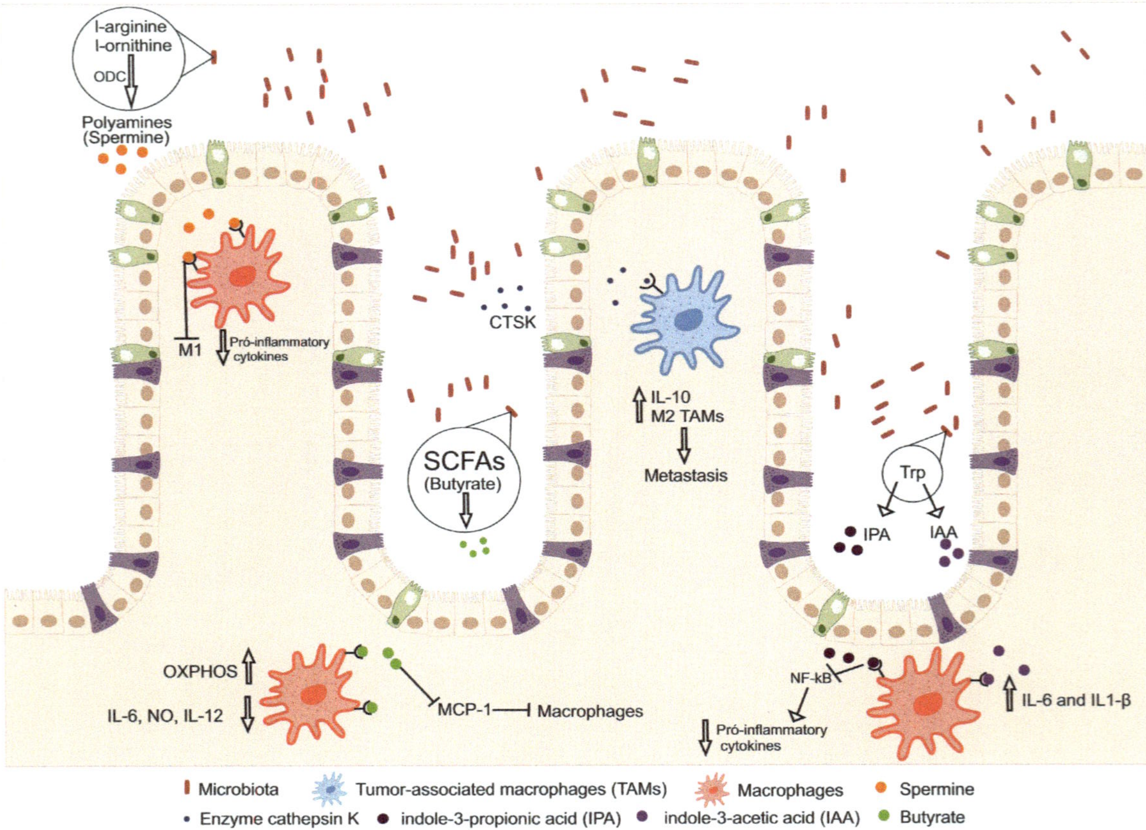

Fig. 2 Microbial metabolites change the metabolism and polarization of intestinal macrophages. Polyamines reduce levels of pro-inflammatory cytokines by reducing the polarizing to the M1 profile. Butyrate, a SCFAs blocks MCP-1, blocking macrophages as well as acting on the metabolism of these cells by increasing OXPHOS and reducing pro-inflammatory cytokines. The enzyme CTSK stimulated by the intestinal microbiota in the neoplastic context increases IL-10 and the polarization of TAMs in M2 leading to tumor metastasis. Tryptophan metabolites, IPA and IAA block NFkB and increase expression of IL-6 and IL1b, respectively. *SCFAs*, short-chain fatty acid; *MCP-1*, monocyte chemoattractant protein-1; *CTSK*, enzyme cathepsin K; *IPA*, indole-3-propionic acid; *IAA*, indole-3acetic acid; *IL1β*, interleukin-1; *Il-6*, interleukin-6; *Il-10*, interleukin-10.

Short-chain fatty acids

Short-chain fatty acids (SCFA) can be ingested by diet and are derived from the bacterial fermentation of the normal intestinal flora [62]. There are several SCFAs produced by enteric bacteria (SCFA are secreted by bacteria from the phylum Bacteroidetes and Firmicutes) such as butyrate, acetate, and propionate the most abundant SCFAs in the gastrointestinal tract (≥95%) [63,64]. Butyrate is mainly produced by phylum Firmicutes, while acetate and propionate are mainly produced by Bacteroidetes [65,66]. SCFA are mainly active in the stimulation of G protein-coupled receptors, these receptors are present in several specific cells and tissues, such as macrophages in the intestine [67]. The G protein-coupled receptor 43 (GPR43) is a receptor for SCFAs (acetate, propionate, and butyrate) also known as free fatty acid receptor 2 (FFAR2) is the focus of studies related to immune cell modulation [68,69]. In macrophages, GPR43 is expressed mainly in M2 profiles in adipose tissue, inducing TNF-α [70]. Also, macrophages from animals deficient in GPR43 stimulated with LPS and with crystals of ATP or monosodium urate monohydrate (MSU) decrease the secretion of IL1-β and a decrease in the production of ROS [71], demonstrating that there is no concentrated evidence of a modulation of the GPR43 receptor in macrophages, varying according to the tissue.

For example, butyrate (BT), a short-chain fatty acid, a metabolite produced by bacteria from the intestinal microbiota, has a potential intestinal modulator of macrophages metabolism [72,73]. In the gut, the potential of butyrate as a modulator of intestinal macrophage metabolism has been demonstrated, positively regulating OXPHOS, which leads to a possible modulation for an anti-inflammatory phenotype [74], evidencing the importance of the microbiota as a phenotype modulator and macrophage metabolism. Interestingly, BT seems to exert the same function in macrophages in peripheral tissue for example by reducing monocyte chemoattractant protein-1 (MCP-1) levels, which is a chemokine that regulates the migration and infiltration of macrophages in the intestine, causing a decrease in the recruitment and migration of macrophages [72]. In addition, the treatment of intestinal macrophages with butyrate led to the negative regulation of LPS-induced pro-inflammatory mediators, including NO, IL-6, and IL-12, by inhibiting histone deacetylase [75]. It is important to mention that patients with IBD have reduced levels of SCFAs-producing bacteria in the mucosa compared to healthy, underscoring their influence on the modulation of intestinal inflammation [63,76].

Axis arginine-polyamines

The intestinal microbiota is considered to be the main responsible for the level of polyamines in the lower part of the intestine. In human cells, spermine, spermidine, and putrescine are the main polyamines [77], they induced protein synthesis, cell growth, and differentiation [78]. For example, spermine inhibits the synthesis of pro-inflammatory cytokines in human mononuclear cells, restricting the immune response [79]. The synthesis of polyamine by bacteria has been gaining emphasis on the possibility of modulation of the microbiota for an increase in polyamine production with the potential for reduced intestinal inflammation [80,81].

The amino acids L-lysine, L-ornithine, L-arginine, and L-aspartic are the precursors of polyamine biosynthesis by bacteria [82]. Arginine is metabolized by arginase to produce ornithine and urea. Ornithine is the precursor of polyamine production via ornithine decarboxylase (ODC), also known as the rate-limiting enzyme required for polyamine biosynthesis [83,84]. The arginine-polyamine axis has been the target of studies on its potential regulation of inflammation, as well as the polarization of macrophages [81,85]. In macrophages, baseline polyamine levels are low and increase with stimulation of IL-4 or IL-13. In addition, they inhibit the expression of pro-inflammatory genes, indicating that polyamines can regulate the function of macrophages [85].

Also, the enzyme spermine oxidase (SMO) is responsible for the conversion of sperm into spermidine. In UC, there is a decrease in sperm levels and an increase in spermidine, and SMO expression is positively regulated in UC tissues, increased levels in mononuclear inflammatory cells [81]. But the effects of the spermidine-spermine axis are somewhat contradictory, and studies show that in a DSS-induced colitis model SMO enzyme depletion aggravates colitis. However, in a model induced by *Citrobacter rodentium*, attenuation of this inflammation occurs, in addition to a decrease in the production of G-CSF [86], which is a granulocyte colony-stimulating factor that plays an important role in the number of macrophages in the gut during inflammation [87]. Suggesting that the increase in polyamine is harmful in *C. rodentium*-induced colitis, while spermidine protects mice during DSS-induced colitis, with a decrease and increase, respectively, in the production of granulocyte/macrophage stimulation factor, G-CSF [86]. Still, on the spermine-spermidine axis, a study showed its potential for the polarization of macrophages, in which the spermine may inhibit the polarization and activation of M1 macrophages by suppressing the enzyme ornithine decarboxylase (ODC) and also suppressing the synthesis of pro-inflammatory cytokines [88]. The ODC regulates the activation of M1 during *Helicobacter pylori* and *C. rodentium* infection. ODC suppression in macrophages resulted in increased inflammation, with increased M1 response and decreased bacterial persistence in mouse models, demonstrating that ODC directly regulates the activation and polarization of macrophages [84]. Data show that patients with active UC and CAC have increased ODC expression in macrophages, such as CSF2 (GM-CSF), TNFα, IL-1α, and IL-1β, confirming that loss of Odc enhances macrophage polarization toward M1 also in the colorectal cancer model associated with colitis [89]. Together, the studies demonstrate the potential of the ODC enzyme, a polyamine precursor, in the polarization of macrophages in inflammatory bowel disease and also in colorectal cancer associated with colitis [84,89].

Tryptophan

Tryptophan (Trp) is known to be an amino acid produced by the bacterial microbiota, such as *Escherichia coli*, however, there is no substantial amount of data to conclude the significant contribution of bacteria-derived Trp in human physiology [90]. The most abundant source of it is through food, while intestinal microflora metabolizes this amino acid by the pathways of the indole/AhR [91], kynurenine/IDO [92], and serotonin [93].

The pathway indole/AhR is thus known, as many metabolites of Trp, such as nature and its derivatives are ligands of AhR such including indole-3-lactic acid (ILA), indole-3-propionic acid (IPA), indole-3-acetic acid (IAA), and indole-3-aldehyde (IAld) and play an important role for host homeostasis. It is known from the vast literature that innate immune cells express AhR, for example, the macrophages [90]. These metabolites, such as IPA inhibit intestinal dysbiosis and decrease the production of pro-inflammatory cytokines in macrophages by inhibiting NF-kβ activity [94]. In addition, the metabolite IAA suppresses inflammatory stress with a decrease in proinflammatory cytokines, IL-1β and IL-6 in RAW264.7 macrophages cell line, evidencing the potential of a metabolite derived from tryptophan metabolism by the microbiota in the modulation of the inflammatory response LPS-induced in macrophages [95].

Conclusion

Macrophages have a wide gradient of phenotypes, which can be changed according to the tissue studied. The resident macrophages of the intestine have classic markers, being CX3CR1hi, F4/80+, CD11b+, CD11c+, CD64+ [96].

Factors such as intestinal microbiota and its metabolites, the integrity of the epithelial barrier, infections, and inflammation can alter the profile of these cells as well as their quantity [23,32,59]. Therefore, it is extremely important to balance these factors, as well as the macrophages present in the intestine to maintain intestinal homeostasis. Studies focusing on macrophages resident in the intestine have great potential for possible therapies related to inflammatory bowel diseases, CRC, and against infections, since these cells react directly to changes in the intestinal microenvironment and are responsible for part of the maintenance of homeostasis.

Understanding the profile of intestinal macrophages, as well as their role in the intestine and their modulation before alterations, will provide a range of possibilities for manipulating these cells for therapeutic benefit, in infectious and inflammatory diseases.

Acknowledgments

This study was supported by Fundação de Amparo à Pesquisa do Estado de São Paulo (FAPESP): grant number: 2019/14755-0. Vinicius Andrade-Oliveira is also a fellow of Pew Latin American Fellow program of the Pew Foundation.

References

[1] Choy MC, Visvanathan K, De Cruz P. An overview of the innate and adaptive immune system in inflammatory bowel disease. Inflamm Bowel Dis 2017;23(1):2–13.
[2] Yatim KM, Lakkis FG. A brief journey through the immune system. Clin J Am Soc Nephrol 2015;10(7):1274–81.
[3] Iwasaki A, Medzhitov R. Control of adaptive immunity by the innate immune system. Nat Immunol 2015;16(4):343–53.
[4] Flajnik MF, Kasahara M. Origin and evolution of the adaptive immune system: genetic events and selective pressures. Nat Rev Genet 2010;11(1):47–59.
[5] Davies LC, Taylor PR. Tissue-resident macrophages: then and now. Immunology 2015;144(4):541–8.
[6] Oishi Y, Manabe I. Macrophages in inflammation, repair and regeneration. Int Immunol 2018;30(11):511–28.
[7] Martinez FO, Gordon S. The M1 and M2 paradigm of macrophage activation: time for reassessment. F1000Prime Rep 2014;6:13.
[8] Murray PJ, et al. Macrophage activation and polarization: nomenclature and experimental guidelines. Immunity 2014;41(1):14–20.
[9] Cani PD, Jordan BF. Gut microbiota-mediated inflammation in obesity: a link with gastrointestinal cancer. Nat Rev Gastroenterol Hepatol 2018;15(11):671–82.
[10] Lucas C, Barnich N, Nguyen HTT. Microbiota, inflammation and colorectal cancer. Int J Mol Sci 2017;18(6):1–27.
[11] Bain CC, Mowat AM. Macrophages in intestinal homeostasis and inflammation. Immunol Rev 2014;260(1):102–17.
[12] Gren ST, Grip O. Role of monocytes and intestinal macrophages in Crohn's disease and ulcerative colitis. Inflamm Bowel Dis 2016;22(8):1992–8.

[13] Bain CC, et al. Resident and pro-inflammatory macrophages in the colon represent alternative context-dependent fates of the same Ly6Chi monocyte precursors. Mucosal Immunol 2013;6(3):498–510.
[14] Thiesen S, et al. CD14(hi)HLA-DR(dim) macrophages, with a resemblance to classical blood monocytes, dominate inflamed mucosa in Crohn's disease. J Leukoc Biol 2014;95(3):531–41.
[15] Becker L, et al. Age-related changes in gut microbiota alter phenotype of muscularis macrophages and disrupt gastrointestinal motility. Cell Mol Gastroenterol Hepatol 2019;7(1):243–245 e2.
[16] Grainger JR, et al. Macrophages in gastrointestinal homeostasis and inflammation. Pflugers Arch 2017;469(3–4):527–39.
[17] Cipriani G, et al. Intrinsic gastrointestinal macrophages: their phenotype and role in gastrointestinal motility. Cell Mol Gastroenterol Hepatol 2016;2(2):120–130 e1.
[18] Hettinger J, et al. Origin of monocytes and macrophages in a committed progenitor. Nat Immunol 2013;14(8):821–30.
[19] Geissmann F, Jung S, Littman DR. Blood monocytes consist of two principal subsets with distinct migratory properties. Immunity 2003;19(1):71–82.
[20] Cerovic V, et al. Intestinal macrophages and dendritic cells: what's the difference? Trends Immunol 2014;35(6):270–7.
[21] Smith PD, et al. Intestinal macrophages and response to microbial encroachment. Mucosal Immunol 2011;4(1):31–42.
[22] Hashimoto D, et al. Tissue-resident macrophages self-maintain locally throughout adult life with minimal contribution from circulating monocytes. Immunity 2013;38(4):792–804.
[23] Mantovani A, et al. The chemokine system in diverse forms of macrophage activation and polarization. Trends Immunol 2004;25(12):677–86.
[24] Kang B, et al. Commensal microbiota drive the functional diversification of colon macrophages. Mucosal Immunol 2020;13(2):216–29.
[25] Soucy G, et al. Clinical and pathological analysis of colonic Crohn's disease, including a subgroup with ulcerative colitis-like features. Mod Pathol 2012;25(2):295–307.
[26] Ince MN, Elliott DE. Immunologic and molecular mechanisms in inflammatory bowel disease. Surg Clin North Am 2007;87(3):681–96.
[27] Hindryckx P, Jairath V, D'Haens G. Acute severe ulcerative colitis: from pathophysiology to clinical management. Nat Rev Gastroenterol Hepatol 2016;13(11):654–64.
[28] Miehlke S, et al. Microscopic colitis: pathophysiology and clinical management. Lancet Gastroenterol Hepatol 2019;4(4):305–14.
[29] Mikkelsen HB, Rumessen JJ. Characterization of macrophage-like cells in the external layers of human small and large intestine. Cell Tissue Res 1992;270(2):273–9.
[30] Wang S, et al. Functions of macrophages in the maintenance of intestinal homeostasis. J Immunol Res 2019;2019:1512969.
[31] Sehgal A, et al. The role of CSF1R-dependent macrophages in control of the intestinal stem-cell niche. Nat Commun 2018;9(1):1272.
[32] Ma TY, et al. TNF-alpha-induced increase in intestinal epithelial tight junction permeability requires NF-kappa B activation. Am J Physiol Gastrointest Liver Physiol 2004;286(3):G367–76.
[33] Matheis F, et al. Adrenergic signaling in muscularis macrophages limits infection-induced neuronal loss. Cell 2020;180(1):64–78.e16.
[34] Mantovani A, Marchesi F. IL-10 and macrophages orchestrate gut homeostasis. Immunity 2014;40(5):637–9.
[35] Edin S, et al. Phenotypic skewing of macrophages in vitro by secreted factors from colorectal cancer cells. PLoS One 2013;8(9):e74982.
[36] Ruder B, Atreya R, Becker C. Tumour necrosis factor alpha in intestinal homeostasis and gut related diseases. Int J Mol Sci 2019;20(8).

[37] Weber B, Saurer L, Mueller C. Intestinal macrophages: differentiation and involvement in intestinal immunopathologies. Semin Immunopathol 2009;31(2):171–84.
[38] Morhardt TL, et al. IL-10 produced by macrophages regulates epithelial integrity in the small intestine. Sci Rep 2019;9(1):1223.
[39] Isidro RA, Appleyard CB. Colonic macrophage polarization in homeostasis, inflammation, and cancer. Am J Physiol Gastrointest Liver Physiol 2016;311(1):G59–73.
[40] Kamada N, et al. Unique CD14 intestinal macrophages contribute to the pathogenesis of Crohn disease via IL-23/IFN-gamma axis. J Clin Invest 2008;118(6):2269–80.
[41] Rogler G, et al. Isolation and phenotypic characterization of colonic macrophages. Clin Exp Immunol 1998;112(2):205–15.
[42] Ogino T, et al. Increased Th17-inducing activity of CD14+ CD163 low myeloid cells in intestinal lamina propria of patients with Crohn's disease. Gastroenterology 2013;145(6):1380–1391.e1.
[43] Yahaya MAF, et al. Tumour-associated macrophages (TAMs) in colon cancer and how to reeducate them. J Immunol Res 2019;2019:2368249.
[44] Zhong X, Chen B, Yang Z. The role of tumor-associated macrophages in colorectal carcinoma progression. Cell Physiol Biochem 2018;45(1):356–65.
[45] Lee C, et al. Targeting of M2-like tumor-associated macrophages with a melittin-based pro-apoptotic peptide. J Immunother Cancer 2019;7(1):147.
[46] Kikuchi T, et al. Characterization of tumor-infiltrating immune cells in relation to microbiota in colorectal cancers. Cancer Immunol Immunother 2020;69(1):23–32.
[47] Li R, et al. Gut microbiota-stimulated cathepsin K secretion mediates TLR4-dependent M2 macrophage polarization and promotes tumor metastasis in colorectal cancer. Cell Death Differ 2019;26(11):2447–63.
[48] Zhang L, Li S. Lactic acid promotes macrophage polarization through MCT-HIF1alpha signaling in gastric cancer. Exp Cell Res 2020;388(2):111846.
[49] Ohashi T, et al. M2-like macrophage polarization in high lactic acid-producing head and neck cancer. Cancer Sci 2017;108(6):1128–34.
[50] Edin S, et al. The distribution of macrophages with a M1 or M2 phenotype in relation to prognosis and the molecular characteristics of colorectal cancer. PLoS One 2012;7(10):e47045.
[51] McLean MH, et al. The inflammatory microenvironment in colorectal neoplasia. PLoS One 2011;6(1):e15366.
[52] Galvan-Pena S, O'Neill LA. Metabolic reprograming in macrophage polarization. Front Immunol 2014;5:420.
[53] Biswas SK, Mantovani A. Orchestration of metabolism by macrophages. Cell Metab 2012;15(4):432–7.
[54] Alves RW, Doretto-Silva L, da Silva EM, et al. The non-canonical role of metabolic enzymes in immune cells and its impact on diseases. Curr Tissue Microenviron 2020;Rep. **1**:221–37.
[55] Viola A, et al. The metabolic signature of macrophage responses. Front Immunol 2019;10:1462.
[56] Banerjee S, et al. ROS-associated immune response and metabolism: a mechanistic approach with implication of various diseases. Arch Toxicol 2020;94(7):2293–317.
[57] Benmoussa K, Garaude J, Acin-Perez R. How mitochondrial metabolism contributes to macrophage phenotype and functions. J Mol Biol 2018;430(21):3906–21.
[58] West AP, et al. TLR signalling augments macrophage bactericidal activity through mitochondrial ROS. Nature 2011;472(7344):476–80.
[59] Postler TS, Ghosh S. Understanding the holobiont: how microbial metabolites affect human health and shape the immune system. Cell Metab 2017;26(1):110–30.
[60] Macpherson AJ, Uhr T. Induction of protective IgA by intestinal dendritic cells carrying commensal bacteria. Science 2004;303(5664):1662–5.

[61] Khosravi A, et al. Gut microbiota promote hematopoiesis to control bacterial infection. Cell Host Microbe 2014;15(3):374–81.
[62] Kim M, et al. Gut microbial metabolites fuel host antibody responses. Cell Host Microbe 2016;20(2):202–14.
[63] Takahashi K, et al. Reduced abundance of butyrate-producing bacteria species in the fecal microbial community in Crohn's disease. Digestion 2016;93(1):59–65.
[64] Feng W, Ao H, Peng C. Gut microbiota, short-chain fatty acids, and herbal medicines. Front Pharmacol 2018;9:1354.
[65] Macfarlane S, Macfarlane GT. Regulation of short-chain fatty acid production. Proc Nutr Soc 2003;62(1):67–72.
[66] Hoverstad T, Midtvedt T. Short-chain fatty acids in germfree mice and rats. J Nutr 1986;116(9):1772–6.
[67] Bilotta AJ, Cong Y. Gut microbiota metabolite regulation of host defenses at mucosal surfaces: implication in precision medicine. Precis Clin Med 2019;2(2):110–9.
[68] Chun E, et al. Metabolite-sensing receptor Ffar2 regulates colonic group 3 innate lymphoid cells and gut immunity. Immunity 2019;51(5):871–884.e6.
[69] Lavoie S, et al. Expression of free fatty acid receptor 2 by dendritic cells prevents their expression of interleukin 27 and is required for maintenance of mucosal barrier and immune response against colorectal tumors in mice. Gastroenterology 2020;158(5):1359–1372.e9.
[70] Nakajima A, et al. The short chain fatty acid receptor GPR43 regulates inflammatory signals in adipose tissue M2-type macrophages. PLoS One 2017;12(7):1–18.
[71] Vieira AT, et al. A role for gut microbiota and the metabolite-sensing receptor GPR43 in a murine model of gout. Arthritis Rheumatol 2015;67(6):1646–56.
[72] Aguilar EC, et al. Butyrate impairs atherogenesis by reducing plaque inflammation and vulnerability and decreasing NFkappaB activation. Nutr Metab Cardiovasc Dis 2014;24(6):606–13.
[73] Ji J, et al. Microbial metabolite butyrate facilitates M2 macrophage polarization and function. Sci Rep 2016;6:24838.
[74] Scott NA, et al. Antibiotics induce sustained dysregulation of intestinal T cell immunity by perturbing macrophage homeostasis. Sci Transl Med 2018;10(464):1–15.
[75] Chang PV, et al. The microbial metabolite butyrate regulates intestinal macrophage function via histone deacetylase inhibition. Proc Natl Acad Sci USA 2014;111(6):2247–52.
[76] Pascal V, et al. A microbial signature for Crohn's disease. Gut 2017;66(5):813–22.
[77] Tofalo R, Cocchi S, Suzzi G. Polyamines and gut microbiota. Front Nutr 2019;6:16.
[78] Nakamura A, Ooga T, Matsumoto M. Intestinal luminal putrescine is produced by collective biosynthetic pathways of the commensal microbiome. Gut Microbes 2019;10(2):159–71.
[79] Zhang M, et al. Spermine inhibits proinflammatory cytokine synthesis in human mononuclear cells: a counterregulatory mechanism that restrains the immune response. J Exp Med 1997;185(10):1759–68.
[80] Kibe R, et al. Upregulation of colonic luminal polyamines produced by intestinal microbiota delays senescence in mice. Sci Rep 2014;4:4548.
[81] Hong SK, et al. Increased expression and cellular localization of spermine oxidase in ulcerative colitis and relationship to disease activity. Inflamm Bowel Dis 2010;16(9):1557–66.
[82] Schneider J, Wendisch VF. Biotechnological production of polyamines by bacteria: recent achievements and future perspectives. Appl Microbiol Biotechnol 2011;91(1):17–30.
[83] Rodriguez PC, Ochoa AC, Al-Khami AA. Arginine metabolism in myeloid cells shapes innate and adaptive immunity. Front Immunol 2017;8:93.
[84] Hardbower DM, et al. Ornithine decarboxylase regulates M1 macrophage activation and mucosal inflammation via histone modifications. Proc Natl Acad Sci USA 2017;114(5):E751–60.

[85] Van den Bossche J, et al. Pivotal advance: arginase-1-independent polyamine production stimulates the expression of IL-4-induced alternatively activated macrophage markers while inhibiting LPS-induced expression of inflammatory genes. J Leukoc Biol 2012;91(5):685–99.
[86] Na YR, et al. GM-CSF induces inflammatory macrophages by regulating glycolysis and lipid metabolism. J Immunol 2016;197(10):4101–9.
[87] Dabritz J. Granulocyte macrophage colony-stimulating factor and the intestinal innate immune cell homeostasis in Crohn's disease. Am J Physiol Gastrointest Liver Physiol 2014;306(6):G455–65.
[88] Zhang M, Wang H, Tracey KJ. Regulation of macrophage activation and inflammation by spermine: a new chapter in an old story. Crit Care Med 2000;28(4 Suppl):N60–6.
[89] Singh K, et al. Ornithine decarboxylase in macrophages exacerbates colitis and promotes colitis-associated colon carcinogenesis by impairing M1 immune responses. Cancer Res 2018;78(15):4303–15.
[90] Agus A, Planchais J, Sokol H. Gut microbiota regulation of tryptophan metabolism in health and disease. Cell Host Microbe 2018;23(6):716–24.
[91] Zelante T, et al. Tryptophan catabolites from microbiota engage aryl hydrocarbon receptor and balance mucosal reactivity via interleukin-22. Immunity 2013;39(2):372–85.
[92] Clarke G, et al. The microbiome-gut-brain axis during early life regulates the hippocampal serotonergic system in a sex-dependent manner. Mol Psychiatry 2013;18(6):666–73.
[93] Yano JM, et al. Indigenous bacteria from the gut microbiota regulate host serotonin biosynthesis. Cell 2015;161(2):264–76.
[94] Zhao ZH, et al. Indole-3-propionic acid inhibits gut dysbiosis and endotoxin leakage to attenuate steatohepatitis in rats. Exp Mol Med 2019;51(9):1–14.
[95] Ji Y, et al. Anti-inflammatory and anti-oxidative activity of Indole-3-acetic acid involves induction of HO-1 and neutralization of free radicals in RAW264.7 cells. Int J Mol Sci 2020;21(5):1–14.
[96] Davies LC, et al. Tissue-resident macrophages. Nat Immunol 2013;14(10):986–95.

CHAPTER 10

Lungs—Inflammatory and respiratory system

Lucas D. Faustino
Division of Rheumatology, Allergy and Immunology, Center for Immunology and Inflammatory Diseases, Massachusetts General Hospital, Harvard Medical School, Boston, MA, United States

The lungs are the primary organs of the respiratory system that continuously supply oxygen to the metabolism of cells throughout the body through a dynamic process of oxygen and carbon dioxide exchange with the blood. This dynamic process of ventilation moves the air through the highly branched conducting airways of the lungs to the alveoli, where epithelial cells are in close opposition to a pulmonary capillary network that facilitate the gas exchange between the respiratory system and the bloodstream. While the ventilation process delivers life-requiring oxygen to the systemic circulation, particles, toxins, allergens, and microbes are also brought into the lungs, where they encounter resident immune cells that protect the host from infection and tissue injury. Macrophages are one of the most abundant immune cells in barrier tissues, including the lungs. They provide the first immune defense to environmental insults and serve as guardians between the host and the outside world.

Macrophages are mononuclear phagocytes found essentially in all tissues of the body that carry out important roles in innate immunity and inflammation, but are also essential for maintenance of normal healthy tissues. They were first described based on their ability to phagocytize microorganisms and scavenger apoptotic cells and cellular debris and are classically identified by the expression of the surface markers F4/80 and CD68 in mice and humans, respectively [1,2].

Despite these common signatures, multi-parameter flow cytometric analyses and genome-wide transcriptional profiling have revealed a remarkable heterogeneity in macrophages throughout the body [3]. In parallel, macrophages have been classified in many ways by researchers on the basis of their developmental origins, organ of residence, and functional properties. Until recently, macrophages were divided into two major subsets known as "M1-like" and "M2-like" based on their pro-inflammatory and

Macrophages in the Human Body
https://doi.org/10.1016/B978-0-12-821385-8.00009-3

anti-inflammatory properties in analogy to the T helper type 1 (Th1) and Th2 lymphocytes, respectively [4]. However, studies over the past few years have shown that this approach is oversimplistic since macrophages possess specialized properties adapted to individual tissues that go beyond their immunological function [5]. It is now becoming clear that individual populations of macrophages are strongly adapted to the needs of their tissue of residency and fulfill roles specific of their anatomical niche. Although the differences in macrophages across the body might reflect their distinct developmental origins, local factors in the environment are determinant for the macrophage identity and their function under both homeostatic and inflammatory conditions. Here, we discuss how this new understanding of macrophage biology provides insights into their function in the lung, an immunologically active barrier site that is continuously exposed to the external environment, and is of crucial physiological importance.

Lung-resident macrophage ontogeny and phenotype

Tissue macrophages were classified as part of the "mononuclear phagocyte system" along with monocytes and their progenitors in the adult bone marrow (BM) by van Furth & Cohn in the mid-1970s [6,7]. The mononuclear phagocyte concept held that committed bone marrow precursors develop into blood monocytes from which tissue macrophages derive [7]. This concept was the prevailing view for decades, despite evidence of the presence of macrophage populations in the yolk sac before embryonic hematopoiesis [8] and macrophage proliferation and persistence in adult tissues irrespective of circulating blood monocytes [9–11]. In recent years, however, a series of more definitive studies have revised our understanding of macrophage ontogeny, demonstrating that many tissue resident macrophages are established during embryonic development, which persist into adulthood regardless of blood monocyte input [12–17]. Although still primarily based on mouse data, the revised concept of the mononuclear phagocyte system now accommodates the existence of tissue macrophages from two distinct origins: those established during embryonic development, which comprise most macrophages that reside in healthy tissues [14,15,17], and those developed in adulthood from circulating blood monocytes [12,17].

The lungs harbor at least two types of macrophages—alveolar macrophages (AMs) and interstitial macrophages (IMs)—that differ in their developmental origin and the anatomical niche they occupy. AMs develop from fetal liver monocytes that populate the lungs around embryonic day 15 (E15) [12,15,18–20]. In the absence of macrophage growth and differentiation factor GM-CSF, there is a deficiency in the AM differentiation from embryonic progenitors in the fetal lungs around E17, indicating that GM-CSF is instrumental for licensing AM development [15,18]. Shortly before and after birth, fetal lung monocytes expand substantially, accompanied by a profound change in the expression of surface markers. These phenotypic changes include the specific upregulation of CD11c integrin and sialic acid-biding immunoglobulin (Ig)-like lectin F (Siglec-F) and concomitant downregulation of CD11b integrin until they reach the mature AM stage.

AMs inhabit the air space of the alveoli where they represent ~90%–95% of the leukocytes in homeostatic conditions. AMs are long-lived and capable of self-renewal without substantial contribution from bone marrow (BM)-derived monocytes [13–15,21], as suggested by the intact AM compartment in $Myb^{-/-}$ mice depleted of monocyte precursors [12] and in leukemic patients with decreased or absent blood monocytes [22]. Despite the well-known embryonic origin of AMs, it has been speculated that under inflammatory conditions and tissue injury, circulating blood monocytes may enter the lungs and differentiate into AMs in response to cues provided by the local microenvironment [23], although the molecular basis for the differentiation of monocytes into alveolar macrophages is incompletely understood.

IMs reside the parenchyma space (interstitium) between the alveoli and capillary beds and can be distinguished from AMs by their distinct surface phenotype, including the expression of CD11b, CX3CR1, and absence of Siglec-F [3,24–27]. IMs can also express variable levels of CD11c and the major histocompatibility complex class II (MHCII) [3,25,28] that lead to their misclassification as dendritic cells (DCs) [29], therefore, limiting their identification in the lungs and the understanding of their role during homeostatic and inflammatory conditions. Although the developmental origin of IMs is not completely understood, they have been shown to require replacement by BM-derived monocytes, which generally display limited lifespans [16,27,30,31]. All these data suggest that different anatomical niches in the lungs—alveoli and parenchyma—use distinct mechanisms to maintain their macrophage populations.

Lung macrophages in homeostatic conditions

Macrophages are strongly adapted to their tissue of residence and perform specific functions that are essential for organ homeostasis [3,5,32,33]. CSF2, also known as GM-CSF, is produced by alveolar epithelial cells and controls much of the unique phenotypic and functional identity of AMs, including lipid catabolism and cytokine production through the induction of peroxisome proliferator-activated receptor-γ (PPAR-γ) [3,18,34,35]. Indeed, mice and humans in which the CSF2-CSF2 receptor axis has been disrupted, a severe pulmonary alveolar proteinosis (PAP) develops owing to excessive surfactant accumulation in the alveoli due to defective AM development [15,36,37]. CSF2 induces the expression of PPAR-γ in AMs [18,35] and its expression is lost in CSF2-deficient mice [38]. In patients with PAP, AMs have reduced expression of PPAR-γ and excessive surfactant lipid accumulation, but systemic CFS2 therapy is effective in rescuing PPAR-γ expression and lipid catabolism in AMs [39]. In addition, mice lacking PPARγ in macrophages develop spontaneous inflammation in the lungs [40], highlighting the important role of the lipid metabolism by AMs in maintaining pulmonary homeostasis.

AMs interact with alveolar epithelial cells through a variety of surface molecules known to dampen the immune responses to injury and facilitate the non-inflammatory removal of apoptotic cells and particulates [41,42]. This state of hyporesponsiveness ensures AMs to clear pathogens and environmental particles without inducing an inflammatory response that might damage the alveolar epithelium and disrupt gas exchange. Examples of anti-inflammatory surface molecules expressed by AMs include the CD200 receptor, which binds to its ligand CD200 on alveolar epithelial cells to inhibit inflammatory responses induced by Toll-like receptor (TLR) ligands and other stimuli [43]. Deficiency in CD200 or CD200 receptor results in spontaneous AM activation and uncontrolled inflammatory responses in the lungs to infection, respectively [43]. Transforming grow factor-β (TGF-β) receptor is another inhibitory surface molecule expressed on AMs. Because members of the TGF-β family are secreted in an inactive (latent) form, they require enzymatic cleavage or interaction with non-proteolytic molecules, such as integrins, for its activation into a bioactive TGF-β [44,45]. In the lungs, the integrin αVβ6 expressed on bronchial and alveolar epithelial cells is critical for the activation of latent TGF-β into bioactive TGF-β [45,46]. The conformational change induced in latent TGF-β by interacting with αVβ6 integrin on alveolar epithelial cells facilitates access of TGF-β receptor on AMs to the bioactive TGF-β, which inhibits pro-inflammatory activities of AMs [45]. Indeed, deficiency in αVβ6 on epithelial cells observed in *Itgb6*$^{-/-}$ mice results in spontaneous pulmonary emphysema due to unrestrained production of the matrix metalloproteinase MMP12 by AMs [47]. AMs from *Itgb6*$^{-/-}$ mice also show a defect in surfactant lipid catabolism similar to that seen in CSF2-deficient mice or patients with PAP [48], and

Itgb6$^{-/-}$ mice develop inflammation in the lungs (and skin) [49,50], which can be reversed by the administration of exogenous active TGF-β1 or by the targeted expression of functional αvβ6 integrin in the lung epithelium [47]. Interleukin (IL)-10 is also critical for maintaining homeostasis in barrier sites. In the lungs, IL-10 has been found to bind the IL-10 receptor on AMs to restrict inflammatory responses through induction of negative regulators, such as the suppressor of cytokine signaling 3 (SOCS3) and the microRNA miR-146b, which blocks the expression of pro-inflammatory cytokines and directly inhibits TLR4 expression and signaling, respectively [51,52].

Pulmonary surfactant-associated proteins are abundant in the alveolar fluid and are produced in the lower airways by type II alveolar epithelial cells [53]. The binding of epithelial-derived surfactant proteins, of which SP-A and SP-D are the most abundant, to the signal regulatory protein α (SIRPα) and TLRs, can also modify phagocytosis, cytokine production, and TLR responsiveness of AMs [54]. For example, SP-A and SP-D can block TLR2 and TLR4 interactions with their respective ligands, as well as their interactions with the TLR co-receptors MD2 and CD14, which prevents the activation of nuclear factor-κB (NF-κB) and the initiation of the inflammatory response [54–56]. Although binding of SP-A and -SP-D to SIRPα inhibits AM-mediated phagocytosis [56], specific surfactant-associated proteins can bind to pathogens, apoptotic cells, and allergens to facilitate their neutralization and elimination without inducing AM activation [57]. AMs may also maintain lung homeostasis by supporting the differentiation of forkhead box P3 (Foxp3) expressing regulatory T cells [58,59].

The role of IMs in pulmonary homeostasis is poorly understood, in part due to the divergent criteria used by researchers to define these lung resident cells [60–62]. However, a prominent feature of IMs is their constitutive production of the anti-inflammatory cytokine IL-10 [62], which can influence the activity of immune cells that reside in the lungs, for example, by reducing the immunogenicity of dendritic cells in the lung [61].

Lung macrophages in inflammatory conditions

Pulmonary macrophages are conditioned by the requirements of the tissue during homeostatic and pathological processes, and their role in maintaining organ homeostasis is equally as important as their participation in immunological responses. This functional plasticity of resident macrophages ensures that inflammatory responses to tissue debris and innocuous antigens are inhibited, while effective immune responses to pathogens are not compromised. It has been speculated that changes in the AM activity during pathological events result from damage to the airway epithelium and loss of exposure to regulatory ligands expressed by epithelial cells that cause AMs to respond to pathogens and other inhaled antigens in a pro-inflammatory, as opposed to an anti-inflammatory, manner.

Given their positioning in the airways, high phagocytic capacity, and expression of a wide range of pattern-recognition receptors, lectins, and scavenger receptors [41], AMs serve as the first line of defense against bacterial, fungal, and viral infections. AMs have the ability to engulf and destroy bacteria, such as *Streptococcus pneumonia, Mycobacteria tuberculosis, Pseudomonas aeruginosa* [63–67], and fungi, such as *Pneumocystis carinii* [68–70]. Examples on how AMs remove these pathogens from the alveoli include the expression of the macrophage receptor with collagenous structure (MARCO), a class A scavenger receptor that bind to structures found on the surface of pathogens [71]. Although MARCO does not directly trigger an inflammatory response from AMs, it helps the binding of other pattern recognition receptors expressed on AMs to pathogen-associated molecules, which can initiate inflammation [72]. AMs also orchestrate the recruitment of neutrophils and effector monocytes to the lungs following infection through the release of inflammatory cytokines, such as IL-1β and type I interferons, which induce the production of chemokines from the respiratory epithelium [73,74].

Among respiratory viral infections, influenza virus is the best studied example. In the absence of AMs, as observed in mice that devoid of CSF2 or its receptor, there is an impairment in virus clearance and an increase in mortality in response to influenza infection [75–78]. Conversely, pulmonary overexpression of CSF2 in mice prevents AM apoptosis and provides resistance to lethal doses of influenza virus [79]. AMs infected by influenza virus can also upregulate the production of type I interferons [80], which restrict viral spread [81] and ameliorate morbidity, for example, through induction of the interferon-inducible antiviral protein IFITM3 [82,83]. The type I interferon-dependent host defense mechanism appears to be impaired in humans infected with the severe acute respiratory syndrome coronavirus 2 (SARS-CoV-2), the virus responsible for the ongoing coronavirus disease 2019 (COVID-19) pandemic, as low levels of type I interferons have been associated with a persistent viral load and an exacerbated inflammatory response in COVID-19 patients [84–86].

AMs have also been implicated in the development and progression of asthma, although it remains uncertain whether they have a pathogenic or protective role [87]. For example, some studies have demonstrated increased IL-10 production by AMs from asthmatic patients related to their counterparts from healthy lungs [88]. In addition, depletion of AMs in mice worsens allergen-induced airway inflammation while adoptive transfer of normal AMs can protect sensitized lungs from damage [89]. On the other hand, other studies found that AMs from patients with asthma express higher levels of costimulatory molecules such as CD80 [90] and, therefore, might be able to stimulate Th2 cell responses.

IMs have also been shown to confer protection against asthma and to reduce responses to inhaled allergens by producing IL-10 [61,91]. However, the role of IMs in pulmonary infection and inflammation, remains unclear. Lung macrophages are known to promote fibrogenesis, matrix remodeling and restoration of the integrity of the respiratory epithelium after lung injury. However, exacerbated tissue injury, in some cases, leads to aberrant wound healing and ultimately fibrosis that is driven, at least in part, by IMs, which have been implicated in the pathogenesis of interstitial lung diseases, such as idiopathic pulmonary fibrosis (IPF) [92]. The relative contribution of tissue-resident IMs and monocyte-derived infiltrating macrophages in this condition is still poorly understood, but some studies have suggested an involvement of newly recruited macrophages on this process [93,94]. Whether the origin of macrophages—tissue-resident versus infiltrating—determines the fate of lung diseases is still in debate.

Conclusion remarks

The lungs harbor two populations of macrophages, AMs and IMs, that represent distinct lineages with specific origins, phenotypes, and functions. Although AMs and IMs likely share functions in lung homeostasis and immunity, some of their properties are highly specific to the anatomical niche they occupy and likely imprinted locally, which ensures precise adaptation to the demands of their environment. These central roles in homeostasis, immunity, and tissue physiology offer clear possibilities for targeted therapeutic intervention. However, for this to be achieve, a better understanding of the AM and IM specialization and maintenance of these cells in their specific lung niches is still needed. For example, it is crucial to identify the factors that control the longevity and self-renewal of AMs and the differentiation of infiltrating monocytes in IMs. In addition, it is still unclear whether macrophages derived from newly recruited monocytes after lung injury and inflammation can truly converge into functionally equivalent IMs (and AMs) or whether they possess different functions and features. Determining the relative contribution of resident and "infiltrating" macrophages to inflammation and tissue injury would create options for preventing their activation or recruitment, with the aim of restoring lung homeostasis. The translation of our knowledge of pulmonary macrophage biology to humans is still a challenge, but of fundamental importance for the treatment of lung diseases.

References

[1] Austyn JM, Gordon S. F4/80, a monoclonal antibody directed specifically against the mouse macrophage. Eur J Immunol 1981;11(10):805–15.
[2] Holness CL, Simmons DL. Molecular cloning of CD68, a human macrophage marker related to lysosomal glycoproteins. Blood 1993;81(6):1607–13.
[3] Gautier EL, et al. Gene-expression profiles and transcriptional regulatory pathways that underlie the identity and diversity of mouse tissue macrophages. Nat Immunol 2012;13(11):1118–28.
[4] Mills CD, et al. M-1/M-2 macrophages and the Th1/Th2 paradigm. J Immunol 2000;164(12):6166–73.
[5] Okabe Y, Medzhitov R. Tissue biology perspective on macrophages. Nat Immunol 2016;17(1):9–17.
[6] van Furth R, Cohn ZA. The origin and kinetics of mononuclear phagocytes. J Exp Med 1968;128(3):415–35.
[7] van Furth R, et al. The mononuclear phagocyte system: a new classification of macrophages, monocytes, and their precursor cells. Bull World Health Organ 1972;46(6):845–52.
[8] Naito M, et al. Development, differentiation, and phenotypic heterogeneity of murine tissue macrophages. J Leukoc Biol 1996;59(2):133–8.
[9] Sawyer RT, Strausbauch PH, Volkman A. Resident macrophage proliferation in mice depleted of blood monocytes by strontium-89. Lab Investig 1982;46(2):165–70.
[10] Volkman A, et al. Differential effects of chronic monocyte depletion on macrophage populations. Lab Investig 1983;49(3):291–8.
[11] Melnicoff MJ, et al. Maintenance of peritoneal macrophages in the steady state. J Leukoc Biol 1988;44(5):367–75.
[12] Schulz C, et al. A lineage of myeloid cells independent of Myb and hematopoietic stem cells. Science 2012;336(6077):86–90.
[13] Yona S, et al. Fate mapping reveals origins and dynamics of monocytes and tissue macrophages under homeostasis. Immunity 2013;38(1):79–91.
[14] Hashimoto D, et al. Tissue-resident macrophages self-maintain locally throughout adult life with minimal contribution from circulating monocytes. Immunity 2013;38(4):792–804.
[15] Guilliams M, et al. Alveolar macrophages develop from fetal monocytes that differentiate into long-lived cells in the first week of life via GM-CSF. J Exp Med 2013;210(10):1977–92.
[16] Jakubzick C, et al. Minimal differentiation of classical monocytes as they survey steady-state tissues and transport antigen to lymph nodes. Immunity 2013;39(3):599–610.
[17] Ginhoux F, Guilliams M. Tissue-resident macrophage ontogeny and homeostasis. Immunity 2016;44(3):439–49.
[18] Schneider C, et al. Induction of the nuclear receptor PPAR-gamma by the cytokine GM-CSF is critical for the differentiation of fetal monocytes into alveolar macrophages. Nat Immunol 2014;15(11):1026–37.

[19] Naito M, Takahashi K, Nishikawa S. Development, differentiation, and maturation of macrophages in the fetal mouse liver. J Leukoc Biol 1990;48(1):27–37.
[20] Orkin SH, Zon LI. Hematopoiesis: an evolving paradigm for stem cell biology. Cell 2008;132(4):631–44.
[21] Murphy J, et al. The prolonged life-span of alveolar macrophages. Am J Respir Cell Mol Biol 2008;38(4):380–5.
[22] Golde DW, Finley TN, Cline MJ. The pulmonary macrophage in acute leukemia. N Engl J Med 1974;290(16):875–8.
[23] Morales-Nebreda L, et al. The heterogeneity of lung macrophages in the susceptibility to disease. Eur Respir Rev 2015;24(137):505–9.
[24] Misharin AV, et al. Flow cytometric analysis of macrophages and dendritic cell subsets in the mouse lung. Am J Respir Cell Mol Biol 2013;49(4):503–10.
[25] Bharat A, et al. Flow cytometry reveals similarities between lung macrophages in humans and mice. Am J Respir Cell Mol Biol 2016;54(1):147–9.
[26] Chakarov S, et al. Two distinct interstitial macrophage populations coexist across tissues in specific subtissular niches. Science 2019;363(6432):1190.
[27] Gibbings SL, et al. Three unique interstitial macrophages in the murine lung at steady state. Am J Respir Cell Mol Biol 2017;57(1):66–76.
[28] Becher B, et al. High-dimensional analysis of the murine myeloid cell system. Nat Immunol 2014;15(12):1181–9.
[29] Plantinga M, et al. Conventional and monocyte-derived CD11b(+) dendritic cells initiate and maintain T helper 2 cell-mediated immunity to house dust mite allergen. Immunity 2013;38(2):322–35.
[30] Bain CC, et al. Long-lived self-renewing bone marrow-derived macrophages displace embryo-derived cells to inhabit adult serous cavities. Nat Commun 2016;7:11852.
[31] Gomez Perdiguero E, et al. Tissue-resident macrophages originate from yolk-sac-derived erythro-myeloid progenitors. Nature 2015;518(7540):547–51.
[32] Gosselin D, et al. Environment drives selection and function of enhancers controlling tissue-specific macrophage identities. Cell 2014;159(6):1327–40.
[33] Lavin Y, et al. Tissue-resident macrophage enhancer landscapes are shaped by the local microenvironment. Cell 2014;159(6):1312–26.
[34] Asada K, et al. Antiinflammatory roles of peroxisome proliferator-activated receptor gamma in human alveolar macrophages. Am J Respir Crit Care Med 2004;169(2):195–200.
[35] Yu X, et al. The cytokine TGF-beta promotes the development and homeostasis of alveolar macrophages. Immunity 2017;47(5):903–912.e4.
[36] Trapnell BC, Whitsett JA, Nakata K. Pulmonary alveolar proteinosis. N Engl J Med 2003;349(26):2527–39.
[37] Nishinakamura R, et al. The pulmonary alveolar proteinosis in granulocyte macrophage colony-stimulating factor/interleukins 3/5 beta c receptor-deficient mice is reversed by bone marrow transplantation. J Exp Med 1996;183(6):2657–62.
[38] Malur A, et al. Restoration of PPARgamma reverses lipid accumulation in alveolar macrophages of GM-CSF knockout mice. Am J Physiol Lung Cell Mol Physiol 2011;300(1):L73–80.
[39] Bonfield TL, et al. Peroxisome proliferator-activated receptor-gamma is deficient in alveolar macrophages from patients with alveolar proteinosis. Am J Respir Cell Mol Biol 2003;29(6):677–82.
[40] Gautier EL, et al. Systemic analysis of PPARgamma in mouse macrophage populations reveals marked diversity in expression with critical roles in resolution of inflammation and airway immunity. J Immunol 2012;189(5):2614–24.
[41] Hussell T, Bell TJ. Alveolar macrophages: plasticity in a tissue-specific context. Nat Rev Immunol 2014;14(2):81–93.
[42] Westphalen K, et al. Sessile alveolar macrophages communicate with alveolar epithelium to modulate immunity. Nature 2014;506(7489):503–6.

[43] Snelgrove RJ, et al. A critical function for CD200 in lung immune homeostasis and the severity of influenza infection. Nat Immunol 2008;9(9):1074–83.
[44] Yehualaeshet T, et al. Activation of rat alveolar macrophage-derived latent transforming growth factor beta-1 by plasmin requires interaction with thrombospondin-1 and its cell surface receptor, CD36. Am J Pathol 1999;155(3):841–51.
[45] Munger JS, et al. The integrin alpha v beta 6 binds and activates latent TGF beta 1: a mechanism for regulating pulmonary inflammation and fibrosis. Cell 1999;96(3):319–28.
[46] Coker RK, et al. Diverse cellular TGF-beta 1 and TGF-beta 3 gene expression in normal human and murine lung. Eur Respir J 1996;9(12):2501–7.
[47] Morris DG, et al. Loss of integrin alpha(v)beta6-mediated TGF-beta activation causes Mmp12-dependent emphysema. Nature 2003;422(6928):169–73.
[48] Koth LL, et al. Integrin beta6 mediates phospholipid and collectin homeostasis by activation of latent TGF-beta1. Am J Respir Cell Mol Biol 2007;37(6):651–9.
[49] Huang X, et al. Expression of the human integrin beta6 subunit in alveolar type II cells and bronchiolar epithelial cells reverses lung inflammation in beta6 knockout mice. Am J Respir Cell Mol Biol 1998;19(4):636–42.
[50] Huang XZ, et al. Inactivation of the integrin beta 6 subunit gene reveals a role of epithelial integrins in regulating inflammation in the lung and skin. J Cell Biol 1996;133(4):921–8.
[51] Bourdonnay E, et al. Transcellular delivery of vesicular SOCS proteins from macrophages to epithelial cells blunts inflammatory signaling. J Exp Med 2015;212(5):729–42.
[52] Curtale G, et al. Negative regulation of toll-like receptor 4 signaling by IL-10-dependent microRNA-146b. Proc Natl Acad Sci USA 2013;110(28):11499–504.
[53] Khoor A, et al. Developmental expression of SP-A and SP-A mRNA in the proximal and distal respiratory epithelium in the human fetus and newborn. J Histochem Cytochem 1993;41(9):1311–9.
[54] Haczku A. Protective role of the lung collectins surfactant protein A and surfactant protein D in airway inflammation. J Allergy Clin Immunol 2008;122(5):861–79 [quiz 880-1].
[55] Yamada C, et al. Surfactant protein A directly interacts with TLR4 and MD-2 and regulates inflammatory cellular response. Importance of supratrimeric oligomerization. J Biol Chem 2006;281(31):21771–80.
[56] Sano H, et al. Pulmonary surfactant protein A modulates the cellular response to smooth and rough lipopolysaccharides by interaction with CD14. J Immunol 1999;163(1):387–95.
[57] Forbes LR, Haczku A. SP-D and regulation of the pulmonary innate immune system in allergic airway changes. Clin Exp Allergy 2010;40(4):547–62.
[58] Soroosh P, et al. Lung-resident tissue macrophages generate Foxp3+ regulatory T cells and promote airway tolerance. J Exp Med 2013;210(4):775–88.
[59] Coleman MM, et al. Alveolar macrophages contribute to respiratory tolerance by inducing FoxP3 expression in naive T cells. Am J Respir Cell Mol Biol 2013;48(6):773–80.
[60] Landsman L, Jung S. Lung macrophages serve as obligatory intermediate between blood monocytes and alveolar macrophages. J Immunol 2007;179(6):3488–94.
[61] Bedoret D, et al. Lung interstitial macrophages alter dendritic cell functions to prevent airway allergy in mice. J Clin Invest 2009;119(12):3723–38.
[62] Kawano H, et al. IL-10-producing lung interstitial macrophages prevent neutrophilic asthma. Int Immunol 2016;28(10):489–501.
[63] Aberdein JD, et al. Alveolar macrophages in pulmonary host defence the unrecognized role of apoptosis as a mechanism of intracellular bacterial killing. Clin Exp Immunol 2013;174(2):193–202.

[64] Gordon SB, et al. Intracellular trafficking and killing of *Streptococcus pneumoniae* by human alveolar macrophages are influenced by opsonins. Infect Immun 2000;68(4):2286–93.
[65] Ballinger MN, et al. Role of granulocyte macrophage colony-stimulating factor during gram-negative lung infection with *Pseudomonas aeruginosa*. Am J Respir Cell Mol Biol 2006;34(6):766–74.
[66] Gonzalez-Juarrero M, et al. Disruption of granulocyte macrophage-colony stimulating factor production in the lungs severely affects the ability of mice to control *Mycobacterium tuberculosis* infection. J Leukoc Biol 2005;77(6):914–22.
[67] LeVine AM, et al. GM-CSF-deficient mice are susceptible to pulmonary group B streptococcal infection. J Clin Invest 1999;103(4):563–9.
[68] Paine 3rd R, et al. Granulocyte-macrophage colony-stimulating factor in the innate immune response to *Pneumocystis carinii* pneumonia in mice. J Immunol 2000;164(5):2602–9.
[69] Chen W, Mills JW, Harmsen AG. Development and resolution of *Pneumocystis carinii* pneumonia in severe combined immunodeficient mice: a morphological study of host inflammatory responses. Int J Exp Pathol 1992;73(6):709–20.
[70] Fleury-Feith J, et al. Bronchoalveolar lavage eosinophilia associated with *Pneumocystis carinii* pneumonitis in AIDS patients. Comparative study with non-AIDS patients. Chest 1989;95(6):1198–201.
[71] Arredouani M, et al. The scavenger receptor MARCO is required for lung defense against pneumococcal pneumonia and inhaled particles. J Exp Med 2004;200(2):267–72.
[72] Novakowski KE, et al. A naturally occurring transcript variant of MARCO reveals the SRCR domain is critical for function. Immunol Cell Biol 2016;94(7):646–55.
[73] Marriott HM, et al. Interleukin-1beta regulates CXCL8 release and influences disease outcome in response to *Streptococcus pneumoniae*, defining intercellular cooperation between pulmonary epithelial cells and macrophages. Infect Immun 2012;80(3):1140–9.
[74] Goritzka M, et al. Alveolar macrophage-derived type I interferons orchestrate innate immunity to RSV through recruitment of antiviral monocytes. J Exp Med 2015;212(5):699–714.
[75] Schneider C, et al. Alveolar macrophages are essential for protection from respiratory failure and associated morbidity following influenza virus infection. PLoS Pathog 2014;10(4):e1004053.
[76] Tumpey TM, et al. Pathogenicity of influenza viruses with genes from the 1918 pandemic virus: functional roles of alveolar macrophages and neutrophils in limiting virus replication and mortality in mice. J Virol 2005;79(23):14933–44.
[77] Tate MD, et al. Critical role of airway macrophages in modulating disease severity during influenza virus infection of mice. J Virol 2010;84(15):7569–80.
[78] Purnama C, et al. Transient ablation of alveolar macrophages leads to massive pathology of influenza infection without affecting cellular adaptive immunity. Eur J Immunol 2014;44(7):2003–12.
[79] Huang FF, et al. GM-CSF in the lung protects against lethal influenza infection. Am J Respir Crit Care Med 2011;184(2):259–68.
[80] Kumagai Y, et al. Alveolar macrophages are the primary interferon-alpha producer in pulmonary infection with RNA viruses. Immunity 2007;27(2):240–52.
[81] Seo SU, et al. Type I interferon signaling regulates Ly6C(hi) monocytes and neutrophils during acute viral pneumonia in mice. PLoS Pathog 2011;7(2):e1001304.
[82] Everitt AR, et al. IFITM3 restricts the morbidity and mortality associated with influenza. Nature 2012;484(7395):519–23.
[83] Bailey CC, et al. Ifitm3 limits the severity of acute influenza in mice. PLoS Pathog 2012;8(9):e1002909.
[84] Blanco-Melo D, et al. Imbalanced host response to SARS-CoV-2 drives development of COVID-19. Cell 2020;181(5):1036–1045.e9.

[85] Hadjadj J, et al. Impaired type I interferon activity and inflammatory responses in severe COVID-19 patients. Science 2020;369(6504):718–24.
[86] Park A, Iwasaki A. Type I and type III interferons—induction, signaling, evasion, and application to combat COVID-19. Cell Host Microbe 2020;27(6):870–8.
[87] Balhara J, Gounni AS. The alveolar macrophages in asthma: a double-edged sword. Mucosal Immunol 2012;5(6):605–9.
[88] Magnan A, et al. Alveolar macrophage interleukin (IL)-10 and IL-12 production in atopic asthma. Allergy 1998;53(11):1092–5.
[89] Bang BR, et al. Alveolar macrophages modulate allergic inflammation in a murine model of asthma. Exp Mol Med 2011;43(5):275–80.
[90] Burastero SE, et al. Increased expression of the CD80 accessory molecule by alveolar macrophages in asthmatic subjects and its functional involvement in allergen presentation to autologous TH2 lymphocytes. J Allergy Clin Immunol 1999;103(6):1136–42.
[91] Sabatel C, et al. Exposure to bacterial CpG DNA protects from airway allergic inflammation by expanding regulatory lung interstitial macrophages. Immunity 2017;46(3):457–73.
[92] Wynn TA, Chawla A, Pollard JW. Macrophage biology in development, homeostasis and disease. Nature 2013;496(7446):445–55.
[93] Osterholzer JJ, et al. Implicating exudate macrophages and Ly-6C(high) monocytes in CCR2-dependent lung fibrosis following gene-targeted alveolar injury. J Immunol 2013;190(7):3447–57.
[94] Misharin AV, et al. Monocyte-derived alveolar macrophages drive lung fibrosis and persist in the lung over the life span. J Exp Med 2017;214(8):2387–404.

CHAPTER 11

Macrophages in the immune-endocrine milieu of reproductive tissues

Maria Christina W. Avellar[a] and Emiliano Barreto[b]
[a]Department of Pharmacology, Federal University of São Paulo, Paulista School of Medicine, São Paulo, São Paulo, Brazil
[b]Laboratory of Cellular Biology, Biological Sciences and Health Institute, Federal University of Alagoas, Maceio, Alagoas, Brazil

Abbreviations

CCR2	chemokine receptor for chemokine ligand 2 (CCL2)
CD	cluster of differentiation
CD206	mannose receptor
CDFR1	CSF1 receptor
CSF1	macrophage colony-stimulating factor 1
CX3CL1	C-X3-C motif chemokine ligand 1
CX3CR1	CX3CL1 receptor 1
CXCL9	C-X-C motif chemokine ligand 9
Gene expression level	(high, Hi; intermediate, Int; low, Lo)
IFN-γ	interferon-γ
IL-10	interleukin-10
IL-1β	interleukin (IL)-1β
LY6C	lymphocyte antigen 6C
NO	nitric oxide
TNF-α	tumor associated factor-alpha

Introduction

In mammals, the regulation of reproductive function is tightly coordinated through the interplay of the hypothalamic-pituitary axis and the gonads (testis or ovaries). Presently, simultaneous appreciation of both local mucosal immune responses and systemic immunity has grown with respect to their roles in the array of functional tasks in the reproductive tract. The interface between the reproductive and immune systems in both males and females has adapted to meet the challenging demands of successful reproduction ensuring continuity of species, while simultaneously supporting internal protection against infectious threats and non-infectious tissue injuries.

Macrophages in the Human Body
https://doi.org/10.1016/B978-0-12-821385-8.00002-0

Unlike other peripheral tissues, these physiological events occur primarily against the hormonal background of sex steroid hormones, which confer more complex dynamics than those found in other peripheral tissues.

In fact, endocrine influences are critical for the appropriate tissue and cellular milieu that dictates the levels of inflammatory mediators and immunomodulators (such as cytokines and antimicrobial proteins), the quality of antigen presentation, the transport of immunoglobulins and the distribution of certain immune cell populations throughout the set of reproductive tissues. Therefore, deciphering the cellular and molecular mechanisms involved in the endocrine-immune interactions are vital to understanding reproductive health, as well as any tissue dysfunctions that may ultimately lead to reproductive diseases and infertility. These are clinical conditions with relevance not only in humans, but also for veterinary medicine in regards to livestock and protection of wild and captive or domesticated mammalian species.

A growing body of evidence demonstrates that macrophages are critical for male and female reproductive health. Indeed, an abundant population of macrophages is found in the gonads (ovary and/or testis); part of this immune cell population is in intimate association with steroidogenic cells in the testis (Leydig cells) and ovaries (granulosa cells), influencing their ability to produce sex steroid hormones (steroidogenesis). Also, macrophages are observed in close association with germ cell development, as well as with fertilizing gametes both in males (spermatogenesis in the testis) and females (folliculogenesis in the ovary). The density and phenotype subsets of macrophages in the gonads and other reproductive organs depend on the developmental and physiological tissue milieu; their developmental roles are recognized as essential for normal reproductive functioning (Fig. 1).

The understanding of their differential numbers, phenotypes, transcriptional/protein marker signatures and roles throughout the lifespan in males and females are now beginning to emerge. Here we review recent progress on the role of macrophages on the complex endocrine-immune regulatory network (i) in the gonads (testis and ovary), the origin tissues of germ cells (spermatozoa and eggs) and (ii) in the downstream reproductive tissue that transport germ cells in the males (epididymis) and females (uterine tube, also known as the oviduct or fallopian tube) after they leave the testicular and ovarian tissue, respectively; and also (iii) in the relationship of testicular and ovarian macrophages with steroidogenic cells in the gonads and sex steroid hormone production (mainly androgens, estrogens, and progesterone) will be highlighted.

Fig.1.12

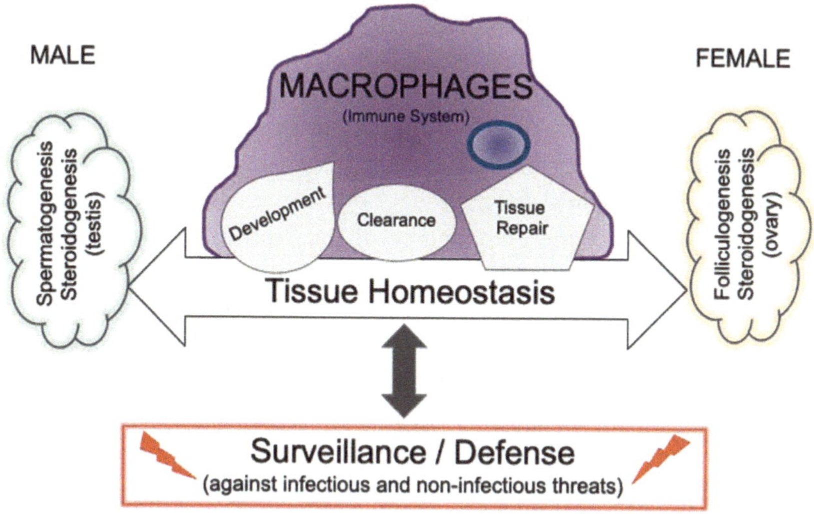

Fig. 1 Overview of macrophages in the milieu of the reproductive tract in males and females and their roles modulating homeostasis during tissue morphogenesis, tissue development and maintenance during lifespan through diverse activities including clearance, tissue/cell repair, steroidogenesis (testis and ovaries), spermatogenesis (testis), folliculogenesis (ovaries), among other activities. These act as sentinels for surveillance and tissue defense against infectious and non-infectious threats.

Macrophages in the male reproductive tract: Focus on the testis and epididymis

The male reproductive system is composed of the testis, the efferent ductules, epididymis, vas deferens, and the accessory glands, which includes the seminal vesicle and the prostate. These components serve to produce, transport, mature, store, and deliver viable spermatozoa to the female reproductive tract [1,2].

Testis

The testis is structured into two compartments.

One is the *seminiferous tubules* that contain columnar Sertoli cells, which are surrounded by a basement membrane. In addition to providing essential nutrients and growth factors, these Sertoli cells have inherent immunosuppressive capabilities and act as supporting cells to warrant a specialized environment in the seminiferous epithelium within the testicular tubules which assists the morphological progress of the immature germ cell (the least differentiated of which is the spermatogonium) into a spermatocyte (spermatogenesis).

The other testicular compartment is the *interstitial space* that contains the Leydig cells, which under the pulsatile control of LH (luteinizing hormone secreted by the pituitary gland into the peripheral circulation), are responsible for testicular steroidogenesis, i.e., production and secretion of testosterone and other steroid hormones, including dihydrotestosterone (DHT) and estradiol, that are essential for male development and fertility. In the adult male, Leydig cells are the chief source of total serum testosterone levels; their steroidogenic function is essential for spermatogenesis and male fertility. The testicular *interstitial space* also contains peritubular myoid cells that surround the seminiferous tubule and support its integrity by laying down the basement membrane and maintaining interaction with other cells, such as Sertoli cells. By secreting growth factors, cytokines, and extracellular matrix, these myoid cells are critical components for testis development and spermatogenesis [2].

Overall, macrophages are by far the most abundant somatic cell population in the postnatal and adult testis of various species; they represent approximately 20%–25% of all interstitial cells present in this tissue. Spatially, testicular macrophages are located in the: (i) interstitium, where they are intimately associated with Leydig cells in the estimated number of 1 to 1.6 macrophages/Leydig cell in different species [3], and in the (ii) peritubular compartment, where they are lying on the surface of the seminiferous tubules in the proximity of the spermatogonial stem cells. Compared to those found in other tissues, testicular macrophages exhibit relatively low inflammatory responses and high immunosuppressive properties; they are also less prompt to respond to an infectious/inflammatory stimulus, and display a continuing synthesis and secretion of anti-inflammatory cytokines [4–8]. The profile of mediators that are relevant for the immunosuppressive phenotype presented by the testicular macrophages is beginning to be unraveled (Fig. 2).

Both steroidogenesis and spermatogenesis are biological functions that involve complex cell-to-cell communication networks, i.e., intra- and intercellular processes between adjacent testicular cells and among the somatic and germ cells that coexist with macrophages in the testicular environment. Part of their critical role in normal male reproductive function is due to their ability to regulate the expression and function of spermatogonial proliferation- and differentiation-inducing factors that assist spermatogenesis, such as CSF1 (macrophage colony-stimulating factor 1) and enzymes involved

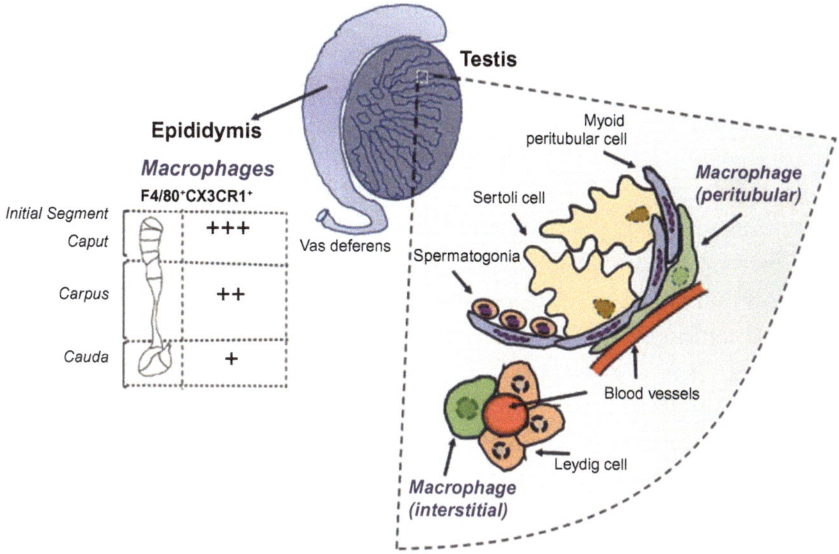

Fig. 2 Schematic representation of a rodent (murine) adult male reproductive tract with focus on the testis (right) and epididymis (left). *Right panel:* Seminiferous tubule epithelial and interstitial compartments illustrating key cell types. Each seminiferous tubule is surrounded by one (rodent) or multiple (human) layer(s) of myoid peritubular cells. Peritubular and interstitial macrophages are shown. Testicular immune privilege is maintained by several factors—blood testicular barrier, immunosuppressive molecules secreted by various testicular cells, a high intratesticular concentration of testosterone produced by Leydig cells, and an alternative M2 macrophage phenotype (expression of surface marker CD163 and high production of IL-10 and 25HC). *Left panel*: epididymis is represented in its four distinct anatomical regions (initial segment, caput, corpus and cauda epididymis). The differential relative number of macrophages along this tissue are associated with their role in immune tolerance in the epididymal proximal regions, and immune defense in the epididymal distal regions.

in retinoic acid biosynthesis. Additionally, it is known that the synthesis and secretion of certain inflammatory mediators (such as IL-1β and TNF-α) by testicular macrophages induce modulation of the transcriptional regulation of key genes involved in steroidogenesis, for example, that will lead to feedback suppression of Leydig cell function [4,5,7].

In fact, the crosstalk between macrophages and Leydig cells during steroidogenesis, and between macrophages and somatic/germ cells during spermatogenesis involves the function of CSF1 on CSF1 receptors (CDFR1) that are differentially expressed in different testicular cells, including Sertoli cells and germ cells [9]. CSF1 can also act as an extrinsic factor that promotes self-renewal of spermatogonial stem cells in vitro [6,10]. Indeed, significant reductions in mating capability, low sperm count, and plasma testosterone (90% lower) are outcomes observed in the *Csf1* mutant mice [11,12]. Whether the deleterious effects of macrophage depletion (and *Csf1* loss) upon reproductive function is the result of local (i.e., testis-specific) and/or systemic loss of macrophages still requires further investigation.

Macrophages are also an important component of the singular testicular immune-privileged environment that prevents antigens (molecular triggers) present on the developing and maturing spermatozoa from inciting an autoimmune response against them (i.e., immune tolerance [7,13]). The temporal setting of this testicular self-tolerance precedes the onset of spermatogenesis that takes place in this tissue later at puberty. This tolerogenic environment results from: (a) its physical structure (blood-testis barrier/Sertoli cell barrier); (b) a local immunosuppressive milieu created by synthesis and secretion of different immunomodulators by different tubular and interstitial cells; and (c) systemic immune tolerance. All these immune-related events are taking place in an environment that is primarily dependent on Leydig cells-directed steroidogenesis [6,14]. The resulting steady-state interactions between the immune system and meiotic germ cell antigens during these events are starting to be teased apart and examined. Overall, testicular macrophages (among other immune cell types) ensure host defense against pathogens, as well as surveillance and response to inflammatory conditions following tissue injuries [4,14,15].

Concerning their immunological properties, testicular macrophages are capable of both producing and secreting inflammatory mediators (such as cytokines), various antimicrobial proteins (such as lysozyme and beta-defensins), and also serving as antigen presenting cells during an immune response. When compared with macrophages found in other tissues, testicular macrophages present a reduced capacity for producing inflammatory factors, and instead display increased production of anti-inflammatory mediators [4,16].

In rodents, for example, testicular macrophages typically display a polarization towards immunoregulatory and immunotolerant features that is associated with the immunosuppressive M2 phenotype; however, this state is highly dependent upon and modulated by local microenvironmental signals that are beginning to be unveiled [14,17]. Upon stimulation by lipopolysaccharide from *E. coli* (a ligand for Toll-like receptor 4, TLR4), testicular macrophages produce and secrete high levels of the anti-inflammatory cytokine IL-10, and low levels of the pro-inflammatory mediators such as NO and TNF-α [5,8]. In addition, these cells not only suppress T-cell proliferation and activation, but also stimulate differentiation of naive T lymphocytes towards an immunosuppressive regulatory phenotype (Treg); this anti-inflammatory profile serves to protect developing germ cells from the deleterious effects of an injurious inflammatory microenvironment [4,8,18].

Macrophages are present and well equipped to perform testis-specific functions as early as prenatal development. During prenatal life, macrophages, in particular CX3CR1-positive cells, are found near the developing vascular plexus in the bipotential gonad of embryos; during this developmental period, they assist in cell differentiation and tissue vascularization that take place during testicular morphogenesis. In the mouse, fetal-liver derived progenitors ($CD115^- CD206^- MHCII^-$) have been identified as the infiltrating cells that give rise to interstitial ($CD206^+ MHCII^-$) and peritubular ($CD206^- MHCII^+$) macrophages. They are found in the fetal testis beginning at embryonic ages E14.5-E17.5 [7,19]. The development and viability of these fetal-derived testicular macrophages are critical for the establishment of a healthy and functional macrophage population in the testis throughout the male reproductive lifespan.

From birth until puberty, macrophages increase in both size and number, staying mainly within the testicular interstitium; in this tissue compartment, some of these macrophages are free, and others closely associate with Leydig cells. Throughout adulthood, testicular macrophages form intimate membrane associations (interdigitations) with these steroidogenic cells, which strengthens the morphofunctional interactions required for Leydig cell proliferation, differentiation and function. For example, the synthesis and secretion of 25-hydroxycholesterol (an intermediate in the testosterone biosynthetic pathway) by macrophages is a required component for steroid biogenesis by Leydig cells [6,20]. Macrophages are also found near the vasculature and within the peritubular myoid cell layer in the adult testis [2,7]. How these macrophages respond to testis microenvironment and are maintained throughout the lifespan are important topics to continue investigating.

At the molecular level, testicular macrophages have been characterized by their differential expression of macrophage surface markers such as F4/80, CD11b, CX3CR1, CD68, and CD163. In the adult, the majority of the resident macrophages (approximately 80%) correspond to $CD163^+$ cells, while the remainder are composed mainly of $CD163^- CD68^+$ cells that represent newly recruited monocytes from the circulation (also known as ED1 cells). During pathological conditions, such as the inflammation associated with bacterial or viral infections (e.g., orchitis), the relative balance and overall function of these macrophage populations in the testis are disrupted [4,8,18]. The increased number of macrophages ($CD68^+$ cells) in the testis from males diagnosed with idiopathic infertility and impaired spermatogenesis is one clinical example of the association between dysbalanced testicular macrophage number/function and male fertility [14,21].

Finally, it is important to consider that androgens, mainly testosterone released by Leydig cells, are an essential components in the regulation of the immunosuppressive phenotype of testicular macrophages [7,22,23]. Prostaglandins and corticosterone present in the testicular interstitial fluid are also important factors that modulate the molecular and functional signatures of these cells in the male gonadal environment [7,18].

Epididymis

The epididymis is a single, long, and highly convoluted tubule that connects the efferent ductules to the vas deferens. Based on anatomical and morphological features, it is divided (from proximal to distal) into an initial segment (most evident in rodent epididymis), caput, corpus and cauda regions. The epidydimal epithelium, composed of various cell types (principal, clear, narrow and basal cells), is in close association with a complex, dynamic network of immune cells. It is also surrounded by smooth muscle and interstitial tissues rich in blood vessels. Throughout its length, this epididymal epithelium tightly regulates sequential changes in the luminal fluid composition (e.g., proteins, ions, solutes, and lipids) and forms a barrier that gives rise to a unique luminal microenvironment crucial for the maturation of the transiting sperm. The microenvironments established from proximal to distal epidydimal regions allow the mammalian immotile spermatozoa (upon leaving the testis) to undergo their maturation process and acquire motility and fertility as they migrate to the distal region of the epididymis where they are stored until ejaculation [1,2].

During their epididymal journey, spermatozoa are protected in this organ against both autoimmunity, and invading pathogens. The blood-epididymis-barrier represents the first line of defense; it selectively restricts the types of compounds that can enter and exit the epididymal lumen and establishes a tolerogenic environment for the continuous flow of autoantigenic spermatozoa which are protected from the organism's own immune system. Macrophages and other immune cell types (such as dendritic cells, mast cells, T and B lymphocytes) are active players in this tolerogenic environment. Along the epididymis, immunoregulatory gene products (such as cytokines, antimicrobial proteins) secreted into the lumen by immune and non-immune cells have roles in establishing the protective epididymal environment in which spermatozoa mature and are stored [2,24] (Fig. 2).

Macrophages constitute the majority of resident immune cells in the epididymis from various mammalian species examined so far. Cell-to-cell interaction between macrophages and epididymal epithelial cells are the chief surveillance components in sensing and responding to changes in the proximal epididymis tissue environment; in this manner, immune tolerance is strictly limited to prevent autoimmunity against antigens of the post-testicular spermatozoa. In the distal epididymal regions, macrophages act primarily as players in tissue surveillance, host defense and immune response against infectious and non-infectious threats. These events are now known as key determinants in maintaining epididymal function, and consequently, male fertility [25,26]. How macrophages prevent circulating pathogens from adversely affecting spermatozoa by monitoring epididymal capillaries in normal steady state conditions and during pathological immune-related infertility is still under investigation.

Phenotypically, a dense network of mature phagocytic cells expressing macrophage and dendritic cell markers (such as F4/80 and CX3CR1, respectively) is present along the epididymis with the highest abundance in the initial segment [27,28]. These cells assist with the phagocytotic removal of senescent sperm [29,30]. More recently, the profile and behavior of CX3CR1$^+$-expressing cells from CX3CR1-GFP transgenic mice revealed shifts in the trajectories of their specific functions along the epididymis (for example, phagocytosis versus antigen-presentation) [27]. Within the epididymal initial segment, the interstitial CX3CR1$^+$ cells project their membranes towards the epithelium tight junctions and establish an intimate connection with epithelial cells; in normal conditions this would allow these cells to rapidly clear damaged epithelial cells and abnormal spermatozoa. In the caput region, a small minority of these interstitial CX3CR1$^+$ cells provide short intraluminal projections. More distally, in the cauda region of the epididymis, the CX3CR1$^+$ cells are found covering the base of the epithelium [27]. Macrophages in this epididymal area would be recruited to resolve inflammation arising as a result of luminal pathogens following infection [25,27,31]. Immunomodulatory molecules that are highly expressed and secreted by epithelial cells along the epididymis, such as certain cytokines and other immunomodulators (including antimicrobials such as β-defensins [24,26,32]), have not yet been demonstrated in epididymal macrophages.

Both peritubular and interstitial mononuclear phagocytes can capture and process circulatory antigens in the proximal epididymal regions that are more highly vascularized; only the interstitial cells exhibit these properties in the cauda epididymal region where vascularization is significantly reduced. Functionally, caput intraepithelial macrophages have been associated with epididymal immune tolerance, while cauda epididymal interstitial macrophages seem to preferentially serve in preventing and combating infectious and non-infectious threats [5,27].

Macrophages in the female reproductive tract

The female internal reproductive organs are composed of the ovaries, the uterine tubes, the uterus and the cervix. The proper development and functioning of the female reproductive tract are essential for it to fulfill its ultimate goal of reproduction that includes the physiological processes of oocyte maturation, fertilization, implantation, fetal growth, and parturition. These are events orchestrated by an interplay of endocrine-immune components that direct estrogens and estrogen-regulated mediators to promote the environmental milieu that contributes to immune tolerance and tissue remodeling in the female reproductive tract. Soluble factors and immune cells undergo periodic changes across the estrous cycle, which are orchestrated by the sex hormones estradiol and progesterone; i.e., estrous, the stage of the reproductive cycle when estradiol levels are known to be high, relative to those seen at diestrous, when estrogen levels are low and progesterone is the predominant hormone [33,34]. Unlike in testis and epididymis, the macrophage population is phenotypically less well characterized in ovary and uterine tubes.

Ovary

The ovary is composed of two parts, i.e., the central medulla, surrounded by the cortex. Its structure contains germ cells (oocytes) and somatic cells (granulosa, theca and stromal cells). The interaction and cross talk of these cells control the formation of oocyte containing follicles, their growth and development, ovulation and the formation of corpus luteum. Depending on the point during ovarian cycle, the adult ovary contains growing/atretic follicles, and the developing/regressing corpora lutea. In mammals, the ovaries are responsible for the cyclical maturation and release of mature germ cells (oocytes) into the uterine tubes. They are also the site for the synthesis and secretion of sex steroids (estrogen and progesterone). Under the influence of gonadotropins (FSH and LH), the ovarian follicles grow and ultimately the oocyte is released from the ruptured ovarian surface epithelium at ovulation.

The released oocyte is then collected by ciliated cells in the fimbria of the uterine tubes; post ovulation, the ovarian surface epithelium proliferates to efficiently repair the wound [35].

Macrophages are abundant in the ovarian interstitium, with population numbers and surface biomarkers highly dependent on the stage of the estrous or menstrual cycle [36–38]. In general, they release cytokines, proteases, and free radicals that converge to generate a cascade of proteolytic and remodeling events that lead to controlled degradation of the ovarian stromal matrix. In fact, during oocyte development, macrophages are recruited into the thecal layers of the follicle, being found in their highest number in the ovary just before ovulation; they are responsible for secreting mediators (such as cytokines, proteases) that act in a paracrine fashion to induce cumulus cells to synthesize components of its extracellular matrix and promote tissue remodeling.

Macrophages are also present in the cortical surface of pre-ovulatory follicles, where they may assist in the process of follicular rupture and/or pre-luteal repair of the ovarian wall [37,39]. Once the oocyte is released, macrophages present in higher numbers relative to the cells in the corpus luteum [11,40]. As the follicle matures, the ovarian granulosa cells express CSF1 and, at least in humans. The CSF1 concentration level in the follicular fluid exceeds that found in the serum [41]. For this reason, follicle-synthesized CSF1 is suggested to play a role in macrophage recruitment to the developing follicle. This is strongly supported by the fact that in the absence of CSF1, ovarian macrophages drastically decrease in number and are not recruited to the developing follicle. Macrophages are also important sources of pro-angiogenic factors (such as VEGF, EGF among others) that are key to support angiogenic processes and the blood supply that are necessary for the different stages of ovarian follicular development; following ovulation, macrophages are conspicuously spread throughout the corpus luteum among the endothelial cells [12,42,43] (Fig. 3).

Two important differences contrast the immune microenvironment of the fetal ovary from that of the fetal testis. Firstly, the ovarian steroidogenic cells remain undifferentiated during the fetal period. Second, unlike their better elucidated male counterparts, fetal ovary immune cells exhibit a tenuous relationship to adult tissue-resident macrophages that is still under debate. There is recent evidence of $CD11b^{Int}F4/80^{Hi}$ cells, consistent with yolk-sac derived progenitors, in mouse fetal ovary at embryonic age E14.5; later, at E16.5, following an increase in the number of $CD11b^{Hi}F4/80^{Int}$ cells that also express the monocyte markers LY6C and CCR2 (leading to a

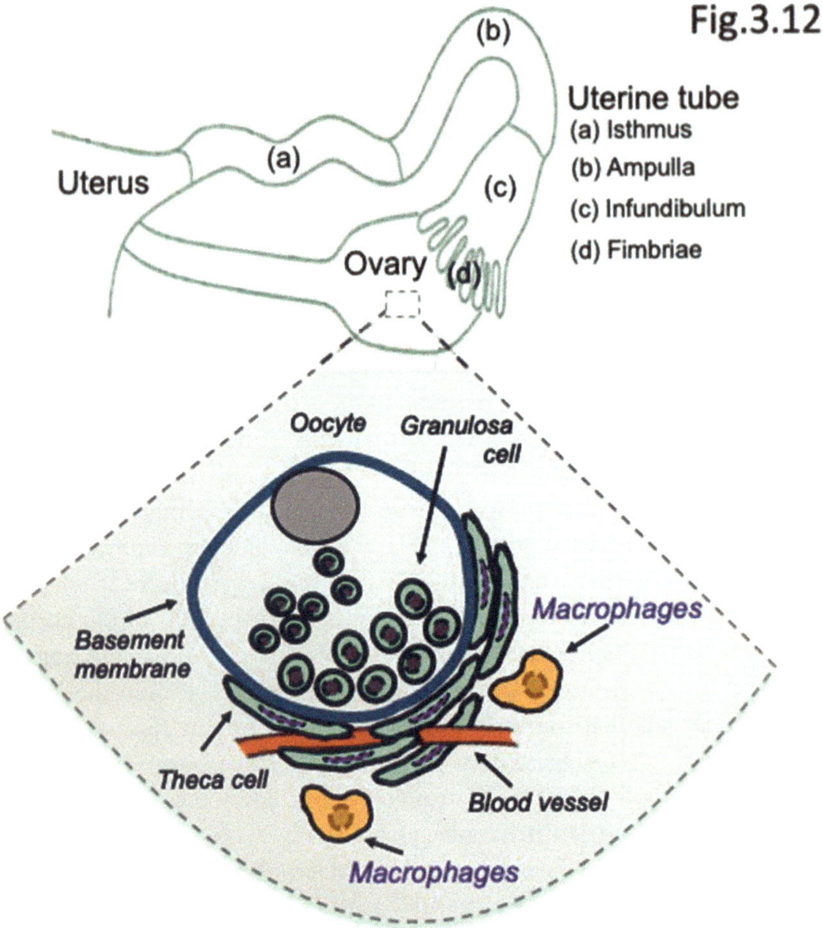

Fig. 3 Schematic representation of a rodent (murine) adult female reproductive tract with focus on the ovary (bottom) and the uterine tube (top) and its distinct anatomical regions (isthmus, ampulla, infundibulum and fimbriae). Ovarian compartments and key cell types are shown. Macrophages are very abundant in the ovary, and they display interstitial localization.

phenotype consistent with resident macrophages). Both cell lineages appear to populate the adult ovarian tissue [44,45].

The expression of the M2 macrophage marker CD206 by the $CD11b^{Int}F4/80^{Hi0}$ macrophages, in the absence of the M1 marker MHCII, indicate that fetal ovarian macrophages exhibit an M2-like phenotype [44]. More recently, macrophage populations have been identified by single cell RNA sequencing in mouse, monkey, and human fetal ovaries [46–48]. $CD14^{+}$ primitive monocytes and activated $MHCII^{+}$ monocyte-derived cells were present in human fetal ovaries [49], supporting reports of macrophage progenitors in the fetal human ovary. In adult ovaries, both embryonic and bone marrow-derived macrophages have the capacity to give rise to distinct macrophage subtypes [44,45].

Macrophages are also the most abundant ovarian immune cell population and are widely distributed in the thecal, interstitial, and luteal regions, as well as in the granulosa cell layer of atretic follicles in the adult mouse ovary [50]. In humans, CD68-positive macrophages are mainly localized near vascular cells and the theca-lutein cell layer of the corpus luteum, while MHCII-positive macrophages reside within corpora lutea [49,51,52]. Other evidence points to macrophages which express $MHCII^{Hi}CD11b^{+}F4/80^{+}$ as the most abundant phagocytes in the ovary, and commonly display flattened cell bodies with dendritic morphology, consistent with an activated macrophage phenotype [44,53]. A small $MHCII^{Lo}CD206^{+}$ macrophage population arises at birth and maintains a stable number during postnatal development [44]. Since monocyte recruitment and macrophage alternative activation (M2) increase in ovaries with aging, an aging-associated chronic inflammation has been suggested to cause a shift in ovarian macrophage ontogeny and polarization [45].

In ovaries destitute of macrophages, either from depletion induced by treatment with clodronate-containing liposomes [50] or by loss-of-function mutation of the *Csf1* gene [12], a negative impact on the length of estrous cycle, ovulation rate and follicle development is observed; these findings support the role of macrophages in regulating normal ovarian function, most notably ovulation and folliculogenesis. The hypothesis is that CSF1, as a chemoattractant factor for macrophages, recruits macrophages which assist follicle growth and development. The presence of another macrophage growth factor/chemoattractant synthesized by the luteal cells has been also considered since, as the corpus luteum matures after copulation, the macrophage number in CSF1-null mutation mouse increases until it reaches around one-third of the wild-type number [12,54].

Macrophages are also found in the granulosa cell layer at advanced stages of ovary follicle atresia [55] and may participate in the removal of cell debris created during granulosa cell apoptosis [51]. An active role for macrophages in the production of factors promoting follicular atresia has also been suggested. Conversely, ovarian macrophages might play a role in the regulation of progesterone production during pregnancy [56] and influence ovulation by producing many factors that may be involved in these events such as cytokines, proteases, and prostaglandins by mechanisms that are awaiting further investigation [55,57].

Uterine tubes

The uterine tubes are essential for the transport of the oocyte between the ovary and the uterus, in addition to providing an environment where spermatozoa capacitation/fertilization and early embryonic development and transport take place. The uterine tube cells maintain the fertilizing capacity of spermatozoa, whereas the follicular fluid facilitates the fertilization process of tubal spermatozoa. Thus, the uterine tubes have an intimate role in fertility and reproduction.

Anatomically, the uterine tubes can be subdivided into four regions, namely uterine, isthmus, ampulla (the most common site of fertilization) and the distal region called infundibulum with the fimbriated ends bordering the ovary which catch the released oocyte during each ovarian cycle. It is composed of three layers (mucosa, seromuscular, and serosa), with the mucosa layer consisting of the lamina propria (where a network of fibroblast-mesenchymal cells with interspersed immune cells is present) and the epithelium (composed of ciliated, secretory, and intercalary cells). The ciliated cells capture the oocyte when it is released at ovulation, guide it down into the ampullary-isthmus junction for fertilization by sperm and then continue to guide the zygote into the uterus [58,59]. Propulsion of gametes and embryos is achieved by complex interaction between the epithelial ciliary activity, muscle contractions, and the flow of tubal secretions from their functional epithelium. The coordination of these events is a prerequisite for successful spontaneous pregnancy. The entire length of the uterine tube epithelium undergoes cyclical changes in morphology and ciliary activity in response to ovarian hormones; estradiol and progesterone mediate changes in tubal morphology across the ovarian cycle [60–62]. The height and numbers of these epithelial cells, as well as their mitotic activity, vary significantly across the cycle. Circulating ovarian sex steroids may regulate these cyclic changes; the mechanism of action of these steroids may involve direct or indirect processes of growth factors and cytokines.

The secretory cells comprise approximately 60% of the epithelial cells found in the uterine tubes, and secrete cytokines, growth factors, antimicrobial proteins, and other immunomodulators into the tubal fluid. The uterine tubes also have a major role in mounting local immune responses against infectious and non-infectious threats by using a complex network of lymphocytes, macrophages, and dendritic cells to control invading pathogens [63]. The endocrine changes affecting this tissue may also modulate and impact the immune responses that are mounted, but the detailed molecular players of these events are still poorly understood.

Macrophages ($CD68^+$-expressing cells) are present all along the uterine tubes, specifically among the epithelial cells and within the connective tissue of the lamina propria [64–67]. They are known to establish important morphological and functional paracrine interactions with the epithelial cells, inserting into the oviduct epithelium and contributing to the epithelial remodeling that takes place during the different phases of the estrous cycle [67,68]. In fact, tubal macrophages display cyclic changes in their numbers, morphology and anatomical location in the tubal tissue during the stages of the menstrual cycle, and in the postmenopausal period [61,65,69]. These temporal-spatial changes in the macrophage invasion of the uterine tubal epithelium at specific stages of the cycle, mainly during the early-mid-luteal phases, occur under the influence of hormonal-driven paracrine loops that operate in the tubal epithelium [67] (Fig. 3).

By secreting cytokines and establishing regulatory paracrine loops, the tubal macrophages contribute to the tubal environment that is an important component of normal uterine tube function, including fertilization. The uterine tube, as well as other parts of the female reproductive system, is immunologically unique in its requirements for tolerance to non-self-cells (sperm) and partially non-self-cells (developing embryo), while responding to an array of sexually transmitted pathogens [59,69,70]. Tubal secretory factors are also important as paracrine signals which influence the smooth muscles of the uterine tubes [71,72].

Although uterine tubes are considered sterile sites, microorganisms can ascend from the lower reproductive tract and cause pelvic inflammatory disease, a relevant cause of reproductive alterations such as tubal infertility and ectopic pregnancy [70,73]. In this context, the presence of abundant macrophages undergoing cyclic changes during the ovarian cycle, reported herein, suggests that these cells play relevant roles in the preventive physiology of the uterine tubes. Following this approach, a study using single-cell transcriptomic profiling provided recent results presenting a cellular and molecular compendium of the human uterine tube that sheds light onto our understanding of uterine tube physiopathology [74].

Macrophages under regulation of sex hormones

Members of the receptor superfamily that are activated by steroid hormones, which include the androgen receptor (AR), estrogen receptor (ER), and progesterone receptors (PR), are expressed on macrophages. The activation of each these receptors can trigger canonical and non-canonical regulatory actions that direct cellular transcriptional programs and signaling cascades that shape the phenotype and immunomodulatory function of the macrophage. Furthermore, involvement of co-modulators and the allosteric modification of the steroid hormone-receptor complex by endogenous ligands are an important aspect of this regulation under the endocrine-dependent environment of the gonads and reproductive tissues [22,75]. Conceptually, macrophages can be seen as hormone-responsive immune surveillance cells in both male and female reproductive tract. Therefore, mechanistic understanding of their phenotype and stimulus-induced activation is highly dependent on their milieu which reflects the temporal changes of sex steroid hormones throughout the lifespan (Fig. 4).

It is worth mentioning that the glucocorticoid receptor (GR) signaling cascade is also present and functional in macrophages, acting as an important regulator of gene-specific mechanisms that direct GR-driven repression of inflammatory response genes. Moreover, differentiation of macrophages to an immunosuppressive/tolerogenic state phenotype that is seen in the male and female gonads and reproductive tissues such as epididymis and uterine tube can be managed by the modulation of GR signaling pathways.

Another important point to highlight is the ability of macrophages (and other immune cells) to synthesize, modify, and metabolize their own sex

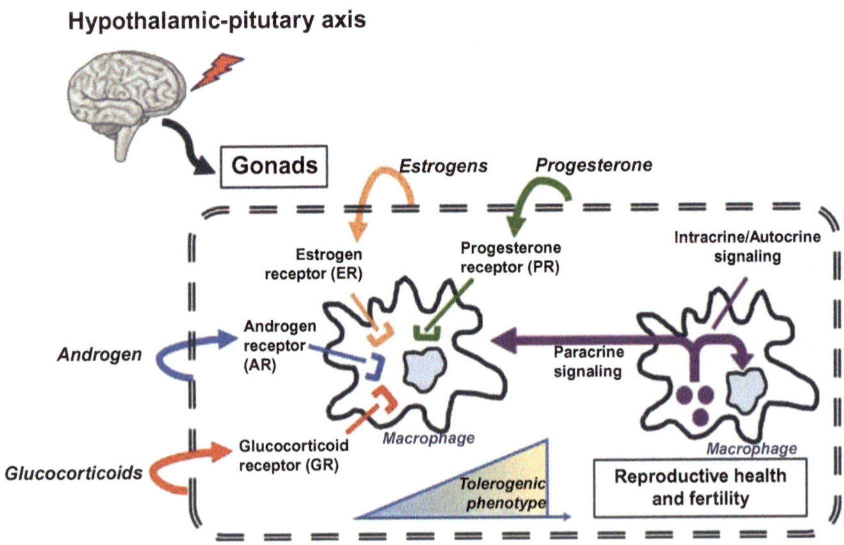

Fig. 4 Schematic representation depicting macrophages as hormone-responsive immune surveillance cells in the gonads and other tissues of the adult male and female reproductive tract.

steroids independent of gonadal sex steroid production. The existence of steroidogenesis and steroid signaling within this immune cell would impose an endogenous steroid-regulatory circuit. The production of sex steroids, i.e., testosterone and estrogen from steroid precursors, by these mechanisms would confer intracrine, autocrine and paracrine roles with signaling effects in both in the cell itself and in neighboring cells [76,77]. The impact and regulation of these events on the physiological and pathological roles of macrophages is still unclear. This is an important scenario to be unraveled since it can allow elucidation of the mechanisms through which sex steroids and endocrine-immune mechanisms control macrophage function and reproductive health. One example of intracrine production of steroids in an immune cell is in the testis, as corticosterone has been shown to be constitutively produced by testicular macrophages [18,77]. By paracrine or autocrine activity, steroids present in the testicular environment could shape the cell's immunosuppressive function, one of the mechanisms by which a steroid hormone could establish and maintain the immune privileged environment of the testis. Thus, investigation addressing specific mechanisms by which sex and other steroids affect macrophages performing physiological surveillance may aid in uncovering new therapeutic targets for controlling endocrine-immune processes that lead to activation or disruption of regulatory cascades. Further elucidation of these transduction events may contribute to developing actionable intervention strategies when assessing clinical conditions that impact reproductive health and fertility in both males and females.

Conclusions/perspectives

The most recent data pinpoint immune-inflammatory processes in the male and female gonads and reproductive organs with local inflammatory consequences that could not only disrupt the tissues themselves and germ cells, but can also be a primary risk for systemic inflammation and a more extensive chronic condition. Whatever the case, these progressions can lead to several deleterious health consequences, including downstream infertility. The reproductive tissues can be seen as an active primary "source" participating in the individual's health and well-being. In males, for example, more accurate urological and andrological markers are needed to serve as key components to better develop diagnostic and therapeutic strategies for refining the treatment of male infertility. This sustains the notion that gaps in male reproductive health-related research are simultaneously basic, translational, and clinical. Targeting normal or activated macrophage phenotypes and interfering with aberrant cross-talk of these immune cells with non-immune cells from the reproductive tract may provide novel strategies to resolve acute and chronic inflammatory conditions associated with reproductive diseases.

In a wider sense, aspects of the complex intercellular interactions (macrophages, germ, and other host cells) in the context of steroid hormones and their receptors (androgens/androgen receptor versus estrogens/estrogen receptor) and inflammation (pro- versus anti-inflammatory conditions) have greatly contributed to highlighting the multiaxial scenario of macrophages within the physiological milieus of the reproductive tissues. Accumulating evidence in this emerging area is underscoring the importance of new technologies at the molecular, cellular, and tissue levels. We are increasingly able to better understand macrophages as potential targets for future therapies in human and veterinary reproductive biology and medicine. In this reproductive scenario, macrophages can be flagged as hormone-responsive immune surveillance cells. The knowledge gained about macrophages in the gonads and downstream reproductive tissues will prove useful to understanding and treating medical conditions of the reproductive system, especially infertility.

References

[1] Avellar MCW, Hinton HT. Epididymis. In: Huhtaniemi I, editor. Encyclopedia of endocrine diseases. 2nd ed, vol. 1. Oxford: Elsevier Science & Technology Books; 2018. https://doi.org/10.1016/B978-0-12-801238-3.65180-2. ISBN: 978-0-12-8122.

[2] Robaire B, Hinton BT. The epididymis. In: Plant TM, Zeleznik AJ, editors. Knobil and Neill's physiology of reproduction. 4th ed. San Diego: Academic Press; 2015. p. 691–771.

[3] Oliveira CFA, Lara NLEM, Cardoso BRL, de França LR, de Avelar GF. Comparative testis structure and function in three representative mice strains. Cell Tissue Res 2020;382:391–404.

[4] Bhushan S, Theas MS, Guazzone VA, Jacobo P, Wang M, Fijak M, Meinhardt A, Lustig L. Immune cell subtypes and their function in the testis. Front Immunol 2020;11:2558.

[5] Bhushan S, Tchatalbachev S, Lu Y, Frohlich S, Fijak M, Vijayan V, et al. Differential activation of inflammatory pathways in testicular macrophages provides a rationale for their subdued inflammatory capacity. J Immunol 2015;194:5455–64.

[6] DeFalco T, Potter SJ, Williams AV, Waller B, Kan MJ, Capel B. Macrophages contribute to the spermatogonial niche in the adult testis. Cell Rep 2015;12:1107–9.

[7] Gu X, Li S-Y, DeFalco T. Immune and vascular contributions to organogenesis of the testis and ovary. FEBS J 2021. https://doi.org/10.1111/febs.15848.

[8] Winnall WR, Muir JA, Hedger MP. Rat resident testicular macrophages have an alternatively activated phenotype and constitutively produce interleukin-10 in vitro. J Leukoc Biol 2011;90:133–43.

[9] Sawaied A, Arazi E, AbuElhija A, Lunenfeld E, Huleihel M. The presence of colony-stimulating factor-1 and its receptor in different cells of the testis; it involved in the development of spermatogenesis in vitro. Int J Mol Sci 2021;22:2325.
[10] Oatley JM, Oatley MJ, Avarbock MR, Tobias JW, Brinster RL. Colony stimulating factor 1 is an extrinsic stimulator of mouse spermatogonial stem cell self-renewal. Development 2009;136:1191–9.
[11] Cohen PE, Hardy MP, Pollard JW. Colony-stimulating factor-1 plays a major role in the development of reproductive function in male mice. Mol Endocrinol 1997;11:1636–50.
[12] Cohen PE, Zhu L, Pollard JW. Absence of colony stimulating factor-1 in osteopetrotic ($csfm^{op}/csfm^{op}$) mice disrupts estrous cycles and ovulation. Biol Reprod 1997;56:110–8.
[13] Fijak M, Meinhardt A. The testis in immune privilege. Immunol Rev 2006;213:66–81.
[14] Meinhardt A, Wang M, Schulz C, Bhushan S. Microenvironmental signals govern the cellular identity of testicular macrophages. J Leukoc Biol 2018;104:757–66.
[15] Redgrove KA, McLaughlin EA. The role of the immune response in chlamydia trachomatis infection of the male genital tract: a double-edged sword. Front Immunol 2014;5:534.
[16] Mossadegh-Keller N, Gentek R, Gimenez G, Bigot S, Mailfert S, Sieweke MH. Developmental origin and maintenance of distinct testicular macrophage populations. J Exp Med 2017;214:2829–41.
[17] Lavin Y, Winter D, Blecher-Gonen R, David E, Keren-Shaul H, Merad M, Jung S, Amit I. Tissue-resident macrophage enhancer landscapes are shaped by the local microenvironment. Cell 2014;159:1312–26.
[18] Wang M, Fijak M, Hossain H, Markmann M, Nusing RM, Lochnit G, et al. Characterization of the microenvironment of the testis that shapes the phenotype and function of testicular macrophages. J Immunol 2017;198:4327–40.
[19] Lokka E, Lintukorpi L, Cisneros-Montalvo S, Mäkelä JA, Tyystjärvi S, Ojasalo V, Gerke H, Toppari J, Rantakari P, Salmi M. Generation, localization and functions of macrophages during the development of testis. Nat Commun 2020;11:4375.
[20] Hales DB. Testicular macrophage modulation of Leydig cell steroidogenesis. J Reprod Immunol 2002;57:3–18.
[21] Frungieri MB, Calandra RS, Lustig L, Meineke V, Köhn FM, Vogt HJ, Mayerhofer A. Number, distribution pattern, and identification of macrophages in the testes of infertile men. Fertil Steril 2002;78:298–306.
[22] Bupp MRG, Jørgensen TN. Androgen-induced immunossupression. Front Immunol 2018;9:794.
[23] Rettew JA, Huet-Hudson YM, Marriott I. Testosterone reduces macrophage expression in the mouse of toll-like receptor 4, a trigger for inflammation and innate immunity. Biol Reprod 2008;78:432–7.
[24] Avellar MCW, Ribeiro CM, Dias-da-Silva MR, Silva EJR. In search of new paradigms for epididymal health and disease: innate immunity, inflammatory mediators, and steroid hormones. Andrology 2019;7:690–702.
[25] Da Silva N, Smith TB. Exploring the role of mononuclear phagocytes in the epididymis. Asian J Androl 2015;17:591–6.
[26] Voisin A, Whitfield M, Damon-Soubeyrand C, Goubely C, Henry-Berger J, Saez F, et al. Comprehensive overview of murine epididymal mononuclear phagocytes and lymphocytes: unexpected populations arise. J Reprod Immunol 2018;126:11–7.
[27] Battistone MA, Mendelsohn AC, Spallanzani RG, Brown D, Nair AV, Breton S. Region-specific transcriptomic and functional signatures of mononuclear phagocytes in the epididymis. Mol Hum Reprod 2020;26:14–29.
[28] Da Silva N, Barton CR. Macrophages and dendritic cells in the post-testicular environment. Cell Tissue Res 2016;363:97–104.

[29] Cooper TG, Yeung CH, Jones R, Orgebin-Crist M-C, Robaire B. Rebuttal of a role for the epididymis in sperm quality control by phagocytosis of defective sperm. J Cell Sci 2002;115:5–7.
[30] Tomlinson MJ, White A, Barratt CL, Bolton AE, Cooke ID. The removal of morphologically abnormal sperm forms by phagocytes: a positive role for seminal leukocytes? Hum Reprod 1992;7:517–22.
[31] Smith TB, Cortez-Retamozo V, Grigoryeva LS, Hill E, Pittet MJ, Da Silva N. Mononuclear phagocytes rapidly clear apoptotic epithelial cells in the proximal epididymis. Andrology 2014;2:755–62.
[32] Ribeiro CM, Silva EJR, Hinton BT, Avellar MCW. β-Defensins and the epididymis: contrasting influences of prenatal, postnatal, and adult scenarios. Asian J Androl 2016;18:323–8.
[33] Pepe G, Braga D, Renzi TA, Villa A, Bolego C, D'Avila F, Barlassina C, Maggi A, Locati M, Vegeto E. Self-renewal and phenotypic conversion are the main physiological responses of macrophages to the endogenous estrogen surge. Sci Rep 2017;7:44270.
[34] Pollard JW, Dominguez MG, Mocci S, Cohen PE, Stanley ER. Effect of colony stimulating factor-1 (CSF-1) null mutation, osteopetrotic (*csfmop*), on the distribution of macrophages in the male reproductive tract. Biol Reprod 1997;56:1290–300.
[35] Auersperg N, Wong AS, Choi KC, Kang SK, Leung PC. Ovarian surface epithelium: biology, endocrinology, and pathology. Endocr Rev 2001;22:255–88.
[36] Kaushic C, Frauendorf E, Rossoll RM, Richardson JM, Wira CR. Influence of the estrous cycle on the presence and distribution of immune cells in the rat reproductive tract. Am J Reprod Immunol 1998;39:209–16.
[37] Takaya R, Fukaya T, Sasano H, Suzuki T, Tamura M, Yajima A. Macrophages in normal cycling human ovaries; immunohistochemical localization and characterization. Hum Reprod 1997;12:1508–18.
[38] White HD, Prabhala RH, Humphrey SL, Crassi KM, Richardson JM, Wira CR. A method for the dispersal and characterization of leukocytes from the human female reproductive tract. Am J Reprod Immunol 2000;44:96–103.
[39] Brännström M, Mayrhofer G, Robertson SA. Localization of leukocyte subsets in the rat ovary during the periovulatory period. Biol Reprod 1993;48:277–86.
[40] Brännström M, Pascoe V, Norman RJ, McClure N. Localization of leukocyte subsets in the follicle wall and in the corpus luteum throughout the human menstrual cycle. Fertil Steril 1994;61:488–95.
[41] Cohen PE, Nishimura K, Zhu L, Pollard JW. Macrophages: important accessory cells for reproductive function. J Leukoc Biol 1999;66:765–72.
[42] Araki M, Fukumatsu Y, Katabuchi H, Schultz LD, Takahashi K, Okamura H. Follicular development and ovulation in macrophage colony-stimulating factor-deficient mice homozygous for the osteopetrosis (*op*) mutation. Biol Reprod 1996;54:478–84.
[43] Buse E, Zöller M, Van Esch E. The macaque ovary, with special reference to the cynomolgus macaque (*Macaca fascicularis*). Toxicol Pathol 2008;36:24S–66S.
[44] Jokela H, Lokka E, Kiviranta M, Tyystjarvi S, Gerke H, Elima K, Salmi M, Rantakari P. Fetal-derived macrophages persist and sequentially maturate in ovaries after birth in mice. Eur J Immunol 2020;50:1500–14.
[45] Zhang Z, Schlamp F, Huang L, Clark H, Brayboy L. Inflammaging is associated with shifted macrophage ontogeny and polarization in the aging mouse ovary. Reproduction 2020;159:325–37.
[46] Li L, Dong J, Yan L, Yong J, Liu X, Hu Y, Fan X, Wu X, Guo H, Wang X, et al. Single-cell RNA-seq analysis maps development of human germline cells and gonadal niche interactions. Cell Stem Cell 2017;20:858–873.e4.
[47] Niu W, Spradling AC. Two distinct pathways of pregranulosa cell differentiation support follicle formation in the mouse ovary. Proc Natl Acad Sci USA 2020;117:20015–26.

[48] Zhao ZH, Li CY, Meng TG, Wang Y, Liu WB, Li A, Cai YJ, Hou Y, Schatten H, Wang ZB, et al. Single-cell RNA sequencing reveals regulation of fetal ovary development in the monkey (*Macaca fascicularis*). Cell Discov 2020;6:97.
[49] Bukovsky A, Caudle MR, Keenan JA, Wimalasena J, Upadhyaya NB, Van Meter SE. Is corpus luteum regression an immune-mediated event? Localization of immune system components and luteinizing hormone receptor in human corpora lutea. Biol Reprod 1995;53:1373–84.
[50] Van der Hoek KH, Maddocks S, Woodhouse CM, van Rooijen N, Robertson SA, Norman RJ. Intrabursal injection of clodronate liposomes causes macrophage depletion and inhibits ovulation in the mouse ovary. Biol Reprod 2000;62:1059–66.
[51] Gaytán F, Morales C, Bellido C, Aguilar E, Sánchez-Criado JE. Ovarian follicle macrophages: is follicular atresia in the immature rat a macrophage-mediated event? Biol Reprod 1998;58:52–9.
[52] Petrovska M, Sedlak R, Nouza K, Presl J, Kinsky R. Development and distribution of the white blood cells within various structures of the human menstrual corpus luteum examined using an image analysis system. Am J Reprod Immunol 1992;28:77–80.
[53] Carlock C, Wu J, Zhou C, Ross A, Adams H, Lou Y. Ovarian phagocyte subsets and their distinct tissue distribution patterns. Reproduction 2013;146:491–500.
[54] Bowen JM, Keyes PL, Warren JS, Townson DH. Prolactin-induced regression of the rat corpus luteum: expression of monocyte chemoattractant protein-1 and invasion of macrophages. Biol Reprod 1996;54:1120–7.
[55] Wu R, Van der Hoek KH, Ryan NK, Norman RJ, Robker RL. Macrophage contributions to ovarian function. Hum Reprod Update 2004;10:119–33.
[56] Jasper MJ, Robertson SA, Van der Hoek KH, Bonello N, Brännström M, Norman RJ. Characterization of ovarian function in granulocyte-macrophage colony-stimulating factor-deficient mice. Biol Reprod 2000;62:704–13.
[57] Ono Y, Nagai M, Yoshino O, et al. $CD11c^+$ M1-like macrophages (MΦs) but not $CD206^+$ $M2^-$ like MΦ are involved in folliculogenesis in mice ovary. Sci Rep 2018;8:8171.
[58] Croxatto HB. Physiology of gamete and embryo transport through the fallopian tube. Reprod BioMed Online 2002;4:160–9.
[59] Varga I, Kachlik D, Miko M, Zishova M, Danisovic L. Rediscovery of forgotten structures inside mucosa of uterine tubes—lymphatic lacunae and their possible role in reproduction. FASEB J 2018;32:S1.
[60] Baracat EC, Simões MJ, Novo NF, Juliano Y, de Lima GR, de Lima Filho OA, Kulay Junior L. Morphologic and morphometric aspects of the uterine tube epithelium during the menstrual cycle. Rev Paul Med 1991;109:204–12.
[61] Crow J, Amso NN, Lewin J, Shaw RW. Morphology and ultrastructure of fallopian tube epithelium at different stages of the menstrual cycle and menopause. Hum Reprod 1994;9:2224–33.
[62] Hwang TS, Song J. Morphometrical changes of the human uterine tubes according to aging and menstrual cycle. Ann Anat 2004;186:263–9.
[63] Wira CR, Grant-Tschudy KS, Crane-Godreau MA. Epithelial cells in the female reproductive tract: a central role as sentinels of immune protection. Am J Reprod Immunol 2005;53:65–76.
[64] Palter SF, Mulayim N, Senturk L, Arici A. Interleukin-8 in the human fallopian tube. J Clin Endocrinol Metab 2001;86:2660–7.
[65] Suenaga Y, Katabuchi H, Fukumatsu Y, Okamura H. Distribution and cytological properties of macrophages in human fallopian tubes. Acta Anat 1998;163:10–9.
[66] Varga I, Miko M, Kachlik D, Zishova M, Danihel Jr L, Babl P. How many cell types form the epithelial lining of the human uterine tubes? Revision of the histological nomenclature of the human tubal epithelium. Ann Anat 2019;224:73–80.

[67] Gaytán M, Morales C, Bellido C, Sánchez-Criado JE, Gaytán F. Macrophages in human fallopian tube and ovarian epithelial inclusion cysts. J Reprod Immunol 2007;73:66–73.
[68] Brenner RM, Maslar IA. The primate oviduct and endometrium. In: Knobil E, Neill JD, editors. The physiology of reproduction. New York: Raven Press; 1988. p. 303–30.
[69] Safwat MD, Habib FA, Oweiss NY. Distribution of macrophages in the human fallopian tubes: an immunohistochemical and electron microscopic study. Folia Morphol (Warsz) 2008;67:43–52.
[70] Quayle AJ. The innate and early immune response to pathogen challenge in the female genital tract and the pivotal role of epithelial cells. J Reprod Immunol 2002;57:61–79.
[71] Gelety TJ, Chaudhuri G. Prostaglandins in the ovary and fallopian tube. Baillieres Clin Obstet Gynaecol 1992;6:707–29.
[72] Charles WA, John FA, Paul WB, Grant YC. Effect of the menstrual cycle on immunological parameters in the human female reproductive tract. J Acquir Immune Defic Syndr 2005;38:S34–6.
[73] Serra P, Venditti M. Ovarian infections. In: Serra GB, editor. The ovary. New York: Raven Press; 1983. p. 313–25.
[74] Dinh HQ, Lin X, Abbasi F, Nameki R, Haro M, Olingy CE, Chang H, Hernandez L, Gayther SA, Wright KN, Aspuria PJ, Karlan BY, Corona RI, Li A, Rimel BJ, Siedhoff MT, Medeiros F, Lawrenson K. Single-cell transcriptomics identifies gene expression networks driving differentiation and tumorigenesis in the human fallopian tube. Cell Rep 2021;35:108978.
[75] Lamorte S, Shinde R, McGaha TL. Nuclear receptors, the aryl hydrocarbon receptor, and macrophage function. Mol Asp Med 2021;78:100942.
[76] Rubinow KB. An intracrine view of sex steroids, immunity, and metabolic regulation. Mol Metab 2018;15:92–103.
[77] Chakraborty S, Pramanik J, Mahata B. Revisiting steroidogenesis and its role in immune regulation with the advanced tools and technologies. Genes Immun 2021;22:125–40.

CHAPTER 12

Macrophages at the maternal-fetal interface

Michael Z. Zulu[a], Clive M. Gray[b,c], Siamon Gordon[d,e], and Fernando O. Martinez[f]
[a]Division of Microbiology and Immunology, Yerkes National Primate Research Center, Emory University, Atlanta, GA, United States
[b]Division of Immunology, Institute of Infectious Disease and Molecular Medicine and Department of Pathology, Faculty of Health Sciences, University of Cape Town, Cape Town, South Africa
[c]Division of Molecular Biology and Human Genetics, Department of Biomedical Sciences, Stellenbosch University, Cape Town, South Africa
[d]Graduate Institute of Biomedical Sciences, College of Medicine, Chang Gung University, Taoyuan City, Taiwan
[e]Sir William Dunn School of Pathology, University of Oxford, Oxford, United Kingdom
[f]Faculty of Health and Medical Sciences, University of Surrey, Guildford, United Kingdom

Introduction

Macrophage origin in placenta

Our understanding of the placenta has deepened in the last century, evolving from a mere filter, a waste- or nutrition-related organ by which the mother feeds the fetus, to a fully supportive immune metabolic effector organ responsible for immune tolerance that supports healthy pregnancies. As any other organ, the placenta is home to a variety of cells that belong to the macrophage or mononuclear phagocyte family. Macrophages are crucial leukocytes in both development and immunity [1]. They express a variety of receptors for self-ligands, pathogen-associated molecular patterns (PAMPs), and danger-associated molecular patterns (DAMPs). In addition, cells are metabolically active, synthesizing bioactive lipids, secreting trophic factors, cytokines and lytic enzymes able to digest or activate macromolecules and the matrix important in tissue remodeling. Although the name, macrophage, was based on the cell's primary function in clearance of invading pathogens, cellular debris, and apoptotic cells by phagocytosis [2], placental macrophages, in common with other tissue macrophages, also perform crucial immunological functions such as antigen presentation, coordination of both innate and adaptive immune responses and wound repair [3].

Prominent figures who contributed to our understating of placental macrophages include Prof. Felix Jacob Marchand (1846–1928) and Dr. Theodor Langhans (1839–1915), but it was Dr. J. Isfred Isidore Hofbauer

Macrophages in the Human Body
https://doi.org/10.1016/B978-0-12-821385-8.00006-8

(1879–1961), who went deeper into the definition of placental macrophages, Fig. 1. Hofbauer focused on pathological-anatomical analysis, biological and chemical characterization of the placenta. In seminal experiments of lipid transfer between mother and fetus, he established parallels between intestinal and the placental villi. By studying Sudan red and iron transfer he identified fetal macrophages as 10–30 μm, pleomorphic, vacuolated cells with a granular cytoplasm, which accumulated lipids if examined promptly [4].

The maternal-fetal interface is a unique anatomical site between the uterine mucosa and the extra-embryonic tissue of the fetus [5]. The placenta is of fetal origin, while the decidua and the myometrium are both of maternal origin, Fig. 2. The placenta has multiple functions that are critical for mammalian reproduction [6]. The first main function is transport of nutrients and metabolic precursors to the fetus through diffusion, transporter-mediated mechanisms or through active endocytosis; these include glucose, amino acids, lipids, and macromolecules such as immunoglobulins. The ABC ATP-binding cassette and SLC solute carrier family have been implicated in transport, as their insufficiency has harmful effects on the pregnancy. Placental vascularization and the development of a complex network of vessels is also vital to maintain an active transport system. Placenta integrity avoids the transfer of undesired factors to the fetus and guarantees the desired polarity in the tissues.

Fig. 1 J. Isfred Isidore Hofbauer was an Austrian-born American gynecologist who performed extensive studies of obstetric organ systems and pioneered new treatments with regard to childbirth.

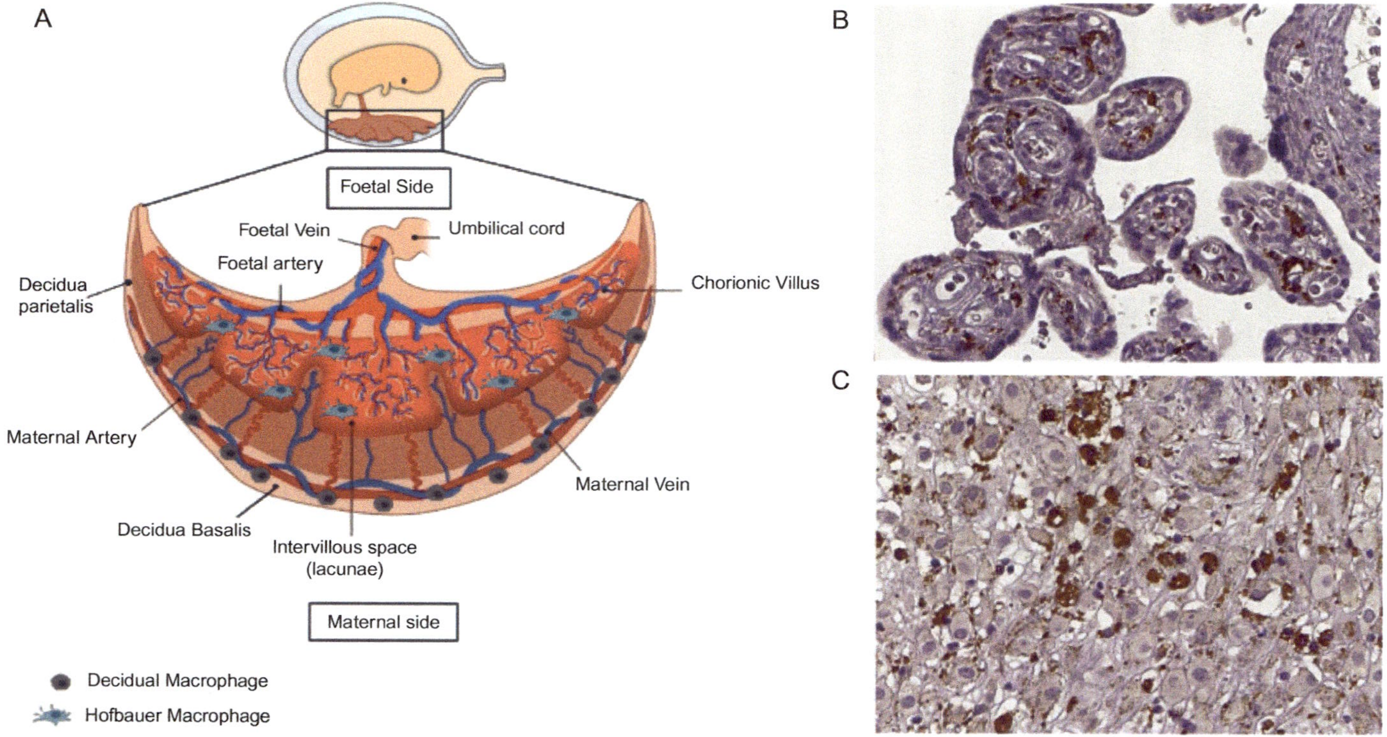

Fig. 2 Macrophage populations in placenta. (A) Following embryo implantation, the decidua divides into two distinct regions: the decidua basalis, which is the portion of the uterus attached to the placenta, and decidua parietalis, which is the layer closest to the endometrium, found between the amnion and the chorion. (B) CD68+ Hofbauer cells and (C) decidual macrophages in term placenta as portrayed in the Human Protein atlas.

Placenta development is an intricate process elegantly summarized in Ref. [7] by Moffet et al. Briefly, the placenta develops from the trophoectoderm or outer layer of the blastocyst in the first days post-fertilization. The undifferentiated blastocyte adheres and actively invades the endometrium which thickens into a large decidua. The endometrium is packed with maternal macrophages which mediate gland development and removal of dead cells as well as debris across the menstruation cycle. During implantations these cells remain, while the endometrium becomes the decidua itself. Cytotrophoblast cells proliferate and send projections that will form primary, and later secondary and tertiary villi. These villi contain a network of canals that empty onto the embryo. Villi also contain fetally-derived macrophages termed Hofbauer cells, which appear d18 post conception, before the vascular connection to the embryo itself is established. This suggests that the first wave of Hofbauer macrophages derives from haemangioblasts in extra-embryonic tissue. Later, when the vascular connection is established in the fetus, we hypothesize that Hofbauer cells could also derive from fetal monocytes or circulating macrophage precursors.

Tissue specialization, phenotype, and functional properties of Hofbauer and decidual macrophages

A central dogma in the macrophage field is that macrophages acquire specific functional programs which are tissue-specific. But is there evidence toward a unique macrophage phenotype or even transcriptome in placenta? Modern technologies provide the tools to study placental macrophages beyond the capabilities of the beginning of the last century. Antibodies and single cell mRNA technology have added to our understanding of the cells and their heterogeneity in placenta.

Placentas are routinely fixed in 10% buffered formalin. Cord, membrane rolls and full-thickness placental samples are then paraffin-embedded and stained with hematoxylin-eosin for routine histological examination. Alternatively, they can also be used in frozen section that preserves epitopes without fixation. CD68 was one of the first antibodies produced that recognized tissue macrophages [8,9]. However, even from early work it became evident that not all antibodies recognizing the same molecule behaved equally. For example, the CD68 clones KP-1, PG-M1, and EBM-11 require different enzymatic processing for detection and stain different cells in placenta.

Histological analysis across pregnancy shows that in the first weeks post implantation the main macrophages in the organ are localized in the decidua, likely remnants of the endometrial macrophage population. This timing has been studied in more detail in the mouse, and the expectation is that the process is similar in humans. There appears to be abundant remodeling of this population, increased influx of circulating monocytes that stay distant to the implantation site and rather accumulate near the decidua basalis area, closer to the myometrium [10].

In humans, during the first weeks of implantation, macrophages are found in high numbers in the maternal decidua and in tissues close in proximity to the placenta. Hofbauer villi macrophages appear later post implantation and represent the only immune population on the fetal side of the placenta in healthy pregnancy. Histological analyses showed their location to be in the stroma of the placental villus, close to fetal capillaries and trophoblasts, making them likely candidates for involvement in regulatory functions during placental development and homeostasis [11,12].

CD68 stains both decidual and Hoffbauer macrophages, pretty much like all macrophages in the body, thus from that point of view there is nothing unique about the cells, Fig. 2. They also express CD14, which is a marker for monocytes and macrophages. The staining of both shows no significant overlap, suggesting either different cell subsets, or different maturation stages for the same cell. CD68 is a late endosomal transmembrane glycoprotein without known function, CD14 instead is part of the LPS lipid binding machinery and mediates the transfer to MD2 and TLR4, for example. Thus CD14+ cells in the placenta could be responsible for the recognition of DAMPs and PAMPs down the Lipid A recognition system. Interestingly, in benign pathology such as obesity, the number of macrophages in the placenta increases [13], where Challier et al. have shown that the numbers of CD68+ and CD14+ cells in obese patients were twice as high of that of lean women. The cells had increased levels of the prototypic TNF, IL-6 and IL-1ß proinflammatory cytokines similar to the profile found in adipose tissue, again failing to highlight a specific placenta phenotype, or a bona fide deactivated phenotype. The literature begins to capture this and the pro-inflammatory role of this wrongly termed M2 macrophages begins to emerge [14].

Macrophages express scavenger receptor and lipid recognition systems important for apoptotic cell removal. In fact, the cells have been described in the proximity of pyknotic cells. For apoptotic cell removal, they will need a machinery of receptors such as CD36, AXL and MERTK. From work in obesity and hyperlipidemia, the expression of lipid receptors emerges as a theme that needs further exploration in the macrophage populations in placenta. AXL, for example, has been studied in Zika virus infection, but only in the endothelium [15]. CD36 together with other fatty acid transporters, has been studied in obesity, but no formal macrophage dedicated study is available. Interspecies differences are ill defined and human studies need developing.

Placental macrophages can also be recognized by their expression of the three subtypes of IgG Fc receptors (CD16, CD32 and CD64). These markers are widely distributed on human leukocytes and classical monocyte/macrophages; however, there is a level of specialization in monocytes, a fraction of circulating monocytes have high expression of CD16, for example. The receptors have been implicated in transfer of immunoglobulins between the mother and the fetus [16,17]. We hypothesize this will also contribute to curb, eliminate, and modulate responses to immunocomplexes. A recent analysis of Fc receptor expression in human macrophages ex vivo shows widespread expression of CD32, and more selective expression of CD16 and CD64, low for example, in bone marrow, but high in spleen and liver [18]. Thus, perhaps the high expression of these three receptors in placenta has both a phenotypic link and functional consequences. Parallel studies between placenta and other organs can clarify this element.

Another immunoglobulin receptor expressed by placental macrophages [19], is the neonatal Fc receptor FcRn, also FCGRT-Fc fragment of IgG receptor and transporter. FCGRT associates with B2 microglobulin, forming a heterodimer with strong similarities to MHC I molecules. Borghi et al. suggest that FcRn is in great extent, uniquely responsible for the IgG transport between the mother and the fetus [20]. The receptor is not neonatal per se, nor it is only an IgG receptor. FCGRT is responsible for the binding and recycling of immunoglobulins in general and albumin, extending their half-life, while providing cell entry for smaller proteins carried by these macromolecules [21].

Phenotypically, placental macrophages have also been shown to express CD163, CD206 and CD209. CD163 is a hemoglobin-haptoglobin complex receptor, CD206 and CD209, also mannose receptor and DC-SIGN, bind to self-mannosylated *N*-glycans which are also PAMPs found on viruses, bacteria, and fungi [22,23]. These receptors are widely expressed in macrophages and DCs in the body, thus their expression does not support a placental macrophage unique proteome. Nota bene, these receptors are usually associated with M2 or alternative activation, while in reality they are also upregulated in macrophage maturation. The fact that the cells can mount inflammatory responses, further discussed below, points to the existence of a basal function, including the ability to respond to DAMPS.

To date there is no unique marker of Decidual macrophages or Hofbauer cells or signature of inflammatory activation/deactivation that sets them apart from the rest of body macrophages. Perhaps at gene or protein signature level, we have more opportunities to define a unique repertoire for the placental macrophages. Before the arrival of single cell RNAseq, there were major efforts to define tissue macrophage transcriptome profiles in all organs, however organs from females were usually not included in these studies. At the single cell level, Li et al. identified seven key manuscripts in this area, summarizing these efforts in a recent review [24]. The studies span from placental blood, to whole placentae and villi versus decidual preparations, and also vary in the use of early to term placenta. This information is vital because both anatomical presence and abundance change in time, and processing and sub-purification can affect the detection of the macrophages and more rare cells such as Dendritic cells. A comprehensive mononuclear phagocyte system of the placenta is still unknown. Of interest, Liu et al. performed a placental single cell RNAseq experiment harvested during the first trimester and could confirm the presence of two different subsets of villi-bound Hofbauer macrophages, distinguishable by their expression of CD74 and HLA-DR [25]. However, these Hofbauer subsets remain ill-defined and need more attention.

Vento-Tormo et al. also studied first trimester placentas and while they focus on Natural Killer (NK) cells and trophoblasts, their data highlight the presence of clear Hofbauer populations, decidual macrophages and interestingly, blood monocytes. Because they digested the tissue, it is difficult to ascertain if the monocytes were already in tissues or simply in the circulation. Suryawanshi et al. defined Hofbauer cells and found high expression of CD163, CD14, CSF1R, and CD68, and although they did not capture clearly the Decidual macrophages, they noticed a population they termed APC (antigen presenting cells), enriched for markers characteristic of dendritic cells. Dendritic cells in placenta need further analysis, and to date they appear restricted to the decidua.

A close approximation to defining phenotypic specialization comes from a recent study by McGovern et al. [26], who focused on the distinction between fetal versus maternal macrophages, and in passing defined gene signatures for the macrophages. The manuscript shows a variety of subsets which add to the emerging theme of heterogeneity the macrophages in the placenta. They identify embryonic and fetal signatures in Hofbauer cell—the markers put forward fall under the categories we have mentioned thus far and do not show a specific pattern yet. The profile resembles that of a mature macrophage with a DC flavor, with a strong HLA component, CD74, HLADR1, HLADPA1, expression of TLR10, LYZ (lysozyme), CD1C, S100A8 amongst others.

Macrophages and their contribution to placental immune privilege in COVID times

Whenever we approach a tissue and find high expression of CD68, CD206, and even Arginase, we term the macrophages as M2. The placenta is no exception and Hofbauer cells have been termed M2-like, with expression of the scavenger receptor CD163 and the pattern recognition receptors CD206 and CD209 and secreting cytokines and chemokines such as IL-10, CCL-2, and CCL-18 [27,28]. Human decidual macrophages were recently characterized as $CD163^{+}$ $CD206^{+}$ $CD209^{+}$ IL-10^{+} $CCL18^{+}$ M2-like cells. The vast literature on macrophages shows that these are widespread markers that can be found in alveolar and colonic macrophages, just to mention a couple of organs that are far from immunologically privileged.

HCs have a phenotype consistent with that of standard macrophages. If stimulated with steroids, for example, Hofbauer cells produce IL-10 and TGF-β [29–31] and over-express CD163, CD206, and CD209, which is very prototypical for macrophages. Demonstrating an M2 profile should include in humans a combination of STAT6 activation and co-expression of transglutaminase 2 and arachidonate 15-lipoxygenase, which are more IL-4/IL-13 selective. An M2 biased macrophage would fail to respond to such challenge.

The reality is more complex and both Hofbauer and decidual macrophages are able to mount classic inflammatory responses, which are behind the many pregnancy complications in viral and bacterial pathologies leading to pregnancy complications such as chorioamnionitis, miscarriage, and preterm delivery [32]. Hofbauer cells have been implicated in vertical transmission of TORCH (Toxoplasmosis, other infections, Rubella, Cytomegalovirus, and Herpes Simplex II virus) infections and Zika virus [33–36], while a study by Quillay and colleagues reported that decidual macrophages are permissive to HIV-1 infection, in vitro [37]. Villitis and Hofbauer cell infiltration in pregnancies complicated by congenital infections such as rubella, varicella, variola, and syphilis, suggest a crucial role of Hofbauer cells as gatekeepers that may either facilitate or prevent in utero transmission of these pathogens [38].

The recent widespread outbreak of corona virus disease (COVID-19) caused by the SARS-CoV-2 opened many questions about pregnancy and risks. Pregnant women are generally more vulnerable to viral infections such as influenza, hepatitis E, varicella, and measles, resulting in increased morbidity and mortality [39,40]. Initial evidence in COVID-19 suggested enhanced vulnerability to SARS-CoV-2 [41]. A recent study from the CDC reported that among women of reproductive-age with SARS-CoV-2 infection, pregnancy was associated with increased risk for hospitalization and intensive care unit admission [42], as well as need for mechanical ventilation and extracorporeal membrane oxygenation [43]. Additional complications of pregnancy that have been associated with SARS-CoV-2 infection include higher rates of stillbirth and preterm delivery [44] as well as maternal deaths [45].

Hetch et al. evaluated the expression of ACE2, TMPRSS2, and SARS-CoV-2, as well as vertical transmission in circa 20 placentas [46]. Transmission is considered rare; the authors showed a polarized expression of ACE2 in the syncytiotrophoblasts, cytotrophoblast and extravillous trophoblast, but did not detect the receptors in the stroma of the villi, the endothelium nor the Hofbauer cells. Others have confirmed ACE2 in cytotrophoblasts, syncytiotrophoblasts and extravillous trophoblasts throughout pregnancy, regardless of COVID-19 status [47]. The topic has expanded significantly [48–51]; cases of placentitis are rare, but they involve fibrin deposition and a strong maternal CD68 infiltrate [52].

Conclusions

The maternal-fetal interface is a unique environment both in cellular composition and cell functions and includes specialized macrophage populations. At this interface, important interactions occur between epithelium, parenchyma and maternal and fetal-derived macrophages. More work is needed to elucidate the role of macrophages in the functional dynamics of the process crucial for pregnancy outcome.

The transitory and dynamic nature of the maternal-fetal interface provides a window of opportunity into studying the malleability of the various macrophage activation subtypes which may play a role in this environment. For example, Decidual macrophages have been implicated in the remodeling required to establish the vessel network, including the prototypic spiral arteries in placenta. These specialized functions may point to unique marker combination for the cells.

Although a role for macrophages in tolerance can be imagined, there is no formal proof that Hofbauer cells or decidual macrophages contribute to immune tolerance, a role which may be more likely to be mediated by Treg and NK cells or the trophoblastic epithelium. The decidua is the anatomic area where the developed immune system of the mother collides with that of the developing embryo; decidual macrophages might play a more important role in this respect due to their interaction with the fetal trophoblasts and maternal lymphocytes. Lack of HLA-A, HLA-B in trophoblasts, and distinct expression and regulation of HLA-C, HLA-E, [53] and HLA-G [54] pinpoint trophoblasts and NK cell interaction as the key determinant for tolerance. In this respect, macrophages and APCs could have a secondary yet important role once a response is activated, and this can lead to functional specialization that needs uncovering.

Due to the fact that Hofbauer cells are not in direct contact with the mothers' immune system, they could play a role in clearing cytokines and inflammatory factors when they cross the trophoblast layer, which in itself is responsible, for example, for chemokine clearance [55]. Expression of decoy and clearance receptors can guide phenotype identification. Macrophages are strong candidates for the elimination of apoptotic cells, and widely occurring genetic errors in specific cell populations in the organ [56]. The Hofbauer cell is not surrounded by immune cells but by fetal fibroblasts. To us, Hofbauer cells could represent a more trophic and structural macrophage, where an interesting report described sprouty protein expression in these cells which can lead to villi branching [57].

Early pathologists already emphasized the need to inspect the placenta in every birth, looking for malformations, vascular accidents and for example, hemoglobin/hemosiderin accumulation in the macrophages and evidence of meconium exposure. Overall, nondigestible material and exogenous pollutants will accumulate in placental macrophages, indicators of hazards to the health of both the mother and child. There is an urgent need for the development and validation of methods to study placenta macrophages in situ and ex vivo. Comparative analysis with other organs will help define placenta specific macrophage functions and phenotypes. Molecular targets and cellular targets for pregnancy complications will emerge from these profiles.

Acknowledgments

We would like to thank Nikoletta Vrikki for her contribution to the manuscript figures.

References

[1] Gordon S. Metchnikoff's legacy: macrophage functions in immunity. Immunology 2008;125:3.

[2] Epelman S, Lavine KJ, Randolph GJ. Origin and functions of tissue macrophages. Immunity 2014;41(1):21–35.

[3] Fujiwara T, et al. Macrophage infiltration predicts a poor prognosis for human Ewing sarcoma. Am J Pathol 2011;179(3):1157–70.

[4] Liepmann W. Zur Biologie der menschlichen placenta. Arch Gynakol 1906;77(1):37–50.

[5] Erlebacher A. Immunology of the maternal-fetal interface. Annu Rev Immunol 2013;31:387–411. https://doi.org/10.1146/annurev-immunol-032712-100003.

[6] Burton GJ, Fowden AL. The placenta: a multifaceted, transient organ. Philos Trans R Soc Lond Ser B Biol Sci 2015;370(1663):20140066.

[7] Turco MY, Moffett A. Development of the human placenta. Development 2019;146:22.

[8] Greywoode GI, McCarthy SP, McGee JO. Labelling of cells of the mononuclear phagocyte system in routinely processed archival biopsy specimens with monoclonal antibody EBM/11. J Clin Pathol 1990;43(12):992–6.

[9] Falini B, et al. PG-M1: a new monoclonal antibody directed against a fixative-resistant epitope on the macrophage-restricted form of the CD68 molecule. Am J Pathol 1993;142(5):1359–72.

[10] Miller L, Hunt JS. Sex steroid hormones and macrophage function. Life Sci 1996;59(1):1–14.

[11] Katabuchi H. The mystery of Hofbauer cells. Placenta 2014;35(10):A2.

[12] Palmeira P, et al. IgG placental transfer in healthy and pathological pregnancies. Clin Dev Immunol 2012;2012:985646.

[13] Challier JC, et al. Obesity in pregnancy stimulates macrophage accumulation and inflammation in the placenta. Placenta 2008;29(3):274–81.

[14] Young OM, et al. Toll-like receptor-mediated responses by placental Hofbauer cells (HBCs): a potential pro-inflammatory role for fetal M2 macrophages. Am J Reprod Immunol 2015;73(1):22–35.

[15] Richard AS, et al. AXL-dependent infection of human fetal endothelial cells distinguishes Zika virus from other pathogenic flaviviruses. Proc Natl Acad Sci USA 2017;114(8):2024–9.

[16] Jensen TS, Matre R. Fc-gamma-receptor activity in the developing human placenta. APMIS 1995;103(6):433–8.

[17] Saji F, Koyama M, Matsuzaki N. Human placental Fc-receptors. Placenta 1994;15(5):453–66.

[18] Bruggeman CW, et al. Tissue-specific expression of IgG receptors by human macrophages ex vivo. PLoS One 2019;14(10):e0223264.

[19] Kiskova T, et al. Expression of the neonatal Fc-receptor in placental-fetal endothelium and in cells of the placental immune system. Placenta 2019;78:36–43.

[20] Borghi S, et al. FcRn, but not FcgammaRs, drives maternal-fetal transplacental transport of human IgG antibodies. Proc Natl Acad Sci USA 2020;117(23):12943–51.

[21] Sand KM, et al. Unraveling the interaction between FcRn and albumin: opportunities for design of albumin-based therapeutics. Front Immunol 2014;5:682.

[22] Svensson-Arvelund J, et al. Human first-trimester trophoblasts promote a tolerant fetal environment by inducing homeostatic macrophages and regulatory T cells through M-CSF, IL10, TGF-β, and TRAIL. J Reprod Immunol 2014;101:29–30.

[23] Sayama S, et al. Human decidual macrophages suppress IFN-gamma production by T cells through costimulatory B7-H1:PD-1 signaling in early pregnancy. J Reprod Immunol 2013;100(2):109–17.

[24] Li H, et al. Single cell transcriptome research in human placenta. Reproduction 2020;160(6):R155–67.
[25] Liu Y, et al. Single-cell RNA-seq reveals the diversity of trophoblast subtypes and patterns of differentiation in the human placenta. Cell Res 2018;28(8):819–32.
[26] Thomas JR, et al. Phenotypic and functional characterization of first-trimester human placental macrophages, Hofbauer cells. J Exp Med 2021;218(1):e20200891.
[27] Mehta RB, et al. Trophoblast cells in immune regulation: modulation of macrophage polarization and production of IL-35. J Reprod Immunol 2011;90(2):165.
[28] Houser BL, et al. Two unique human decidual macrophage populations. J Immunol 2011;186(4):2633–42.
[29] Johnson EL, Chakraborty R. Placental Hofbauer cells limit HIV-1 replication and potentially offset mother to child transmission (MTCT) by induction of immunoregulatory cytokines. Retrovirology 2012;9:101.
[30] Tang ZH, et al. Glucocorticoids enhance CD163 expression in placental Hofbauer cells. Endocrinology 2013;154(1):471–82.
[31] Svensson J, et al. Macrophages at the fetal-maternal Interface express markers of alternative activation and are induced by M-CSF and IL-10. J Immunol 2011;187(7):3671–82.
[32] Tang ZH, et al. Placental Hofbauer cells and complications of pregnancy. Reprod Sci 2011;1221:103–8.
[33] Quicke KM, et al. Zika virus infects human placental macrophages. Cell Host Microbe 2016;20(1):83–90.
[34] Simoni MK, et al. Zika virus infection of Hofbauer cells. Am J Reprod Immunol 2017;77(2), e12613.
[35] Miner JJ, Diamond MS. Zika virus pathogenesis and tissue tropism. Cell Host Microbe 2017;21(2):134–42.
[36] Esser-Nobis K, et al. Comparative analysis of African and Asian lineage-derived Zika virus strains reveals differences in activation of and sensitivity to antiviral innate immunity. J Virol 2019;93(13):e00640-19.
[37] Quillay H, et al. Distinct characteristics of endometrial and decidual macrophages and regulation of their permissivity to HIV-1 infection by SAMHD1. J Virol 2015;89(2):1329–39.
[38] Haram K, et al. Early development of the human placenta and pregnancy complications. J Matern Fetal Neonatal Med 2020;33(20):3538–45.
[39] Jamieson DJ, Theiler RN, Rasmussen SA. Emerging infections and pregnancy. Emerg Infect Dis 2006;12(11):1638–43.
[40] Kourtis AP, Read JS, Jamieson DJ. Pregnancy and infection. N Engl J Med 2014;371(11):1077.
[41] Wang W, Zhang Q, Qu F. The potential benefits of Chinese integrative medicine for pregnancy women during the COVID-19 pandemic. Integr Med Res 2020;9(3), 100461.
[42] Ellington S, et al. Characteristics of women of reproductive age with laboratory-confirmed SARS-CoV-2 infection by pregnancy status—United States, January 22-June 7, 2020. MMWR Morb Mortal Wkly Rep 2020;69(25):769–75.
[43] Wang X, Wang D, He S. The role of a cytokine storm in severe COVID-19 disease in pregnancy. Am J Obstet Gynecol 2020;223(5):780–2.
[44] Kingston EV. High rates of stillbirth and preterm delivery in women with covid-19 and the efficacy of ECMO in pregnancy. BMJ 2020;370:m2921.
[45] Cheng SO, Khan S, Alsafi Z. Maternal death in pregnancy due to COVID-19. Ultrasound Obstet Gynecol 2020;56(1):122.
[46] Hecht JL, et al. SARS-CoV-2 can infect the placenta and is not associated with specific placental histopathology: a series of 19 placentas from COVID-19-positive mothers. Mod Pathol 2020;33(11):2092–103.

[47] Gengler C, et al. SARS-CoV-2 ACE-receptor detection in the placenta throughout pregnancy. Clin Microbiol Infect 2021;27(3):489–90.
[48] Facchetti F, et al. SARS-CoV2 vertical transmission with adverse effects on the newborn revealed through integrated immunohistochemical, electron microscopy and molecular analyses of placenta. EBioMedicine 2020;59:102951.
[49] Smithgall MC, et al. Third-trimester placentas of severe acute respiratory syndrome coronavirus 2 (SARS-CoV-2)-positive women: histomorphology, including viral immunohistochemistry and in-situ hybridization. Histopathology 2020;77(6):994–9.
[50] Taglauer E, et al. Consistent localization of SARS-CoV-2 spike glycoprotein and ACE2 over TMPRSS2 predominance in placental villi of 15 COVID-19 positive maternal-fetal dyads. Placenta 2020;100:69–74.
[51] Sharps MC, et al. A structured review of placental morphology and histopathological lesions associated with SARS-CoV-2 infection. Placenta 2020;101:13–29.
[52] Linehan L, et al. SARS-CoV-2 placentitis: an uncommon complication of maternal COVID-19. Placenta 2021;104:261–6.
[53] Papuchova H, et al. The dual role of HLA-C in tolerance and immunity at the maternal-fetal interface. Front Immunol 2019;10:2730.
[54] Papuchova H, et al. Three types of HLA-G+ extravillous trophoblasts that have distinct immune regulatory properties. Proc Natl Acad Sci USA 2020;117(27):15772–7.
[55] Martinez de la Torre Y, et al. Protection against inflammation- and autoantibody-caused fetal loss by the chemokine decoy receptor D6. Proc Natl Acad Sci USA 2007;104(7):2319–24.
[56] Coorens THH, et al. Inherent mosaicism and extensive mutation of human placentas. Nature 2021;592(7852):80–5.
[57] Anteby EY, et al. Human placental Hofbauer cells express sprouty proteins: a possible modulating mechanism of villous branching. Placenta 2005;26(6):476–83.

CHAPTER 13

Targeting macrophages in systemic diseases

Alexandre Wagner Silva de Souza[a], Wilson de Melo Cruvinel[b], and Luís Eduardo Coelho Andrade[a]
[a]Rheumatology Division, Medicine Department, Escola Paulista de Medicina, Universidade Federal de São Paulo, São Paulo, Brazil
[b]Pontifícia Universidade Católica de Goiás, Goiânia, Brazil

Introduction

Monocytes and macrophages are innate immune cells that play an important role on inflammatory responses, host defense, tissue repair, and immune modulation. Together with dendritic cells and neutrophils, monocytes and macrophages are considered professional phagocytes [1]. Circulating monocytes are bone marrow-derived cells that traffic from the blood stream to tissues in inflammatory conditions, where they differentiate into monocyte-derived macrophages and monocytes-derived dendritic cells (DC) [2,3]. On the other hand, tissue-resident macrophages are different from monocyte-derived macrophages and monocytes-derived DC, since these cells have embryonic origin and are self-renewing cells in their respective tissues. Tissue-resident macrophages are heterogeneous cells and have sentinel function in the majority of the tissues in the body. Examples of tissue-resident macrophages include Langerhans cells, alveolar macrophages, microglial cells, Kupffer cells, and red pulp macrophages in spleen [4] (Fig. 1).

The plasticity of these innate immune cells leads to the development of phenotypically and functionally distinct subsets of monocytes and macrophages that are induced and activated by cytokines, growth factors and chemokines depending on the type and extent of tissue damage or the current tissue requirements [5]. Besides their functions in tissue homeostasis under physiologic conditions, the distinct macrophage and monocyte subsets have been recognized as important players in the pathogenesis of different systemic autoimmune and inflammatory diseases [6]. This chapter aims to review the current evidence of the role of monocytes and macrophages in systemic inflammatory diseases (Fig. 2).

Macrophages in the Human Body
https://doi.org/10.1016/B978-0-12-821385-8.00012-3

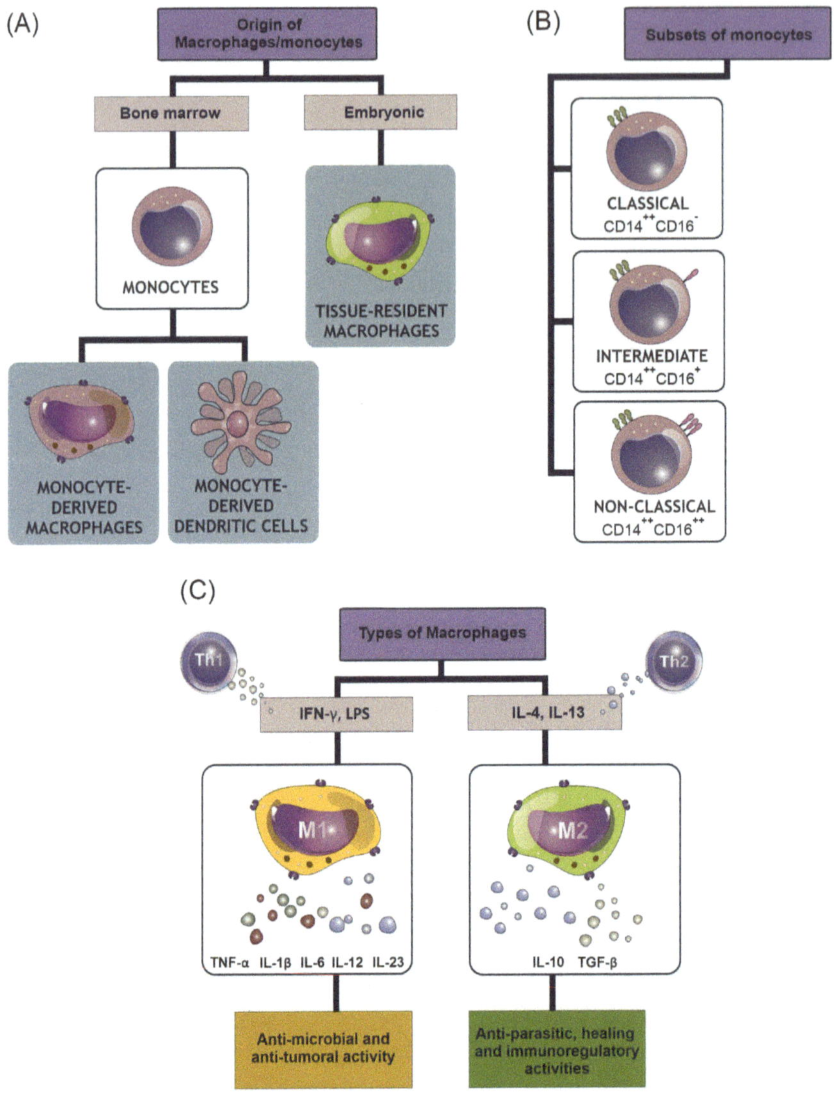

Fig. 1 Origen and subsets of monocytes and macrophages. (A) Circulating monocytes are bone marrow-derived cells that differentiate into monocyte-derived macrophages and monocytes-derived dendritic cells. Tissue-resident macrophages have embryonic origin and are self-renewing cells in different tissues. (B). Three subsets of monocytes are recognized based on the expression of the surface markers CD14 and CD16: classical ($CD14^{++}CD16$), intermediate ($CD14^{++}CD16^{+}$) and non-classical monocytes ($CD14^{+}CD16^{++}$). (C). Polar phenotypes of macrophages: M1 macrophage phenotype is activated by Th1 cytokines, such as interferon (IFN)-γ, and lipopolysaccharide (LPS); M2 macrophages are induced by Th2 cytokines such as IL-4 and IL-13.

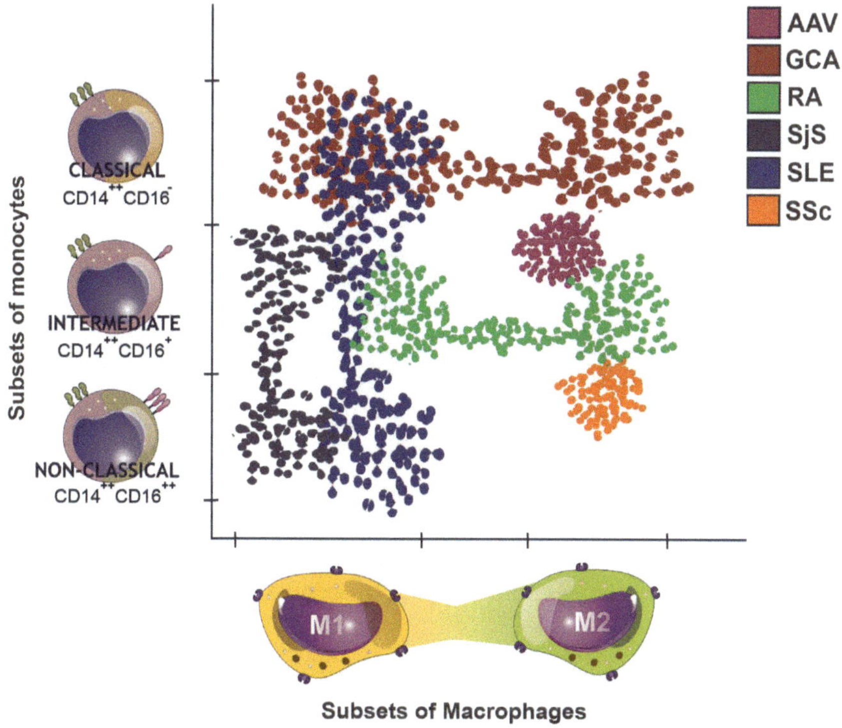

Fig. 2 Two-dimension dot histogram graph of the distribution of macrophage and monocyte subsets in different systemic autoimmune and inflammatory diseases. *Dots* were plotted based on literature reports on the frequency of macrophage and monocyte subsets in tissues and peripheral blood, respectively. For example, intermediate monocytes and M2 macrophages predominate in AAV, whereas classical monocytes and M1 macrophages prevail in SLE. A continuous phenotypic spectrum is observed for macrophages in RA and GCA, as well as for monocytes in SLE and SSj. *AAV*, ANCA-associated vasculitis; *BD*, Behçet disease; *GCA*, giant cell arteritis; *IgAV*, IgA vasculitis; *KD*, Kawasaki disease; *RA*, rheumatoid arthritis; *SjS*, Sjögren syndrome; *SLE*, systemic lupus erythematosus; *SSc*, systemic sclerosis; *TAK*, Takayasu arteritis.

Distribution of monocytes and their subtypes in the peripheral blood

Monocytes are heterogeneous cells that circulate in the blood stream and present high plasticity. Three subsets of monocytes are recognized based on the expression of the surface markers CD14 and CD16: classical ($CD14^{++}CD16^{-}$), intermediate ($CD14^{++}CD16^{+}$) and non-classical monocytes ($CD14^{+}CD16^{++}$) [7]. Classical monocytes comprise 80%–95%

of all blood monocytes and present high phagocytic capacity and scavenger function. Most classical monocytes differentiate in DC in tissues. Between 2% and 11% of blood monocytes have the intermediate phenotype, which is highly pro-inflammatory and contributes to the production of reactive oxygen species, antigen presentation, activation of T cells and angiogenesis. Intermediate monocytes may differentiate either into macrophages or DC in tissues. The percentage of non-classical monocytes ranges between 2% and 8% of all monocytes in the peripheral blood and they predominantly differentiate into macrophages in tissues. These cells have high mobility and patrolling function for endothelial injury, but they also have a role on antigen presentation and T-cell activation [8,9].

The number and percentage of circulating monocytes and their subsets have been assessed in different systemic autoimmune and inflammatory diseases, including systemic lupus erythematosus (SLE), rheumatoid arthritis (RA), systemic sclerosis (SSc), Sjögren syndrome (SjS), and systemic vasculitides. Table 1 depicts disturbances in the distribution of monocyte subtypes in different systemic autoimmune and inflammatory diseases.

Macrophage subtypes in tissues

In patients with systemic autoimmune and inflammatory diseases, monocytes migrate to affected tissues and organs, where they differentiate into macrophages or DC [9]. Different phenotypes of macrophages may be found in inflamed tissues including the classically activated M1 and the alternatively activated M2 subtypes. The latter are further subdivided into M2a, M2b, M2c, and M2d phenotypes [1,5]. Th1 cytokines, such as interferon (IFN)-γ, and lipopolysaccharide (LPS) induce the macrophage M1 phenotype. These cells produce high levels of pro-inflammatory cytokines (e.g., TNF-α, IL-1β, IL-6, IL-12 and IL-23) and produce reactive oxygen species. Therefore, M1 macrophages have anti-microbial and anti-tumoral activity, in addition to promoting inflammation [5,45]. Conversely, M2 macrophage subsets are more diverse than the M1 phenotype, but, in general, M2 macrophages have high phagocytosis capacity and are very efficient in removing debris and apoptotic cells. These cells also promote wound healing as they have pro-fibrotic, pro-angiogenic and anti-inflammatory properties. M2 macrophages are induced by Th2 cytokines such as IL-4 and IL-13, and they produce high amounts of IL-10 and transforming growth factor (TGF)-β. In contrast to the favorable anti-parasitic, immunoregulatory, and healing properties of M2 macrophages, this subpopulation participates in tumor progression [5,45].

Table 1 Monocytes and macrophages in different systemic autoimmune and inflammatory diseases.

Disease	Monocyte subtypes in the peripheral blood	Macrophage polarization in tissues	References
SLE	↓ Classical subset and ↑ non-classical than controls. Correlation between anti-dsDNA titer and the number of classical monocytes.	Indirect evidence of the role of M1 macrophages in the pathogenesis.	[10–12]
RA	↑ $CD14^+CD16^+$ monocytes and correlation with disease activity. ↑ Intermediate monocytes in the peripheral blood and in synovial fluid of arthritic joints.	Although, the number of macrophages in the synovium correlates with levels of M1-related cytokines in the serum and in the synovia, a mix of M1 and M2 macrophages infiltrates synovial tissue.	[13–22]
SSc	$CD16^+$ monocytes were associated with the diffuse form of SSc, with increased modified Rodnan's skin score and with the severity of interstitial lung disease.	↑ Expression of the M2 markers CD163 and CD204 by macrophages in the fibrotic skin. Serum levels of the sCD163 are increased in SSc patients. Increased sCD163 release by PBMC was associated with SSc disease progression.	[23–28]
SjS	↑ Non-classical (mature monocytes) in the peripheral blood and ↑ intermediate monocytes in salivary glands.	No data regarding macrophages polarization in humans with SjS. There is indirect evidence to a shift toward the M1 phenotype.	[6,29,30]
GCA	↑ Classical monocytes prior to therapy. Glucocorticoids normalized the distribution of monocytes, but led to ↓ non-classical monocytes.	M2 macrophages were predominant in sites of tissue destruction while M1 macrophages were observed within the inner intima of arteries with high degree of intimal hyperplasia in temporal arteries and the aorta.	[31,32]
TAK	No evidence available	The expression of the M2 marker CD206 dominated the M1 marker CD86. ↑ M1 macrophages in the aorta from TAK patients and from patients with atherosclerotic disease than in heart transplant donors.	[33]

Continued

Table 1 Monocytes and macrophages in different systemic autoimmune and inflammatory diseases.—cont'd

Disease	Monocyte subtypes in the peripheral blood	Macrophage polarization in tissues	References
KD	No evidence available	↑ M1 macrophages in coronary artery aneurysms of KD patients. ↑ mRNA expression of M1 markers (TLR2 and TLR4) and M2 markers (ARG1, CCR2, TLR1, TLR8, TLR5, MS4A6A, CD36, and MS4A4A) in macrophages.	[34,35]
AAV	↑ Intermediate monocytes than controls and a significant correlation between CD16 and MPO expression. In vitro stimulation of monocytes with anti-MPO antibodies induced higher production of IL-1β, IL-6 and IL-8. This was not observed with anti-PR3 antibodies.	↑ M2 macrophages in early pauci-immune glomerulonephritis and in airway lesions of GPA patients. Association between sCD163 and active glomerulonephritis in AAV. Serum from AAV patients induced the polarization toward the M2c macrophage subtype.	[36–41]
IgAV	No evidence available	↑ Tubulo-interstitial and glomerular M2 macrophages were associated with poor renal outcomes.	[42]
BD	↑ Intermediate monocytes in PBMC. Controversial results regarding the distribution of classical and non-classical monocytes. Colchicine increases non-classical monocytes in the peripheral blood.	No evidence available	[43,44]

AAV, ANCA-associated vasculitis; *BD*, Behçet's disease; *GCA*, giant cell arteritis; *IgAV*, IgA vasculitis; *KD*, Kawasaki disease; *RA*, rheumatoid arthritis; *SjS*, Sjögren's syndrome; *SLE*, systemic lupus erythematosus; *SSc*, systemic sclerosis; *TAK*, Takayasu arteritis.

There is enough evidence that macrophages play an important role as effector cells in the pathogenesis of different systemic autoimmune diseases. Different macrophage subtypes are related to the production of cytokines and chemokines, enhancement of the inflammatory process and ultimately the pathogenic organ effects of these diseases [6].

The role of monocytes and macrophages in systemic autoimmune diseases

Systemic lupus erythematosus

Monocytes in systemic lupus erythematosus

In SLE patients, controversial findings were reported by the literature regarding the distribution of monocyte subsets based on the expression of CD14 and CD16 in the peripheral blood. Early studies showed that $CD14^+ CD16^+$ monocytes (possibly corresponding to intermediate or non-classical monocytes) were increased in SLE patients compared to controls, especially in those presenting lupus nephritis, whereas in another study no differences were found between SLE and controls [10,46,47]. In one study, SLE patients presented fewer circulating $CD14^+ CD16^-$ monocytes (possibly corresponding to classical monocytes) than controls [48]. When monocyte subsets were analyzed (i.e., classical, intermediate and non-classical monocytes), SLE patients presented an increase in non-classical monocytes and a decrease in classical monocytes compared to controls [10,11]. Females with SLE presented more striking abnormalities of monocytes subsets than male patients [12]. Moreover, the frequency of non-classical monocytes in the peripheral blood correlated positively with anti-double stranded DNA antibody titer [10]. On the other hand, the monocyte count does not appear to correlate with disease activity [47]. Interestingly, one study showed that SLE serum is able to induce monocyte apoptosis via C5a [49]. Finally, macrophage depletion attenuated skin and kidney disease severity in a SLE mouse model [50].

There is evidence that monocytes can contribute to the characteristic B-cell hyperactivity observed in SLE via CD40-CD40L interaction. SLE patients present increased frequency of circulating CD40L-positive monocytes [51] and recombinant CD40L stimulated in vitro production of IgG by B cells from SLE patients but not from normal controls [52]. In addition, CD40L overexpression can induce lupus-like manifestations whereas CD40L neutralization decreased the production of autoantibodies in lupus-prone mice [53,54].

Macrophages in systemic lupus erythematosus

In SLE patients, high circulating levels of M1-related cytokines (i.e., IFN-γ, IL-1β, IL-6, and TNF-α) combined with lower TGF-β serum levels in patients with active disease indicate a potential shift to the macrophage M1 phenotype [45,55–57]. As an exception to this M1-macrophage cytokine profile, IL-10, an M2-macrophage associated cytokine, is increased and plays a paradoxically pro-inflammatory role in SLE patients and its production by different cell types, including autoreactive helper T cells, follicular T cells and B-cells, is dependent on large amounts of type I IFN. IFN-α in turn induces a pro-inflammatory gain of function of IL-10, as it enhances STAT-1 activation by IL-10, which increases the expression of CXCL9 and CXCL10 in IFN-α-primed macrophages. These chemokines induce immune cell migration (e.g., Th1 cells, macrophages, NKT cells and cytotoxic $CD8^+$ T cells) to affected tissues. Moreover, IL-10 loses its ability to suppress TNF-α production in IFN-α-primed macrophages. Finally, IL-10 has a role in stimulating autoreactive B cells [58–60]. Macrophages in SLE present increased expression of intercellular adhesion molecule 1 (ICAM-1) and sialic acid-binding Ig-like lectin 1 (Siglec-1, CD169), which is intensely reduced by glucocorticoid therapy [61]. It is relevant that increased Siglec-1-positive macrophages promote activation of T and B lymphocytes, and the frequency of Siglec-1-positive monocytes has been shown to correlate with SLE disease activity [61,62].

Neopterin is a marker of macrophage activation triggered by IFN-γ and is considered a signature of the M1-macrophage phenotype. Neopterin serum levels are increased in SLE patients and this marker correlates positively with the number of circulating apoptotic lymphocytes in the peripheral blood, suggesting that the hyperactivation of M1 macrophages may be associated with an impaired removal of apoptotic cells in SLE [56]. In fact, defective phagocytosis of macrophages in SLE has been demonstrated in several studies, with potential impairment of the clearance of apoptotic cells and debris [63,64]. SLE patients have circulating antibodies against scavenger receptors of macrophages and that might further affect the ability of M2 macrophages to clear apoptotic cells in SLE [65]. In addition, myeloid cells from SLE patients presenting active disease have increased expression of M1-associated genes, including STAT1 and SOCS3, whereas the expression of M2-associated genes (e.g., STAT3, STAT6, and CD163) was not altered [66].

These findings suggest a potential role for the M1/M2 macrophage imbalance in SLE pathogenesis, by favoring the M1 pro-inflammatory profile at the expense of the anti-inflammatory and apoptotic bodies/immunocomplex clearing role of M2 macrophages. Thus, targeting monocytes and macrophages may be a reasonable therapeutic approach in SLE. In fact, adoptive transfer of IL-4-induced M2 has ameliorated the disease in murine SLE [67] and in vitro treatment of human SLE monocytes with pioglitazone, an M2-polarizing drug, has decreased the production of pro-inflammatory cytokines [68]. In addition, pharmacological treatment also has the potential to reverse the M1/M2 polarization toward recovery of M2 profile in SLE. Sodium valproate, a histone deacetylase inhibitor, was shown to modulate in vitro the epigenetic configuration of monocyte-derived macrophages from SLE patients by decreasing their pro-inflammatory phenotype [69]. The aryl hydrocarbon receptor (AhR) agonist indole-3-carbinol induces in vitro differentiation of SLE monocytes to an M2-macrophage phenotype with secretion of anti-inflammatory cytokines [70].

Rheumatoid arthritis

Monocytes in rheumatoid arthritis

In RA, the role of $CD14^{+}CD16^{+}$ monocytes seems to be clearer than in SLE, since these cells are increased in the peripheral blood from RA patients compared to controls and these cells are associated with disease activity. In addition, the expression of CD16 is markedly increased in monocytes from the synovial fluid in RA compared to monocytes from the peripheral blood [13]. When looking into monocyte subsets, all studies show increased frequency of intermediate monocytes in the peripheral blood of RA patients in comparison to controls [14–16], as well as in the synovial fluid of arthritic joints [17,18].

Macrophages in rheumatoid arthritis

Macrophages have a relevant role in the pathogenesis of RA as the number of these cells is increased in the synovial tissue of RA patients and is positively correlated with disease activity and with cartilage and joint destruction [19,71,72]. Serial synovial biopsies in RA patients have shown a decrease in tissue macrophages in parallel with successful response to therapy, regardless of the nature of the therapeutic agent [20]. Experimental models of arthritis have shown that macrophage M1 polarization is associated with higher inflammatory burden in affected joints [21]. In line with these findings, the number of macrophages in the synovium is correlated with circulating levels of pro-inflammatory cytokines such as TNF-α and IL-6 and the expression of M1-related cytokines, including IL-1, TNF-α, IL-6 and IL-12, is increased in the inflamed synovia from RA patients [20–22,72]. In contrast, the immunohistochemistry analysis of macrophages in the synovial tissue from RA patients shows a mix of M1 and M2 markers such as folate receptor β (M1 marker) and CD163 (M2 marker) [15]. Interestingly, some studies show that synovial macrophages in RA produce more IL-10 and TGF-β than TNF-α and inducible nitric oxide synthetase (iNOS), indicating that in some instances the resolution process counterbalances the inflammatory process [21,73–75].

Targeting macrophages is a promising strategy in RA and the prototype strategy is represented probably by antagonizing the key macrophage cytokine TNF-α with various immune-biological drugs, which has proven widely successful [76]. Modulation of macrophage polarization represents a potential therapeutic perspective in RA. SIRT-1, a histone deacetylase with anti-inflammatory properties, has been shown to promote down-regulation of M1-related genes and to induce over-expression of M2-related genes. SIRT-1 transgenic mice subjected to the collagen-induced arthritis model exhibited reversal of M1 macrophage polarization in the synovia and milder clinical and histological arthritis, with less pronounced cartilage and bone destruction as compared to the wild type mice [77]. Another approach has been explored by interfering in the NOTCH-mediated differentiation of M1 macrophages. The NOTCH inhibitor Thapsigargin has been shown to reverse M1 polarization in a TNF-α transgenic arthritis mouse model, increasing the frequency of M2 macrophages and decreasing the severity of arthritis [78]. Furthermore, IL-10 has demonstrated to have a beneficial therapeutic effect in collagen-induced arthritis by inhibiting M1 macrophage polarization [79,80].

Systemic sclerosis

Monocytes in systemic sclerosis

SSc patients have been reported to present increased number of $CD14^+$ monocytes in the peripheral blood mononuclear cells (PBMC) in comparison to controls, but CD16 expression was not analyzed in this study [81]. Another study showed increased circulating $CD16^-$ monocytes in the peripheral blood of SSc patients in general compared to controls, whereas patients with the diffuse form of SSc presented more $CD16^+$ cells than controls and this subpopulation of monocytes was correlated with the modified Rodnan skin score. Furthermore, the number of circulating $CD16^+$ monocytes was associated with the severity of interstitial lung disease in SSc [23].

Macrophages in systemic sclerosis

Macrophage infiltration in the skin of SSc patients has been noted for more than 30years [82]. Macrophage migration inhibitory factor (MIF) is probably involved in the accumulation of macrophages in SSc skin, as MIF has been shown to be hyper expressed at the basal and suprabasal keratinocytes of SSc skin, and MIF serum levels are increased in SSc patients [83]. The pathophysiology of SSc is characterized by vascular abnormalities, immune activation, and excessive deposition of extracellular matrix in the skin and different organs. The increased extracellular matrix in SSc comprises high amounts of structurally normal collagen synthesized by subpopulations of fibroblasts [84]. In this pro-fibrotic context, M2 macrophages play an important role as a regulator of fibrosis [1,5,24]. The shift toward the M2 phenotype of macrophages in SSc is highlighted by the expression of the M2 markers CD163 and CD204 by macrophages in the fibrotic skin from SSc patients [25,81]. Th2-induced cytokines (e.g., IL-4 and IL-13) and transforming growth factor β (TGF-β) play a major role in this process [26,85]. In fact, elevated serum levels of M2-inducing cytokines, i.e., IL-4, IL-13, and IL-10, have been demonstrated in patients with SSc [86,87]. Ample evidence of the M2 activation in SSc is corroborated by in vitro studies, animal models and clinical studies. For example, CD204-deficient mice fail to develop silica-induced fibrosis in animal model [88]. Serum levels of the soluble form of CD163 (sCD163), a surrogate marker of the M2 activation, are increased in SSc patients compared to controls, and sCD163 release by PBMC from SSc patients upon in vitro activation was clearly associated with disease progression [27,28]. However, the representation of macrophages in SSc may be even more complex, as it has been reported that part of the $CD204^{+}$ $CD163^{+}$ $CD206^{+}$ M2 macrophages in SSc co-expresses CD80 and CD86 M1 markers [89].

From a mechanistic and target-therapeutic point of view, evidence has accumulated that M2 macrophages play an important role in activating tissue fibroblasts and inducing fibrosis in SSc by means of promoting increased levels of TGF-β, vascular endothelial growth factor (VEGF), and platelet-derived growth factor (PDGF) [90,91]. It has been shown that cadherin 11 mediates increased TGF-β secretion by macrophages in SSc skin, and blockade of cadherin 11 decreased skin thickness and myofibroblast numbers in a murine fibrosis model [92].

Sjögren syndrome

Monocytes in Sjögren syndrome

In SjS, interesting results were described regarding the proportion of monocyte subsets in the peripheral blood and salivary gland tissue. In one study, an increased frequency of non-classical monocytes was found in the peripheral blood from SjS patients as compared to controls, whereas a mouse model of SjS showed that such non-classical monocytes differentiated into dendritic cells in the salivary glands [29]. On the other hand, another study showed the expansion of the pro-inflammatory (classic) subset of monocytes ($CD14^{high}CD16^{low}$) in the salivary glands of SjS patients in association with increased IL-34 mRNA expression in these glands. IL-34 is a cytokine known to induce the pro-inflammatory phenotype of monocytes [30].

Macrophages in Sjögren syndrome

In mouse models of SjS, the inflammatory infiltration of lacrimal glands is mixed and includes predominantly B cells, followed by macrophages, mature $CD4^+$ T helper cells, $CD8^+$ cytotoxic T cells, NK cells and myeloid immunoregulatory cells [93]. In the salivary glands of non-obese diabetic (NOD) mice (i.e., an experimental model of SjS), there is inflammatory infiltration comprising both phenotypes M1 and M2 macrophages among other cell types, such as B cells and T cells [94]. This inflammatory response results in permanent damage to exocrine tissue and elicits loss of secretory function [95].

Conversely, there are no data regarding macrophages polarization in humans with SjS. The indirect evidence points out to a shift toward the M1 phenotype, as IFN-ɣ, an M1-inducing cytokine, was associated with SjS syndrome in one study [6]. In addition, a higher number of IFN-ɣ-secreting cells in PBMC was observed in SjS patients with Raynaud phenomenon as compared to SjS patients without Raynaud phenomenon and controls [96]. Moreover, studies have shown higher serum levels of M1-related pro-inflammatory cytokines, such as IL-6, IL-12, IL-23, and IL-33, in SjS compared to controls [97–99]. Active disease was related to increased serum IL-12 and lower serum levels of the anti-inflammatory cytokine IL-35 [98]. The salivary concentration of the pro-inflammatory cytokine IL-6 was increased in SjS patients compared to controls, and higher expression of IL-17 and IL-18 as well as higher mRNA expression of IFN-ɣ were all described in salivary glands from SjS patients, indicating a local production of cytokines associated with the M1 phenotype in the affected exocrine tissue [96,100]. Another indirect evidence of the M1 polarization in SjS is the finding of increased salivary production of CXCL10 in either pre-clinical or in established primary SjS patients [101,102]. CXCL10 is an IFN-ɣ-induced chemokine responsible for the migration of monocytes/macrophages to inflamed salivary glands and is associated with the M1-polarization [5].

The role of monocytes and macrophages in systemic vasculitis

Monocytes in systemic vasculitis

The subsets of monocytes in the peripheral blood were also evaluated in patients with systemic vasculitides such as giant cell arteritis (GCA) and antineutrophil cytoplasmic antibody (ANCA)-associated vasculitis (AAV) [31,36]. Patients with newly diagnosed GCA and polymyalgia rheumatica (PMR) prior to any treatment presented monocytosis due to high number of circulating classical monocytes. Glucocorticoid therapy led to normalization of total monocytes in the peripheral blood and their three subsets in PMR patients, while only a decrease in non-classical monocytes was observed in GCA patients [31]. Conversely, $CD16^+$ monocytes, namely intermediate monocytes, are predominant in AAV patients compared to healthy and disease controls [36,37]. In intermediate monocytes, the expression of myeloperoxidase (MPO), but not of proteinase 3 (PR3), was significantly correlated with CD16 expression. In monocytes purified from PBMC in AAV patients, the in vitro stimulation with anti-MPO antibodies led to increased production of pro-inflammatory cytokines such as interleukin (IL)-1β, IL-6 and IL-8, but this effect was not observed with antibodies against PR3 [36,37].

Recently, two studies have approached monocyte subtypes distribution in the peripheral blood from patients with Behçet disease (BD). Li et al. analyzed monocytes in a small number of treatment-naïve BD patients (N=16) and in 15 controls. BD patients presented higher percentage of intermediate monocytes in PBMC, whereas the percentage of non-classical monocytes was lower than controls. In six BD patients, treatment led to a decrease in the percentage of intermediate monocytes with a concomitant increase of non-classical monocytes [43]. The other study approaching monocytes in BD patients analyzed the number of monocyte subsets in the peripheral blood in a larger number of HC (N=39) and BD patients (N=49), most presenting stable disease and under therapy. Intermediate monocytes were increased while classical monocytes were decreased in the peripheral blood from BD patients compared to controls. Moreover, colchicine therapy for BD led to a significant increase in the number of non-classical monocytes in the peripheral blood [44].

Macrophages in systemic vasculitis

To date, the role of macrophages has been addressed in the pathophysiology of different forms of systemic vasculitides, including large-vessel vasculitis such as GCA and Takayasu arteritis (TAK), Kawasaki disease (KD) as medium-vessel vasculitis, and in AAV and IgA vasculitis (IgAV) as small-vessel vasculitis [31–35,38–42,103–107].

In the immunopathogenesis of GCA, dendritic cell activation occurs in the *vasa vasorum* and it is followed by the infiltration of $CD4^+$ T cell and macrophages in the arterial wall [103]. In temporal arteries from GCA patients, infiltrating macrophages express markers of non-classical monocytes ($CD68^+CD16^+CX3CR1^+CCR2^-$) [31]. The activation of macrophages in newly diagnosed GCA is highlighted by the higher concentration of circulating products of macrophage activation in comparison to healthy controls, such as soluble CD163 (i.e., a scavenger receptor found in M2 macrophages) and of YKL-40 (a glycoprotein secreted by activated macrophages that induces the maturation of monocytes into macrophages) [104]. In temporal arteries, macrophages from GCA patients are more acutely activated, since these cells present a higher expression of the acute activation marker (MRP-14—*migration factor-related protein*-14) than the MRP-8, an intermediate/late macrophage activation marker [105].

The balance between M1 and M2 macrophages was evaluated in the aorta and in temporal arteries form GCA patients, using the folate receptor β (FRβ) and the CD206 as M1 and M2 markers, respectively. M2 macrophages were in sites of tissue destruction while M1 macrophages were observed within the inner intima of arteries with high degree of intimal hyperplasia. The same pattern was found for temporal arteries and for the aorta. Furthermore, the in vitro treatment of macrophages with M-CSF led to higher expression of FRβ, whereas GM-CSF induced the expression of CD206 by these cells [32].

In TAK, macrophages are the most abundant cells observed in the aorta, as compared to T cells, B cells, and NK cells. The expression of the M2 marker CD206 dominated the M1 marker CD86. Even then, M1 macrophages were more frequent in the aorta from TAK patients and from patients with atherosclerotic disease than in heart transplant donors. Intriguingly, histological disease activity and daily prednisone dose were correlated with the expression of the pan-T $CD3^+$ marker and not with the macrophage infiltrate [33]. Treatment with leflunomide is beneficial to some TAK patients. Interestingly, leflunomide was shown to inhibit M2 polarization by blocking STAT-6 phosphorylation in PBMC from TAK patients, and leflunomide also induced the apoptosis of M2 macrophages, and inhibited the release of M2-derived CCL22 and the profibrotic cytokine TGF-β [106].

KD is a systemic inflammatory disease that presents with medium-vessel vasculitis and affects children younger than 5years. A study assessed macrophage subtypes in coronary arteries from KD patients who underwent by-pass due to coronary aneurysm comparing to patients with atherosclerotic disease. In this study, M1 macrophages were more frequent in KD while M2 macrophages were more common in atherosclerotic disease [34]. Conversely, another study showed a complex and dynamic representation of macrophage subtypes in KD. In the acute phase of the disease, the mRNA expression of two M1 markers (TLR2 and TLR4) and of eight M2 markers (ARG1, CCR2, TLR1, TLR8, TLR5, MS4A6A, CD36, and MS4A4A) was increased in PBMC, and all these macrophage markers were significantly decreased after successful treatment with intravenous immunoglobulin (IVIG). Moreover, epigenetic changes were associated with this modulation in mRNA expression of M1 and M2 genes. Hypomethylation of the CpG sites in the promoter region of three M1 genes (IL1R1, TLR2, TLR4) and of three M2 markers (IL1R2, TLR5, ARG1) was observed in macrophages from KD patients [35].

AAV is a group of necrotizing small vessel vasculitis associated with few or no immune deposits. ANCA caused by reactivity against myeloperoxidase (MPO) and proteinase 3 (PR3) represent biomarkers of AAV [107]. Macrophage subtypes were evaluated in pauci-immune glomerulonephritis (GN) and in airways lesions of AAV patients. In the early stages of pauci-immune GN, $CD68^+$ and $CD163^+$ macrophages (i.e., M2 macrophages) were the predominant cells at sites of fibrinoid necrosis, exceeding the number of infiltrating neutrophils and T cells [38]. Indeed, as an indirect evidence of the role of M2 macrophages in GN in AAV patients, urinary levels of the soluble form of the M2 marker sCD163 are associated with active renal vasculitis with high sensitivity (83%) and specificity (96%) [39].

In the airway granulomatous inflammatory lesions of patients with granulomatosis with polyangiitis (GPA), macrophages were more common than T cells, B cells and NK cells. The M2 macrophage marker (CD163) had higher expression than the M1 macrophage marker (CD86) in GPA airway lesions. As a mirror of the Th2 response, the predominance of M2 macrophages in GPA patients was associated with the increased expression of the Th2 transcription factor GATA-3. Moreover, M2 macrophages were related to the prescription of immunosuppressive agents and with the daily dose of prednisone [40]. The M2 phenotype of macrophages is further divided into four different subtypes named as M2a, M2b, M2c, and M2d macrophages. Ohlsson et al. analyzed the in vitro stimulation of M0 (non-activated) macrophages by serum samples from AAV patients with active disease and from controls. Serum from AAV patients induced the polarization toward the M2c macrophage subtype, whereas serum from controls and from AAV patients in remission induced predominantly the M0 macrophage subtype [41].

IgA vasculitis is a small vessel vasculitis with immune deposits of predominantly IgA1. IgAV affects mainly children and involves the skin, joints, gastrointestinal tract, and the kidneys [107]. The expression of the M1 and M2 macrophage markers MRP 8/14 and CD163, respectively, was evaluated in kidney biopsy specimens of IgAV patients. Tubulo-interstitial and glomerular counts of M2 macrophages were associated with grades III and IV of the International Study of Kidney Disease in Children (ISKDC) classification and with glomerular crescents.

Macrophage activation syndrome and hyperactive immune syndromes

Macrophage activation syndrome (MAS) or secondary hemophagocytic lymphohistiocytosis (HLH) is a life-threatening condition that may develop in systemic autoimmune diseases, infectious diseases and malignancies. MAS is characterized by hemophagocytosis, which is defined by the engulfment of erythrocytes, leukocytes, and platelets by phagocytes, mainly macrophages. However, hemophagocytosis is neither sensitive nor specific for the diagnosis of MAS. There is no absolute biomarker for MAS and the diagnosis is based on several clinical and laboratory findings, including sustained fever, hyperferritinemia, pancytopenia, fibrinolytic consumptive coagulopathy, and liver dysfunction (Table 2) [108].

There is a host of evidence for the participation of classically activated M1 macrophages in MAS. IFN-ɣ and TNF-α are key elements in M1 polarization and both appear to be crucial in the development of MAS, as hemophagocytic macrophages in the liver of MAS patients express high amounts of TNF-α [109] and primary HLH patients with IFN-ɣ receptor deficiency failed to develop hemophagocytosis [110]. In addition, CD163, a scavenger receptor expressed in classically activated M2 macrophages, is extensively present in MAS macrophages and increased serum levels of soluble CD163 (sCD163) is considered a valuable diagnostic biomarker of the disease [111].

In frank contrast, other series of evidence point to the participation of M2 macrophages in MAS. The M2-inducing cytokine IL-4 has been shown to induce hemophagocytosis by macrophages in vivo, whereas IFN-ɣ blockade did not inhibit hemophagocytosis [112]. IL-10, an anti-inflammatory cytokine, increases the expression of CD163 by macrophages, which seem to contribute an anti-inflammatory role by clearing free hemoglobin by the

Table 2 2004 Hemophagocytic lymphohistiocytosis (HLH) diagnostic criteria [108].

1. Molecular diagnosis consistent with HLH
2. Diagnosis of HLH fulfilled (5 of the 8 criteria below)
Fever
Splenomegaly
Cytopenia (affecting ≥ 2 of 3 lineages in the peripheral blood)
Hemoglobin <90g/L (hemoglobin <100g/L in infants <4weeks)
Platelets $<100\times10^9$/L
Neutrophils $<1.0\times10^9$/L
Hypertriglyceridemia or hyperfibrinogenemia
Fasting triglycerides ≥ 3.0mmol/L (i.e., ≥ 265mg/dL)
Fibrinogen ≥ 1.5g/L
Hemophagocytosis in bone marrow or spleen or lymph nodes. No evidence of malignancy
Low or no NK cell activity (according to local laboratory evidence)
Ferritin ≥ 500µg/L
sCD25 (i.e., soluble IL-2 receptor ≥ 2400U/mL)

A diagnosis of HLH can be stablished if the patient fulfills criterion 1 or 2.

CD163-mediated endocytosis of haptoglobin-bound hemoglobin. It is relevant that hemophagocytosis occurs at late stages of MAS and actually is not documented in up to a third of cases. Taken together, these observations suggest that macrophages play a dual role in MAS, exhibiting a pro-inflammatory M1 phenotype in the hyper-inflammatory stages of the disease and progressing to an M2 anti-inflammatory phenotype at later stages [113].

References

[1] Murray PJ, Wynn TA. Protective and pathogenic functions of macrophage subsets. Nat Rev Immunol 2011;11(11):723–37. https://doi.org/10.1038/nri3073.

[2] Guilliams M, Ginhoux F, Jakubzick C, Naik SH, Onai N, Schraml BU, et al. Dendritic cells, monocytes and macrophages: a unified nomenclature based on ontogeny. Nat Rev Immunol 2014;14:571–8. https://doi.org/10.1038/nri3712.

[3] Shi C, Pamer EG. Monocyte recruitment during infection and inflammation. Nat Rev Immunol 2011;11(11):762–74. https://doi.org/10.1038/nri3070.

[4] Davies LC, Jenkins SJ, Allen JE, Taylor PR. Tissue-resident macrophages. Nat Immunol 2013;14(10):986–95. https://doi.org/10.1038/ni.2705.

[5] Sica A, Mantovani A. Macrophage plasticity and polarization: in vivo veritas. J Clin Invest 2012;122(3):787–95. https://doi.org/10.1172/JCI59643.

[6] Ma WT, Gao F, Gu K, Chen DK. The role of monocytes and macrophages in autoimmune diseases: a comprehensive review. Front Immunol 2019;10:1140. https://doi.org/10.3389/fimmu.2019.01140.

[7] Ziegler-Heitbrock L, Ancuta P, Crowe S, Dalod M, Grau V, Hart DN, et al. Nomenclature of monocytes and dendritic cells in blood. Blood 2010;116(16):e74–80. https://doi.org/10.1182/blood-2010-02-258558.

[8] Wong KL, Tai JJ, Wong WC, Han H, Sem X, Yeap WH, et al. Gene expression profiling reveals the defining features of the classical, intermediate, and nonclassical human monocyte subsets. Blood 2011;118(5):e16–31. https://doi.org/10.1182/blood-2010-12-326355.
[9] Sampath P, Moideen K, Ranganathan UD, Bethunaickan R. Monocyte subsets: phenotypes and function in tuberculosis infection. Front Immunol 2018;9:1726. https://doi.org/10.3389/fimmu.2018.01726.
[10] Zhu H, Hu F, Sun X, Zhang X, Zhu L, Liu X, et al. $CD16^+$ monocyte subset was enriched and functionally exacerbated in driving T-cell activation and B-cell response in systemic lupus erythematosus. Front Immunol 2016;7:512. https://doi.org/10.3389/fimmu.2016.00512.
[11] Mukherjee R, Barman PK, Thatoi PK, Tripathy R, Das BK, Ravindran B. Non-classical monocytes display inflammatory features: validation in sepsis and systemic lupus erythematous. Sci Rep 2015;5:13886. https://doi.org/10.1038/srep13886.
[12] Jiang W, Zhang L, Lang R, Li Z, Gilkeson G. Sex differences in monocyte activation in systemic lupus erythematosus (SLE). PLoS One 2014;9(12). https://doi.org/10.1371/journal.pone.0114589, e114589.
[13] Kawanaka N, Yamamura M, Aita T, Morita Y, Okamoto A, Kawashima M, et al. CD14+, CD16+ blood monocytes and joint inflammation in rheumatoid arthritis. Arthritis Rheum 2002;46:2578–86. https://doi.org/10.1002/art.10545.
[14] Rossol M, Kraus S, Pierer M, Baerwald C, Wagner U. The CD14(bright) CD16+ monocyte subset is expanded in rheumatoid arthritis and promotes expansion of the Th17 cell population. Arthritis Rheum 2012;64:671–7. https://doi.org/10.1002/art.33418.
[15] Tsukamoto M, Seta N, Yoshimoto K, Suzuki K, Yamaoka K, Takeuchi T. $CD14^{bright}$CD16+ intermediate monocytes are induced by interleukin-10 and positively correlate with disease activity in rheumatoid arthritis. Arthritis Res Ther 2017;19:28. https://doi.org/10.1186/s13075-016-1216-6.
[16] Tsukamoto M, Suzuki K, Seta N, Takeuchi T. Increased circulating CD14brightCD16+ intermediate monocytes are regulated by TNF-α and IL-6 axis in accordance with disease activity in patients with rheumatoid arthritis. Clin Exp Rheumatol 2018;36:540–4.
[17] Yoon BR, Yoo SJ, Choi Y, Chung YH, Kim J, Yoo IS, et al. Functional phenotype of synovial monocytes modulating inflammatory T-cell responses in rheumatoid arthritis (RA). PLoS One 2014;9(10). https://doi.org/10.1371/journal.pone.0109775, e109775.
[18] Smiljanovic B, Radzikowska A, Kuca-Warnawin E, Kurowska W, Grün JR, Stuhlmüller B, et al. Monocyte alterations in rheumatoid arthritis are dominated by preterm release from bone marrow and prominent triggering in the joint. Ann Rheum Dis 2018;77:300–8. https://doi.org/10.1136/annrheumdis-2017-211649.
[19] Mulherin D, Fitzgerald O, Bresnihan B. Synovial tissue macrophage populations and articular damage in rheumatoid arthritis. Arthritis Rheum 1996;39:115–24. https://doi.org/10.1002/art.1780390116.
[20] Haringman JJ, Gerlag DM, Zwinderman AH, Smeets TJ, Kraan MC, Baeten D, et al. Synovial tissue macrophages: a sensitive biomarker for response to treatment in patients with rheumatoid arthritis. Ann Rheum Dis 2005;64:834–8. https://doi.org/10.1136/ard.2004.029751.
[21] Tardito S, Martinelli G, Soldano S, Paolino S, Pacini G, Patane M, et al. Macrophage M1/M2 polarization and rheumatoid arthritis: a systematic review. Autoimmun Rev 2019;18. https://doi.org/10.1016/j.autrev.2019.102397, 102397.
[22] Tak PP, Smeets TJ, Daha MR, Kluin PM, Meijers KA, Brand R, et al. Analysis of the synovial cell infiltrate in early rheumatoid synovial tissue in relation to local disease activity. Arthritis Rheum 1997;40:217–25. https://doi.org/10.1002/art.1780400206.

[23] Lescoat A, Lecureur V, Roussel M, Sunnaram BL, Ballerie A, Coiffier G, et al. CD16-positive circulating monocytes and fibrotic manifestations of systemic sclerosis. Clin Rheumatol 2017;36:1649–54. https://doi.org/10.1007/s10067-017-3597-6.
[24] Worrell JC, O'Reilly S. Bi-directional communication: conversations between fibroblasts and immune cells in systemic sclerosis. J Autoimmun 2020;113. https://doi.org/10.1016/j.jaut.2020.102526, 102526.
[25] Higashi-Kuwata N, Makino T, Inoue Y, Takeya M, Ihn H. Alternatively activated macrophages (M2 macrophages) in the skin of patient with localized scleroderma. Exp Dermatol 2009;18:727–9. https://doi.org/10.1111/j.1600-0625.2008.00828.x.
[26] Chia JJ, Lu TT. Update on macrophages and innate immunity in scleroderma. Curr Opin Rheumatol 2015;27:530–6. https://doi.org/10.1097/BOR.0000000000000218.
[27] Nakayama W, Jinnin M, Makino K, Kajihara I, Makino T, Fukushima S, et al. Serum levels of soluble CD163 in patients with systemic sclerosis. Rheumatol Int 2012;32:403–7. https://doi.org/10.1007/s00296-010-1691-z.
[28] Bielecki M, Kowal K, Lapinska A, Chyczewski L, Kowal-Bielecka O. Increased release of soluble CD163 by the peripheral blood mononuclear cells is associated with worse prognosis in patients with systemic sclerosis. Adv Med Sci 2013;58:126–33. https://doi.org/10.2478/v10039-012-0076-9.
[29] Wildenberg ME, Welzen-Coppens JM, van Helden-Meeuwsen CG, Bootsma H, Vissink A, van Rooijen N, et al. Increased frequency of CD16+ monocytes and the presence of activated dendritic cells in salivary glands in primary Sjögren syndrome. Ann Rheum Dis 2009;68:420–6. https://doi.org/10.1136/ard.2008.087874.
[30] Ciccia F, Alessandro R, Rodolico V, Guggino G, Raimondo S, Guarnotta C, et al. IL-34 is overexpressed in the inflamed salivary glands of patients with Sjogren's syndrome and is associated with the local expansion of pro-inflammatory CD14(bright)CD16+ monocytes. Rheumatology (Oxford) 2013;52:1009–17. https://doi.org/10.1093/rheumatology/kes435.
[31] van Sleen Y, Wang Q, van der Geest KSM, Westra J, Abdulahad WH, Heeringa P, et al. Involvement of monocyte subsets in the immunopathology of giant cell arteritis. Sci Rep 2017;7:6553. https://doi.org/10.1038/s41598-017-06826-4.
[32] Jiemy WF, van Sleen Y, van der Geest KS, Ten Berge HA, Abdulahad WH, Sandovici M, et al. Distinct macrophage phenotypes skewed by local granulocyte macrophage colony-stimulating factor (GM-CSF) and macrophage colony-stimulating factor (M-CSF) are associated with tissue destruction and intimal hyperplasia in giant cell arteritis. Clin Transl Immunol 2020;9. https://doi.org/10.1002/cti2.1164, e1164.
[33] Dos Santos JP, Artigiani Neto R, Mangueira CLP, Filippi RZ, Gutierrez PS, Westra J, et al. Associations between clinical features and therapy with macrophage subpopulations and T cells in inflammatory lesions in the aorta from patients with Takayasu arteritis. Clin Exp Immunol 2020;202:384–93. https://doi.org/10.1111/cei.13489.
[34] Ohashi R, Fukazawa R, Shimizu A, Ogawa S, Ochi M, Nitta T, et al. M1 macrophage is the predominant phenotype in coronary artery lesions following Kawasaki disease. Vasc Med 2019;24:484–92. https://doi.org/10.1177/1358863X19878495.
[35] Guo MM, Chang LS, Huang YH, Wang FS, Kuo HC. Epigenetic regulation of macrophage marker expression profiles in Kawasaki disease. Front Pediatr 2020;8:129. https://doi.org/10.3389/fped.2020.00129.
[36] Park J, Lee EB, Song YW. Decreased tumour necrosis factor-α production by monocytes of granulomatosis with polyangiitis. Scand J Rheumatol 2014;43(5):403–8. https://doi.org/10.3109/03009742.2014.894568.
[37] O'Brien EC, Abdulahad WH, Rutgers A, Huitema MG, O'Reilly VP, Coughlan AM, et al. Intermediate monocytes in ANCA vasculitis: increased surface expression of ANCA autoantigens and IL-1β secretion in response to anti-MPO antibodies. Sci Rep 2015;5:11888. https://doi.org/10.1038/srep11888.

[38] Zhao L, David MZ, Hyjek E, Chang A, Meehan SM. M2 macrophage infiltrates in the early stages of ANCA-associated pauci-immune necrotizing GN. Clin J Am Soc Nephrol 2015;10:54–62. https://doi.org/10.2215/CJN.03230314.
[39] O'Reilly VP, Wong L, Kennedy C, Elliot LA, O'Meachair S, Coughlan AM, et al. Urinary soluble CD163 in active renal vasculitis. J Am Soc Nephrol 2016;27(9):2906–16. https://doi.org/10.1681/ASN.2015050511 [Epub 2016 Mar 3. Erratum in: J Am Soc Nephrol. 2018;29:2255].
[40] de Souza AWS, van Timmeren M, Sanders JS, Stegeman C, Heeringa P, Kallenberg CGM, et al. M2 macrophage is the predominant phenotype in airways inflammatory lesions in patients with granulomatosis with polyangiitis. Arthritis Res Ther 2017;18(19):100. https://doi.org/10.1186/s13075-017-1310-4.
[41] Ohlsson SM, Linge CP, Gullstrand B, Lood C, Johansson A, Ohlsson S, et al. Serum from patients with systemic vasculitis induces alternatively activated macrophage M2c polarization. Clin Immunol 2014;152:10–9. https://doi.org/10.1016/j.clim.2014.02.016.
[42] Kim J, Choi SE, Lee KH, Jeong HJ, Shin JI, Lim BJ. Tubulointerstitial infiltration of M2 macrophages in Henoch-Schönlein purpura nephritis indicates the presence of glomerular crescents and bad clinical parameters. Biomed Res Int 2019;2019. https://doi.org/10.1155/2019/8579619, 8579619.
[43] Li C, Liu J, Yu X, Li L, Wang Z, Shi J, et al. Aberrant monocyte subsets in patients with Behçet's disease. Clin Immunol 2021;225. https://doi.org/10.1016/j.clim.2021.108683, 108683.
[44] Gazzito Del Padre TC, Belem JMFM, de Aguiar MF, Torquato HFV, Paredes-Gamero EJ, Abdulahad WH, et al. Distribution of monocytes subpopulations in the peripheral blood from patients with Behçet's disease – impact of disease status and colchicine use. Clin Immunol 2021;231. https://doi.org/10.1016/j.clim.2021.108854, 108854.
[45] Shapouri-Moghaddam A, Mohammadian S, Vazini H, Taghadosi M, Esmaeili SA, Mardani F, et al. Macrophage plasticity, polarization, and function in health and disease. J Cell Physiol 2018;233(9):6425–40. https://doi.org/10.1002/jcp.26429.
[46] Sümegi A, Antal-Szalmás P, Aleksza M, Kovács I, Sipka S, Zeher M, et al. Glucocorticosteroid therapy decreases CD14-expression and CD14-mediated LPS-binding and activation of monocytes in patients suffering from systemic lupus erythematosus. Clin Immunol 2005;117(3):271–9. https://doi.org/10.1016/j.clim.2005.09.002.
[47] Li Y, Lee PY, Sobel ES, Narain S, Satoh M, Segal MS, et al. Increased expression of FcgammaRI/CD64 on circulating monocytes parallels ongoing inflammation and nephritis in lupus. Arthritis Res Ther 2009;11(1):R6. https://doi.org/10.1186/ar2590.
[48] Wu Z, Zhang S, Zhao L, Fei Y, Wang L, Li J, et al. Upregulation of CD16-monocyte subsets in systemic lupus erythematous patients. Clin Rheumatol 2017;36(10):2281–7. https://doi.org/10.1007/s10067-017-3787-2.
[49] Bengtsson AA, Sturfelt G, Gullstrand B, Truedsson L. Induction of apoptosis in monocytes and lymphocytes by serum from patients with systemic lupus erythematosus – an additional mechanism to increased autoantigen load? Clin Exp Immunol 2004;135:535–43. https://doi.org/10.1111/j.1365-2249.2003.02386.x.
[50] Chalmers SA, Chitu V, Herlitz LC, Sahu R, Stanley ER, Putterman C. Macrophage depletion ameliorates nephritis induced by pathogenic antibodies. J Autoimmun 2015;57:42–52. https://doi.org/10.1016/j.jaut.2014.11.007.
[51] Katsiari CG, Liossis SN, Souliotis VL, Dimopoulos AM, Manoussakis MN, Sfikakis PP. Aberrant expression of the costimulatory molecule CD40 ligand on monocytes from patients with systemic lupus erythematosus. Clin Immunol 2002;103(54–62). https://doi.org/10.1006/clim.2001.5172.

[52] Harigai M, Hara M, Fukasawa C, Nakazawa S, Kawaguchi Y, Kamatani N, et al. Responsiveness of peripheral blood B cells to recombinant CD40 ligand in patients with systemic lupus erythematosus. Lupus 1999;8:227–33. https://doi.org/10.1191/096120399678847678.

[53] Higuchi T, Aiba Y, Nomura T, Matsuda J, Mochida K, Suzuki M, et al. Cutting edge: ectopic expression of CD40 ligand on B cells induces lupus-like autoimmune disease. J Immunol 2002;168:9–12. https://doi.org/10.4049/jimmunol.168.1.9.

[54] Wang X, Huang W, Schiffer LE, Mihara M, Akkerman A, Hiromatsu K, et al. Effects of anti-CD154 treatment on B cells in murine systemic lupus erythematosus. Arthritis Rheum 2003;48:495–506. https://doi.org/10.1002/art.10929.

[55] Umare V, Pradhan V, Nadkar M, Rajadhyaksha A, Patwardhan M, Ghosh KK, et al. Effect of proinflammatory cytokines (IL-6, TNF-α, and IL-1β) on clinical manifestations in Indian SLE patients. Mediat Inflamm 2014;2014. https://doi.org/10.1155/2014/385297, 385297.

[56] Jin O, Sun LY, Zhou KX, Zhang XS, Feng XB, Mok MY, et al. Lymphocyte apoptosis and macrophage function: correlation with disease activity in systemic lupus erythematosus. Clin Rheumatol 2005;24(2):107–10. https://doi.org/10.1007/s10067-004-0972-x.

[57] Becker-Merok A, Eilertsen GØ, Nossent JC. Levels of transforming growth factor-beta are low in systemic lupus erythematosus patients with active disease. J Rheumatol 2010;37(10):2039–45. https://doi.org/10.3899/jrheum.100180.

[58] Sharif MN, Tassiulas I, Hu Y, Mecklenbräuker I, Tarakhovsky A, Ivashkiv LB. IFN-alpha priming results in a gain of proinflammatory function by IL-10: implications for systemic lupus erythematosus pathogenesis. J Immunol 2004;172(10):6476–81. https://doi.org/10.4049/jimmunol.172.10.6476.

[59] Geginat J, Vasco M, Gerosa M, Tas SW, Pagani M, Grassi F, et al. IL-10 producing regulatory and helper T-cells in systemic lupus erythematosus. Semin Immunol 2019;44. https://doi.org/10.1016/j.smim.2019.101330, 101330.

[60] Godsell J, Rudloff I, Kandane-Rathnayake R, Hoi A, Nold MF, Morand EF, et al. Clinical associations of IL-10 and IL-37 in systemic lupus erythematosus. Sci Rep 2016;6:34604. https://doi.org/10.1038/srep34604.

[61] Biesen R, Demir C, Barkhudarova F, Grün JR, Steinbrich-Zöllner M, Backhaus M, et al. Sialic acid-binding Ig-like lectin 1 expression in inflammatory and resident monocytes is a potential biomarker for monitoring disease activity and success of therapy in systemic lupus erythematosus. Arthritis Rheum 2008;58:1136–45. https://doi.org/10.1002/art.23404.

[62] Macauley MS, Crocker PR, Paulson JC. Siglec-mediated regulation of immune cell function in disease. Nat Rev Immunol 2014;14(10):653–66. https://doi.org/10.1038/nri3737.

[63] Herrmann M, Voll RE, Zoller OM, Hagenhofer M, Ponner BB, Kalden JR. Impaired phagocytosis of apoptotic cell material by monocyte-derived macrophages from patients with systemic lupus erythematosus. Arthritis Rheum 1998;41:1241–50. https://doi.org/10.1002/1529-0131(199807)41:7<1241::AID-ART15>3.0.CO;2-H.

[64] Baumann I, Kolowos W, Voll RE, Manger B, Gaipl U, Neuhuber WL, et al. Impaired uptake of apoptotic cells into tingible body macrophages in germinal centers of patients with systemic lupus erythematosus. Arthritis Rheum 2002;46:191–201. https://doi.org/10.1002/1529-0131(200201)46:1<191::AID-ART10027>3.0.CO;2-K.

[65] Wermeling F, Chen Y, Pikkarainen T, Scheynius A, Winqvist O, Izui S, et al. Class A scavenger receptors regulate tolerance against apoptotic cells, and autoantibodies against these receptors are predictive of systemic lupus. J Exp Med 2007;204(10):2259–65. https://doi.org/10.1084/jem.20070600.

[66] Labonte AC, Kegerreis B, Geraci NS, Bachali P, Madamanchi S, Robl R, et al. Identification of alterations in macrophage activation associated with disease activity in systemic lupus erythematosus. PLoS One 2018;13. https://doi.org/10.1371/journal.pone.0208132, e0208132.
[67] Li F, Yang Y, Zhu X, Huang L, Xu J. Macrophage polarization modulates development of systemic lupus erythematosus. Cell Physiol Biochem 2015;37:1279–88. https://doi.org/10.1159/000430251.
[68] Mohammadi S, Saghaeian-Jazi M, Sedighi S, Memarian A. Immunomodulation in systemic lupus erythematosus: induction of M2 population in monocyte-derived macrophages by pioglitazone. Lupus 2017;26(12):1318–27. https://doi.org/10.1177/0961203317701842.
[69] Mohammadi S, Saghaeian-Jazi M, Sedighi S, Memarian A. Sodium valproate modulates immune response by alternative activation of monocyte-derived macrophages in systemic lupus erythematosus. Clin Rheumatol 2018;37:719–27. https://doi.org/10.1007/s10067-017-3922-0.
[70] Mohammadi S, Memarian A, Sedighi S, Behnampour N, Yazdani Y. Immunoregulatory effects of indole-3-carbinol on monocyte-derived macrophages in systemic lupus erythematosus: a crucial role for aryl hydrocarbon receptor. Autoimmunity 2018;51:199–209. https://doi.org/10.1080/08916934.2018.1494161.
[71] Sack U, Stiehl P, Geiler G. Distribution of macrophages in rheumatoid synovial membrane and its association with basic activity. Rheumatol Int 1994;13:181–6. https://doi.org/10.1007/BF00390265.
[72] Tu J, Wang X, Gong X, Hong W, Han D, Fang Y, et al. Synovial macrophages in rheumatoid arthritis: the past, present, and future. Mediat Inflamm 2020;2020:1583647. https://doi.org/10.1155/2020/1583647.
[73] Lin NY, Beyer C, Giessl A, Kireva T, Scholtysek C, Uderhardt S, et al. Autophagy regulates TNFα-mediated joint destruction in experimental arthritis. Ann Rheum Dis 2013;72:761–8. https://doi.org/10.1136/annrheumdis-2012-201671.
[74] Tsuneyoshi Y, Tanaka M, Nagai T, Sunahara N, Matsuda T, Sonoda T, et al. Functional folate receptor beta-expressing macrophages in osteoarthritis synovium and their M1/M2 expression profiles. Scand J Rheumatol 2012;41:132–40. https://doi.org/10.3109/03009742.2011.605391.
[75] Ambarus CA, Noordenbos T, de Hair MJ, Tak PP, Baeten DL. Intimal lining layer macrophages but not synovial sublining macrophages display an IL-10 polarized-like phenotype in chronic synovitis. Arthritis Res Ther 2012;14:R74. https://doi.org/10.1186/ar3796.
[76] McInnes IB, Schett G. Cytokines in the pathogenesis of rheumatoid arthritis. Nat Rev Immunol 2007;7(6):429–42. https://doi.org/10.1038/nri2094.
[77] Park SY, Lee SW, Lee SY, Hong KW, Bae SS, Kim K, et al. SIRT1/adenosine monophosphate-activated protein kinase alpha signaling enhances macrophage polarization to an anti-inflammatory phenotype in rheumatoid arthritis. Front Immunol 2017;8:1135. https://doi.org/10.3389/fimmu.2017.01135.
[78] Sun W, Zhang HW, Wang H, Chiu YG, Wang MM, Ritchlin CT, et al. Targeting notch-activated M1 macrophages attenuates joint tissue damage in a mouse model of inflammatory arthritis. J Bone Miner Res 2017;32:1469–80. https://doi.org/10.1002/jbmr.3117.
[79] Ye L, Wen ZY, Li YQ, Chen BN, Yu T, Liu LY, et al. Interleukin-10 attenuation of collagen-induced arthritis is associated with suppression of interleukin-17 and retinoid-related orphan receptor gamma t production in macrophages and repression of classically activated macrophages. Arthritis Res Ther 2014;16(2):R96. https://doi.org/10.1186/ar4544.
[80] Chen S, Chen BN, Wen ZY, Huang Z, Ye L. IL-33/ST2- mediated inflammation in macrophages is directly abrogated by IL-10 during rheumatoid arthritis. Oncotarget 2017;8:32407–18. https://doi.org/10.18632/oncotarget.16299.

[81] Higashi-Kuwata N, Jinnin M, Makino T, Fukushima S, Inoue Y, Muchemwa FC, et al. Characterization of monocyte/macrophage subsets in the skin and peripheral blood derived from patients with systemic sclerosis. Arthritis Res Ther 2010;12:R128. https://doi.org/10.1186/ar3066.

[82] Ishikawa O, Ishikawa H. Macrophage infiltration in the skin of patients with systemic sclerosis. J Rheumatol 1992;19:1202–6.

[83] Selvi E, Tripodi SA, Catenaccio M, Lorenzini S, Chindamo D, Manganelli S, et al. Expression of macrophage migration inhibitory factor in diffuse systemic sclerosis. Ann Rheum Dis 2003;62:460–4. https://doi.org/10.1136/ard.62.5.460.

[84] Cutolo M, Soldano S, Smith V. Pathophysiology of systemic sclerosis: current understanding and new insights. Expert Rev Clin Immunol 2019;15:753–64. https://doi.org/10.1080/1744666X.2019.1614915.

[85] Nguyen JK, Austin E, Huang A, Mamalis A, Jagdeo J. The IL-4/IL-13 axis in skin fibrosis and scarring: mechanistic concepts and therapeutic targets. Arch Dermatol Res 2020;312:81–92. https://doi.org/10.1007/s00403-019-01972-3.

[86] Huang XL, Wang YJ, Yan JW, Wan YN, Chen B, Li BZ, et al. Role of anti-inflammatory cytokines IL-4 and IL-13 in systemic sclerosis. Inflamm Res 2015;64:151–9. https://doi.org/10.1007/s00011-015-0806-0.

[87] Fuschiotti P. Role of IL-13 in systemic sclerosis. Cytokine 2011;56:544–9. https://doi.org/10.1016/j.cyto.2011.08.030.

[88] Beamer CA, Holian A. Scavenger receptor class a type I/II (CD204) null mice fail to develop fibrosis following silica exposure. Am J Phys 2005;289:L186–95. https://doi.org/10.1152/ajplung.00474.2004.

[89] Soldano SP, Trombetta CP, Contini PP, Tomatis VM, Ruaro BM, Brizzolara RP, et al. Increase in circulating cells coexpressing M1 and M2 macrophage surface markers in patients with systemic sclerosis. Ann Rheum Dis 2018;2018. https://doi.org/10.1136/annrheumdis-2018-213648, 213648.

[90] Christmann RB, Sampaio-Barros P, Stifano G, Borges CL, de Carvalho CR, Kairalla R, et al. Association of interferon- and transforming growth factor beta-regulated genes and macrophage activation with systemic sclerosisrelated progressive lung fibrosis. Arthritis Rheumatol 2014;66:714–25. https://doi.org/10.1002/art.38288.

[91] Braga TT, Agudelo JS, Camara NO. Macrophages during the fibrotic process: M2 as friend and foe. Front Immunol 2015;6:602. https://doi.org/10.3389/fimmu.2015.00602.

[92] Wu M, Pedroza M, Lafyatis R, George AT, Mayes MD, Assassi S, et al. Identification of cadherin 11 as a mediator of dermal fibrosis and possible role in systemic sclerosis. Arthritis Rheumatol 2014;66:1010–21. https://doi.org/10.1002/art.38275.

[93] Schenke-Layland K, Xie J, Magnusson M, Angelis E, Li X, Wu K, Reinhardt DP, et al. Lymphocytic infiltration leads to degradation of lacrimal gland extracellular matrix structures in NOD mice exhibiting a Sjögren's syndrome-like exocrinopathy. Exp Eye Res 2010;90:223–37. https://doi.org/10.1016/j.exer.2009.10.008.

[94] Baban B, Liu JY, Abdelsayed R, Mozaffari MS. Reciprocal relation between GADD153 and Del-1 in regulation of salivary gland inflammation in Sjögren syndrome. Exp Mol Pathol 2013;95:288–97. https://doi.org/10.1016/j.yexmp.2013.09.002.

[95] Srivastava A, Makarenkova HP. Innate immunity and biological therapies for the treatment of Sjögren's syndrome. Int J Mol Sci 2020;21:9172. https://doi.org/10.3390/ijms21239172.

[96] Willeke P, Schlüter B, Schotte H, Domschke W, Gaubitz M, Becker H. Interferon-gamma is increased in patients with primary Sjogren's syndrome and Raynaud's phenomenon. Semin Arthritis Rheum 2009;39:197–202. https://doi.org/10.1016/j.semarthrit.2008.04.002.

[97] Tishler M, Yaron I, Shirazi I, Yossipov Y, Yaron M. Increased salivary interleukin-6 levels in patients with primary Sjögren's syndrome. Rheumatol Int 1999;18:125–7. https://doi.org/10.1007/s002960050070.

[98] Fogel O, Rivière E, Seror R, Nocturne G, Boudaoud S, Ly B, et al. Role of the IL-12/IL-35 balance in patients with Sjögren syndrome. J Allergy Clin Immunol 2018;142:258–268.e5. https://doi.org/10.1016/j.jaci.2017.07.041.
[99] Awada A, Nicaise C, Ena S, Schandéné L, Rasschaert J, Popescu I, et al. Potential involvement of the IL-33-ST2 axis in the pathogenesis of primary Sjogren's syndrome. Ann Rheum Dis 2014;73(6):1259–63. https://doi.org/10.1136/annrheumdis-2012-203187.
[100] Sakai A, Sugawara Y, Kuroishi T, Sasano T, Sugawara S. Identification of IL-18 and Th17 cells in salivary glands of patients with Sjögren's syndrome, and amplification of IL-17-mediated secretion of inflammatory cytokines from salivary gland cells by IL-18. J Immunol 2008;181:2898–906. https://doi.org/10.4049/jimmunol.181.4.2898.
[101] Hernández-Molina G, Michel-Peregrina M, Hernández-Ramírez DF, Sánchez-Guerrero J, Llorente L. Chemokine saliva levels in patients with primary Sjögren's syndrome, associated Sjögren's syndrome, pre-clinical Sjögren's syndrome and systemic autoimmune diseases. Rheumatology (Oxford) 2011;50:1288–92. https://doi.org/10.1093/rheumatology/ker019.
[102] Lee YJ, Scofield RH, Hyon JY, Yun PY, Lee HJ, Lee EY, et al. Salivary chemokine levels in patients with primary Sjogren's syndrome. Rheumatology (Oxford) 2010;49:1747–52. https://doi.org/10.1093/rheumatology/keq121.
[103] Weyand CM, Goronzy JJ. Immune mechanisms in medium and large-vessel vasculitis. Nat Rev Rheumatol 2013;9:731–40. https://doi.org/10.1038/nrrheum.2013.161.
[104] van Sleen Y, Sandovici M, Abdulahad WH, Bijzet J, van der Geest KSM, Boots AMH, et al. Markers of angiogenesis and macrophage products for predicting disease course and monitoring vascular inflammation in giant cell arteritis. Rheumatology (Oxford) 2019;58:1383–92. https://doi.org/10.1093/rheumatology/kez034.
[105] Mihm B, Bergmann M, Brück W, Probst-Cousin S. The activation pattern of macrophages in giant cell (temporal) arteritis and primary angiitis of the central nervous system. Neuropathology 2014;34:236–42. https://doi.org/10.1111/neup.12086.
[106] Cui X, Kong X, Chen R, Ma L, Jiang L. The potential role of leflunomide in inhibiting vascular fibrosis by down-regulating type-II macrophages in Takayasu's arteritis. Clin Exp Rheumatol 2020;38(Suppl 124):69–78.
[107] Jennette JC, Falk RJ, Bacon PA, Basu N, Cid MC, Ferrario F, et al. 2012 revised international Chapel Hill consensus conference nomenclature of vasculitides. Arthritis Rheum 2013;65:1–11. https://doi.org/10.1002/art.37715.
[108] Henter JI, Horne A, Aricó M, Egeler RM, Filipovich AH, Imashuku S, et al. HLH-2004: diagnostic and therapeutic guidelines for hemophagocytic lymphohistiocytosis. Pediatr Blood Cancer 2007;48:124–31. https://doi.org/10.1002/pbc.21039.
[109] Billiau AD, Roskams T, Van Damme-Lombaerts R, Matthys P, Wouters C. Macrophage activation syndrome: characteristic findings on liver biopsy illustrating the key role of activated, IFN-gamma-producing lymphocytes and IL-6- and TNF-alpha-producing macrophages. Blood 2005;105:1648–51. https://doi.org/10.1182/blood-2004-08-2997.
[110] Tesi B, Sieni E, Neves C, Romano F, Cetica V, Cordeiro AI, et al. Hemophagocytic lymphohistiocytosis in 2 patients with underlying IFNgamma receptor deficiency. J Allergy Clin Immunol 2015;135:1638–41. https://doi.org/10.1016/j.jaci.2014.11.030.
[111] Sakumura N, Shimizu M, Mizuta M, Inoue N, Nakagishi Y, Yachie A. Soluble CD163, a unique biomarker to evaluate the disease activity, exhibits macrophage activation in systemic juvenile idiopathic arthritis. Cytokine 2018;110:459–65. https://doi.org/10.1016/j.cyto.2018.05.017.
[112] Milner JD, Orekov T, Ward JM, Cheng L, Torres-Velez F, Junttila I, et al. Sustained IL-4 exposure leads to a novel pathway for hemophagocytosis, inflammation, and tissue macrophage accumulation. Blood 2010;116:2476–83. https://doi.org/10.1182/blood-2009-11-255174.
[113] Crayne CB, Albeituni S, Nichols KE, Cron RQ. The immunology of macrophage activation syndrome. Front Immunol 2019;10:119. https://doi.org/10.3389/fimmu.2019.00119.

Index

Note: Page numbers followed by *f* indicate figures and *t* indicate tables.

L

M

FSC
www.fsc.org
MIX
Papier | Fördert gute Waldnutzung
FSC® C083411

Kolibri 360 GmbH
Ferdinand-Jühlke-Straße 7
99095 Erfurt, Deutschland
produktsicherheit@kolibri360.de